ENGINEERING ECONOMICS AND ECONOMIC DESIGN FOR PROCESS ENGINEERS

ENGINEERING ECONOMICS AND ECONOMIC DESIGN FOR PROCESS ENGINEERS

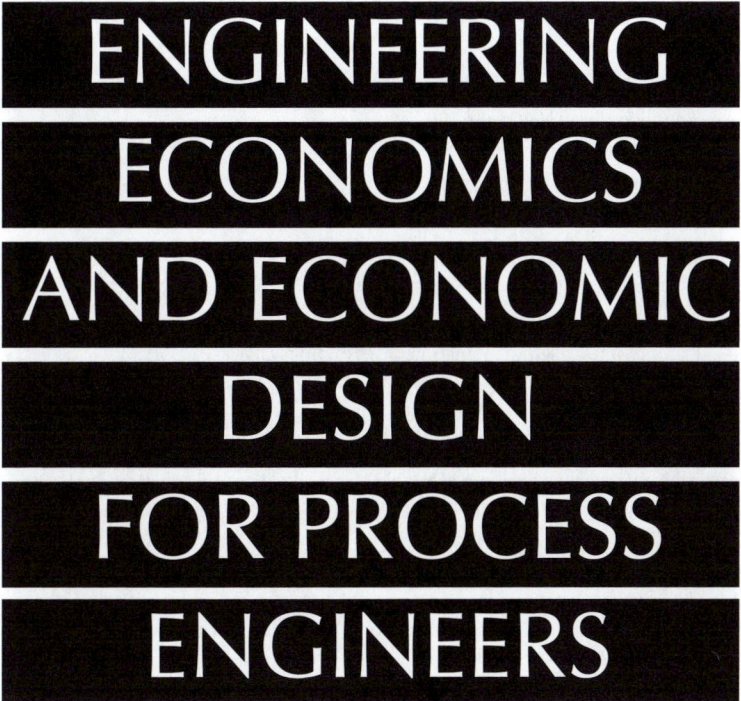

Thane Brown
University of Cincinnati, Ohio

CRC Press
Taylor & Francis Group
Boca Raton   London   New York

CRC Press is an imprint of the
Taylor & Francis Group, an informa business

CRC Press
Taylor & Francis Group
6000 Broken Sound Parkway NW, Suite 300
Boca Raton, FL 33487-2742

© 2007 by Taylor & Francis Group, LLC
CRC Press is an imprint of Taylor & Francis Group, an Informa business

International Standard Book Number-10: 0-8493-8212-2 (Hardcover)
International Standard Book Number-13: 978-0-8493-8212-3 (Hardcover)

Library of Congress Cataloging-in-Publication Data

Brown, Thane.
   Engineering economics and economic design for process engineers / Thane Brown.
      p. cm.
    ISBN-13: 978-0-8493-8212-3 (alk. paper)
    ISBN-10: 0-8493-8212-2 (alk. paper)
    1. Engineering economy. I. Title.

TA177.4.B76 2006
658.15024'62--dc22   2006018328

Visit the Taylor & Francis Web site at
http://www.taylorandfrancis.com

and the CRC Press Web site at
http://www.crcpress.com

About the Author

Thane Brown is presently an adjunct professor at the University of Cincinnati, where he teaches engineering economics. He is a member of the Advisory Council for the Chemical and Materials Engineering Department at the University of Dayton. He gives several seminars each year at both schools.

Prior to teaching, Mr. Brown worked for Procter & Gamble for over 36 years. His last position there was Director of North American Engineering. His organization, comprised of P&G and contractor engineers, designed and built over $1 billion of plant facilities per year. At P&G, he worked in a variety of engineering and manufacturing roles. His engineering work mainly involved the conceptualization, design, construction, and startup of plants in the shortening and oil, coffee, soft drink, juice, feminine products, and papermaking businesses. His manufacturing assignments were in shortening, oil, and juice.

He has written a number of articles on engineering economics, batch pressure filtration, and heat transfer. Mr. Brown is a registered professional engineer in Ohio and has a B.S. in chemical engineering from Oregon State University.

Introduction

This book contains the material I wish I'd known when I started working. Engineers play a big role in how much money a company makes. By bringing economic thinking into all parts of their work, they maximize their impact on profits. In broad terms, they deal with two types of economic decisions:

- Is the project economically justifiable?
- Of the options available, which is best from an economic standpoint?

The book provides the tools and methodology to answer these questions. It will help engineers (and students about to enter the workforce) integrate technical and economic decision making, creating more profit and growth for their companies. It covers two broad topics: engineering economics and economic design methodology. Its methods are simple, fast, and inexpensive to use and apply.

The first section of the book covers the basics of Engineering Economics — time value of money, capital and production cost estimation, economic evaluation methods, and risk analysis. The focus is on the early stages of engineering — process development, feasibility engineering, and conceptual design. This is the time when the economic structure of a project is set and when the engineer's impact on profits is greatest.

Secondly, the book presents a methodology for economically designing a plant or process. During my Procter & Gamble career I observed that there are almost always several technically acceptable answers to any design question. One can use economic factors to select the best of them, and this part of the book is based upon that premise. It shows how to combine technical and economic decision making to create economically optimal designs. Whereas the book focuses on the early project stages, one can use the thought processes and most of the methods in all project phases

ECONOMICS IN ENGINEERING

Companies exist for one reason — to make money (profits). Profits, sales, expenses, and taxes are interrelated:

$$\text{Profit}_{AT} = \text{Sales} + \text{Other Income} - \text{Expenses} - \text{Taxes}$$

For a company making money, profits are the crucial part of its cash flow. They determine whether the company will have enough money to invest in capital, to acquire other companies, acquire other brands, and to pay dividends to its shareholders.

When they create new products or processes and design and build plants and processes, engineers affect both profits and cash flow. Their main impact is in the areas of:

- Production cost spending (This is the main expense in the profit equation.)
- Depreciation (This is a part of production cost.)
- Capital spending (This is one of the uses of profits.)
- Sales (This and production cost spending are the main factors in the profit equation. Engineers affect sales by inventing new or improving products, building new capacity, and reducing the cost of producing products.)

Good engineers develop sound cost structures for their products and projects. These cost structures:

- Enable the company's products to be competitively priced in the market place.
- Enable the company to earn a reasonable profit margin on these products.
- Have financial returns that meet the company's return on investment (ROI) criteria.

When developing projects, engineers continually review their economic attractiveness. Additionally, most companies require formal assessments several times during the life of a project. These assessments check whether a project is still on track to meet its objectives. For example, will the project still deliver its intended business result? Does it meet or exceed the company's ROI criteria (plus any other financial objectives)? Is the projected startup date acceptable? Are the project risks still manageable? These risks include market risks, technical risks, project cost risk, and schedule risk. Of them, market risk is the most significant:

- *Market risk* — The sales volume or selling price is lower than expected. Introductory marketing expenses are higher, and so on.
- *Technical risk* — Yields or operating efficiencies are lower than estimated. Emissions are higher. Equipment does not work correctly, and so on.
- *Project cost risk* — The capital cost or production costs are higher than expected.
- *Schedule risk* — The startup is late or prolonged.

Engineers are responsible for controlling the last three and may get involved assessing all four. They often use sensitivity analysis and decision-tree tools to understand the risks and manage decision making.

ECONOMIC DESIGN

Economic design is one of the ways engineers create profit. It involves knowing the economic impact of engineering choices and finding the economic balance between capital spending and production cost spending. (The balance point is defined by a company's ROI criteria.) It seeks the answers to:

- Is it better to spend more capital and have lower production costs?
- Is it better to spend less capital and have higher production costs?

These two questions are asked over and over again while the process is being developed and the plant or process is being designed. Balancing capital and production cost is what helps ensure the overall project is economically justifiable.

HOW DOES ONE DO ECONOMIC DESIGN — A MODEL

Engineers are continually deciding among alternates or options. The economic optimization model provides a framework — a method — for doing this. It has three phases:

- Defining the business and technical purposes of the design before starting work
- Creating a list of options to be analyzed
- Analyzing the options and selecting the most economic. Three steps are involved:
  - Eliminating the technically inadequate options
  - Economically analyzing those that are acceptable
  - Selecting the option having the best after-tax economics

Every practicing engineer — whether involved in developing processes, designing a process or a plant, or making plant upgrades — can use the tools presented here. The economic design model is shown in Figure I.1.

WHEN DOES ONE DO ECONOMIC DESIGN?

Simply put, economic design is done in every phase of a design. Because the economic framework in a project is set in what this book calls the *Process Development*, *Feasibility*, and *Conceptual* phases of a design, this book will deal only with those. (Companies may subdivide projects differently or use different names for the phases but the work is the same. You will find the inputs to and outputs for each project phase described in Appendix V.) The following briefly describes these phases:

- *Process development*. In this phase, one defines the process steps, operating conditions, and raw and packaging material specs. When this phase is complete, one will have defined a technically feasible process. Some typical options or studies are:
  - What process steps (functions) and operating conditions are feasible?
  - What unit operations are best for each process function?

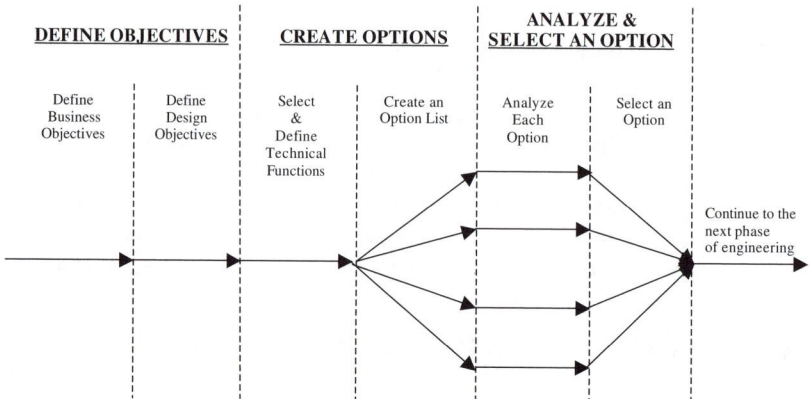

FIGURE I.1 The economic design model.

- From product quality, environmental, safety, and economic standpoints, what are the proper raw materials and packaging materials?
- Should recycle or purge streams be used?
- *Feasibility.* Here one begins to develop design details to decide whether a proposal is economically feasibility. Some typical options or studies are:
  - How should the tradeoffs among the reactor (or dominant unit operation), separation systems, and the heat recovery systems be optimized?
  - How many separate processes or plants should exist? Where should they be located?
  - Should the process be batch or continuous?
  - For health, safety, or environmental reasons, should the plant or process be located in an unpopulated area?
- *Conceptual.* In this phase, one develops the major features of the feasible option or options. Some typical options or studies are:
  - For the major equipment, which of the technically feasible types are the most cost-effective?
  - What is the heat recovery plan? What process streams will exchange energy with each other? About how much energy is economical to exchange?
  - Should a surge exist between unit operations? How much?

HOW THIS BOOK IS ORGANIZED

Section I: Engineering Economics

Chapter 1 sets the stage by explaining the engineer's role in creating economically feasible projects. It then discusses the economics of projects — how they are funded, what kinds of investments are needed, and how revenues, expenses, profits, and risks are interrelated. Last, it covers how cash flows into and out of a company.

Chapter 2 deals with the time value of money. It discusses present and future values, annuities, interest rates, inflation, and inflation indices. These are the precursor materials for Chapter 5.

Chapter 3 details how to create order-of-magnitude and study grade estimates for the investments in a project — capital, startup expenses, and working capital. The estimating methods are ideal for use when there is little design detail.

Chapter 4 concerns itself with production costs. It explains how to make study grade production cost estimates.

Chapter 5 applies the time value basics from Chapter 2 to proposal and project evaluation. It covers net present value, annual cost, ROI, breakeven volume, sensitivity analysis, and decision trees. It also briefly mentions other evaluation methods.

Section II: Economic Design Methodology

Chapter 6 explains the economic design model. This three-phase model — Defining Objectives, Creating Options, and Analyzing/Selecting an Option — describes a process for doing economic design in a project.

Chapter 7 covers the reasons for setting business and technical objectives early in a project. It then explains both business and technical objectives and how to go about setting them.

Chapter 8 and Chapter 9 cover methodology designed to foster the creation of a thorough list of options for study. Today, inadequate option list creation is the major shortcoming in the creation of economic designs. Chapter 8 presents a discipline for flow sheet creation designed to keep options open. This involves specifying the technical function for each process step before selecting unit operations and equipment types. Chapter 9 discusses how to create a list of options for study. It presents several tools that will stimulate one's thinking and help create a complete list.

Chapter 10 explains how to anayze different options and how to select the most economic of the technically feasible options. It covers a general method for economic analysis, the economics of selecting equipment, and the economics of plant siting. As an added topic, it explains how to decide whether or not to install extra capacity for an assumed future need.

Chapter 11 presents a number of common economic design case studies:

- Finding the optimal cooling water outlet temperature in a heat exchanger that is cooling a hot stream
- Finding the optimal catalyst usage in a reactor/filter system
- Finding the optimal amount of heat recovery in a heat exchanger loop
- Determining whether to build a grass-roots plant or whether to expand an existing plant
- Finding the economically optimal number of plants

Appendices: There are five Appendices — Definitions of Key Terms, Inflation Indices, Compound Interest Tables, Equipment Pricing Data, and Project Phase Inputs and Outputs.

This is a practical book for every student and practicing engineer. It presents the tools of engineering economics and economic design. Using them, every engineer can increase their company's profits.

TERMINOLOGY

I need to clarify a few terms. I use three words in their broadest sense — *project*, *design*, and *engineer*. All three include the work done from the process development (or process synthesis) phase through detailed design engineering. A *project* or *design* may thus refer to work being done by a research and development (R&D) engineer defining a process or to the work being done by a plant engineer specifying equipment for plant maintenance. Likewise, the term *engineer* refers to all types of engineers — those in R&D, process design, facilities, a plant, and so forth.

Additionally, I use the convention of K = 1000 and M = 1,000,000.

Table of Contents

Section I

Engineering Economics

1 The Economic Side of Engineering

To create economically viable projects, engineers must create products whose cost is competitive in the marketplace and design plants in which there is an economic balance between capital investment and production cost.

Companies exist for one reason — to make money. The good ones do this honestly, ethically, and legally while demonstrating care and respect for the health and safety of their employees, for the communities in which they operate, and for the environment. This chapter discusses how engineers affect and are affected by economics.

1.1 THE ENGINEER'S ROLE IN PROFIT CREATION

Making money requires that a company turn a profit after it pays taxes. Engineers affect profits in a number of ways:

- They create the opportunity for new sales by developing new or improved products. In addition, they develop the processes that make these products.
- They create the opportunity for reducing expenses by developing processes that reduce costs.
- They design and build or modify plants or processes that will:
  - Produce new or improved products
  - Increase production capacity, enabling increased sales
  - Reduce expenses

1.2 ECONOMIC VIABILITY

When R&D engineers develop products and processes, they must do it so that their company can competitively price the products. Getting the price right requires raw and packaging materials that are not too costly as well as processes simple and efficient enough to have affordable manufacturing costs. Additionally, they must develop their processes so that the capital cost for the process is economically justifiable. In other words, the return on the capital investment must meet or exceed the company's return on investment (ROI) criteria and must meet any cash flow limitations. From an economic standpoint, the development engineer must be able to estimate raw material, packaging material, production, and capital costs. S/he then uses these to calculate the project ROI and determine whether the project is economically viable.

If a development project looks attractive enough, a company will fund the project, authorizing plant design work to begin. At this point, the engineer's job is to design the plant economically. Obviously, the design is the major determiner of capital cost. The design also has a major effect on production costs. Getting both correct requires careful attention to six factors affecting capital and production costs:

- Plant location as it affects raw and packaging material in freight, product shipping, and labor costs
- Product yields and material losses
- Plant staffing levels, which are affected by plant and equipment layouts, equipment selection, and levels of automation
- Plant operating efficiencies
- Utility usage
- The amount of capital needed as it affects depreciation charges

Economic tradeoffs exist between production costs and capital cost. For example, one could spend more capital to increase product yield. Economic design means finding the economic balance between production costs and capital cost. In other words, the engineer must decide whether the ROI of investing more capital to have lower production costs is acceptable. S/he must decide:

- Is it better to spend more capital and have lower production costs?
- Is it better to spend less capital and have higher production costs?

To illustrate, assume you are building capacity to produce a specialty chemical. You will ship 90% of your production to ten locations around the U.S. You must decide how many plants to build and where to locate them. When making this decision, you would consider the following tradeoffs:

- You could minimize shipping costs by building ten plants located next to each of the ten major shipping points.
- However, this would maximize capital costs because one large plant is less expensive to build than ten smaller plants.
- In addition, this would maximize manufacturing costs because one large plant has lower costs than ten smaller plants.

To make the decision, you would look for the number of plants that minimizes the "combination" of shipping, manufacturing, and capital costs. Chapter 5 discusses the methods for "combining" costs. A company's ROI criteria for investments is a key factor in combining costs. Chapter 10 includes a discussion of "The Economics of Plant Siting."

Whether developing products and processes or designing plants, engineers must develop economically viable products. To create economically viable projects, engineers must create products whose cost is competitive in the marketplace and design plants in which there is an economic balance between capital investment and production cost.

1.3 PROJECT AUTHORIZATION

Project authorization is simply getting company agreement to proceed with a project. If the project is large enough, this either will include agreement to spend existing budget funds or, if there are no funds in the budget for the project, will require appropriating new funds. Economics are almost always involved in the authorization decision. ROI and cash flow are the usual economic criteria used when making authorization decisions. There are also legal, ethical and "good citizen" reasons to fund a project. Examples of these are: legal (complying with new environmental regulations); ethical (eliminating personnel safety hazards beyond that required by OSHA); and good citizen (upgrading the outside appearance of a plant).

1.4 TYPES OF PROJECTS

An engineer might work on different types of projects. Categorized by their benefit to a company, they are:

- *New product introductions and product upgrades.* These involve bringing new or improved products to the marketplace. These projects create new volume and sales by meeting unfilled customer needs.
- *Capacity increases.* These allow a company to make more product by increasing production capability. The need for capacity occurs when a company cannot or soon expects it will not be able to make enough product.
- *Cost reduction.* These can involve any number of things — labor productivity increases, yield improvements (loss reductions), utility savings, maintenance cost reductions, in-freight cost savings, or distribution (shipping) cost reductions.
- *Maintenance work.* When replacing equipment, one usually must consider whether to upgrade the equipment. For example, should one replace a corroded carbon steel pump in kind or with a 316 stainless steel unit?
- *Health/safety/environmental improvements.* These upgrade a plant's basic systems to make them safer (for both its employees and the community) or to reduce the type or amount of emissions from the plant. New regulations or a company's desire to better protect its employees and the community provide the driving force for these projects.

1.5 FUNDING

With respect to funding, there are two kinds of engineering work: that funded by an expense budget and that funded by a capital project. Expense budgets usually fund only people and people-related costs such as travel. Capital projects fund people costs plus the purchase and installation of capital facilities such as buildings, process equipment, utilities, and so on.

1.5.1 FUNDING CONSIDERATIONS

A variety of funding considerations exist but they boil down to economic and non-economic. The major economic ones are ROI and cash flow. The non-economic considerations are mainly alignment with company strategy and legal/regulatory requirements.

- Economic:
  - *ROI.* Companies usually define a minimum acceptable ROI for project funding. If a project does not meet this minimum, it is not funded. The minimum ROI criteria may vary dependent upon the type of project. For example, if a company is trying to grow its business via the introduction of new products, it may raise the minimum acceptable ROI for all other kinds of projects. This way it will funnel more money to new product work.
  - *Cash flow.* A company usually limits the amount of capital available for investment to control its cash flow.

  Non-economic:
  - *Strategy.* A company may require most projects to be in alignment with its strategy. This helps ensure that it spends money doing the things it has defined as important.
  - *Legal/regulatory.* Regulatory agencies — at the local, state, and national levels — routinely pass new laws and regulations that affect plant operation and require changes to the plants.

1.5.2 FUNDING DECISIONS

A company often funds preliminary work, such as process development or feasibility studies, from some part of its expense budget. Even when funds are already included in a budget, projects usually go through a step where company management authorizes each project.

For capital spending, most companies have detailed authorization procedures before capital funds can be committed. The level of the person authorizing a project depends on the amount of money involved. For example, a maintenance manager might approve a small maintenance project whereas the plant manager would most likely authorize the larger ones. Multimillion dollar projects are as a rule approved at a high level in a company. At least one Fortune 500 company uses a special appropriations committee to authorize all multimillion dollar projects.

When a company considers funding a project, it will first look at the project's ROI. If the project meets its minimum ROI criteria, the company will most likely compare the project to other requests for funding. Because of cash flow limitations on the amount of capital available for investment, the company will generally select the proposals having the best ROIs. It may also apportion its authorizations — a percentage to new or upgraded products, a percentage to capacity increases, a percentage to cost reductions. There is a least one major exception: health, safety, and environmental work. These projects seldom have a "hard-number" financial return and thus no quantifiable ROI. A company funds them because of its commitment to protect its employees, the community, and the environment or because of enacted laws and regulations.

To calculate the ROI of a project, one must estimate its investments, expenses, revenues, and cash flow timing (Chapter 5 explains how to calculate ROI.)

Example 1. This example illustrates the funding process for a large, new product project. For the purpose of the example, we will assume R&D and Customer Research have developed a new product, a health oil. So that the company can make this product, R&D must develop a new process and raw material specifications. After that, the Plant Design Group will design the process, purchase equipment, build the process, and start it up. Because of the project's marketplace risk and expected size of the project (capital and staffing), the company will do the funding in stages.

For the first stage, R&D estimates how much money they will spend developing the process and specifications. Although the money will come from their expense budget, both R&D and General Management approve projects of this size. Nearing the end of the development work, R&D and the Plant Design Group will study the economic feasibility of the project.

In the second stage (assuming the feasibility was positive), the Plant Design Group will ask for two sums of money. One will fund the engineering work that will lead to a full start appropriation and will provide funds for the purchase of long delivery-time equipment. (Purchasing equipment this early in a project is a way to shorten the schedule without incurring too much risk.) The other request will fund engineering, equipment buying, construction, and startup of a test market plant that can make small amounts of product for customer testing.

When engineering is far enough along to have a fairly accurate fix on the capital cost and production costs for the plant, the Plant Design Group will ask for the third stage, the full-start, appropriation. This will authorize engineering, buying, construction, and startup of all the facilities needed to make the product. (This also signals the Buying Group to begin purchase negotiations for raw and packaging materials.) For very large or very high-risk projects, this may be only a partial authorization.

If the third stage authorization was only partial, the company must make a fourth and final approval. Using four stages helps to ensure the accuracy of capital costs, marketplace data, and project economics before final approval and before signing material purchase contacts.

As a part of each funding decision, the company will review and update the estimates of the key economic data for the project — capital cost, production cost, sales volume/revenues, and cash flow timing. It will also assess whether the new product will take volume away from any other of the company's products (a negative impact on revenues). Using this data, they update the ROI estimate.

1.6 INVESTMENTS, REVENUES, EXPENSES, AND PROFIT

1.6.1 INVESTMENTS

Capital is only one of several investments needed for a project. The others are startup expense, working capital, royalties and licensing, supplier advances, and introductory marketing/advertising/sales expense. Engineers inventing processes and designing plants affect all except the marketing, advertising, and sales costs.

- *Startup expenses* are the expenses above normal due to the startup of a project.
- *Working capital* is mainly the value of inventories, accounts payable, and accounts receivable.
- *Royalties and licensing* are the fees a company pays for the use of another company's patents.
- *Supplier advances* include any prepayments to equipment, raw material, or packaging material suppliers. These advances could include items such as preliminary engineering, early equipment purchases, supplier-tooling changes, or process modifications in a supplier's plant. These are the kind of things a supplier might do so it could make a special product, raw material, or packaging material.
- *Introductory marketing/advertising/sales expenses* are the above-normal costs needed to first get customers to buy a new or upgraded product.

Chapter 3 discusses estimating capital, startup expenses, working capital, royalties/licenses, and supplier advances.

1.6.2 REVENUES, EXPENSES, AND PROFITS

The influence different kinds of projects have on revenues, expenses, and profits varies:

- *New product introduction.* The purpose of a new product is to generate new sales volume and profit. The amount of profit is a function of the amount of new volume, the selling price, and the cost to make the product. The new product may cause sales and profits of present brands to decline (cannibalization). Any loss in present profit decreases the net profit from the new introduction.
- *Product upgrade.* Product upgrades are intended to increase sales or to keep a product competitive. When the purpose is to increase sales, the profit comments are the same as for new products. If the intent is to maintain competitiveness, there may be no profit increase. In fact, if the production costs for the upgraded product are higher, profits would decline.
- *Capacity increase.* When a company cannot make enough product for Sales to sell or when it anticipates missed sales in the near future, it would increase its production capacity. Profit will increase when Sales sells more product. Costs may drop if the company realizes economies of scale from its added capacity.
- *Cost savings.* The profit increase comes from having lower expenses. Also, once costs are lower, Sales may be able to drop the selling price, thereby increasing volume and profits.

- *Maintenance*. Maintenance work is intended to keep a plant running well, so these projects usually do not affect profits.
- *Health, safety, and environmental*. Most often, profits decrease because production costs (expense) increase.

1.6.3 PROJECT RISKS AND RISK ANALYSIS

No matter the kind of project, risk is involved. Nothing is certain until the startup is complete and the plant or process is running as intended. Most engineers are clear that technical risks exist — low yields, low operating efficiencies, more emissions, equipment selection problems, and so on. However, there are other risks: higher capital cost, late startup, higher production cost, lower sales volume, lower selling price, more introductory marketing expenses, and so on. Some are more significant than the technical risk primarily because one can solve most technical problems with a reasonable amount of money or time. In my experience, market risks — those associated with sales volume, selling price, and introductory marketing expense — are the most significant.

The risks come mostly from two sources — estimating variability or error and problems with some part of the project. The kinds of things that go wrong are:

Type of Project	What Might Go Wrong
New product introduction	• Expected sales volume does not materialize. • The actual selling price is less than predicted. • Expenses — including R&D costs and production costs — are higher than anticipated. • Cannibalization of existing products is more than expected. • Investments are higher than predicted.
Product upgrade	• If sales volume is expected to increase, the risk comments are the same as for new products. • If the upgrade was done to maintain competiveness. • Sales volume drops in spite of the upgrade. • The upgraded product cannibalizes volume from other products. • Investments are higher than predicted. • Expenses are higher than estimated.
Capacity increase	• The expected volume increase does not materialize. • Investments are higher than predicted. • Expenses are higher than anticipated.
Cost savings	• The expected cost reductions do not materialize. • Investments are higher than predicted.
Maintenance	• Capital cost is higher than expected.
Health, safety, and environmental	• Investments are higher than predicted. • Production costs are higher than predicted.

These types of risks are a normal part of doing business. The two main tools used to analyze and manage risk are sensitivity analysis and decision trees. One uses sensitivity analysis to understand what could happen when actual events are different from what was estimated. One uses decision trees to better understand how unexpected events could affect the outcome of the project. Chapter 5 discusses both in more detail.

1.7 ECONOMIC WORK AFTER FUNDING

After each stage of project authorization, engineers do process development or plant design work. This is where engineers spend most of their time and this is where they can influence economics. They affect economics by integrating economic and design thinking into everything they do, such as:

- Raw material selection
- Packaging material selection
- Process/flow sheet creation
- Unit operation selection
- Selecting the number of plants and their location
- Equipment sizing and selection
- Plant and equipment layouts
- The amount of automation

The second section of this book is devoted to this topic.

1.8 CORPORATE CASH FLOW

So far, we have discussed ROI and project justification, how engineers create profit, how projects are funded, and how revenues can flow from capital investments. Now we will talk about from where the money for the project will come.

1.8.1 WHAT IS CASH FLOW?

When a company invests capital in a project, the money either can "come out of the company's checkbook" or be borrowed. When deciding whether it can afford to invest money, a company must go through the same thought process that we as individuals go through when making a similar decision. Say you are considering buying a rental condominium property for $150K. Before deciding what to do, you would have to decide several things:

- How much of a down payment can you afford?
- How much can you charge each month for rent?
- How much can you afford to spend monthly for expenses — mortgage payments, maintenance, condominium fees, insurance, utilities, and so on?
- How much tax would you have to pay — income tax on the rent and property tax on the condo?
- How much can you write off each year for depreciation?

- How would you pay your monthly expenses when the condo is unoccupied?
- Will you have to take out a loan to make the down payment and to create an expense reserve? If so, how much interest would you have to pay monthly?

In other words, you would have to sort out whether your expected cash flows — cash on hand, expenses, depreciation, taxes, and rental income — would balance so you could pay the bills when they are due and so you would make money on your investment.

Companies must do exactly the same thing when they invest their money in new products, new capacity, and similar projects. This section explains how cash flows within a company. It shows what different activities compete for the money a company makes — activities such as R&D spending to create new products, expense spending to operate environmental processes, capital spending to build capacity, dividends to shareholders, advertising spending to create interest in purchasing the company's products, interest payments on loans, and so on.

With this knowledge an engineer will able to appreciate:

- Why some projects are funded and others are not
- How important it is to save a dollar of capital or a dollar of production expense
- Why saving a dollar of production cost is more important than saving a dollar of capital
- Why money is spent on advertising rather than on a savings project

This section will convey the concept of cash flows within a company, not teach accounting. Because it is dealing with concepts, it will cover only the major cash flows. The ones not covered are usually smaller in magnitude. Figure 1.1 is a diagram of the cash flows.

Looking at the page in a company's Annual Report entitled "Consolidated Statements of Cash Flows," one sees that total cash flow for a company is comprised of cash flows from Operating Activities, Investing Activities, and Financing Activities. Bajkowski[1] describes the three cash flows as: "The operating cash flow is designed to measure a company's ability to generate cash from day-to-day operations as it provides goods and services to its customers … The investing segment … captures changes in a company's investment in the firm … The financial segment…examines how the company finances its endeavors and how it rewards its shareholders through dividend payments."

$$Cash\ flow_{total} = Cash\ flow_{operating\ activities} + Cash\ flow_{investing\ activities}$$
$$+ Cash\ flow_{financing\ activities} \qquad (1.1)$$

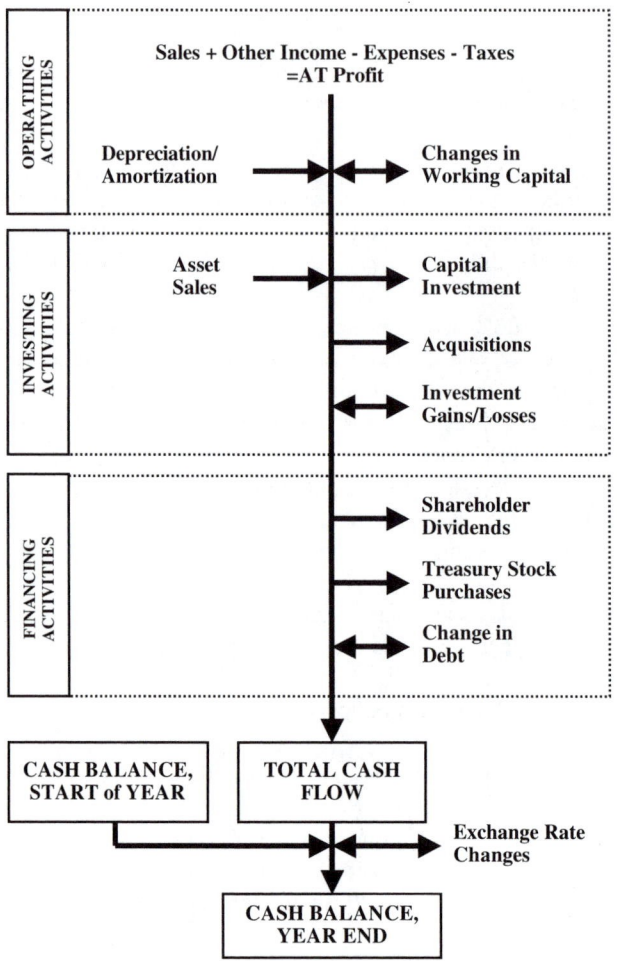

FIGURE 1.1 Simplified corporate cash flow.

1.8.2 OPERATING ACTIVITIES

Operating cash flow is the net financial result of a company's operating strategies and plans and is defined by:

$$\text{Cash flow}_{\text{operating activities}} = \text{Profit}_{\text{after tax}} + \text{Depreciation/amortization} + \text{Working capital changes}$$

1.8.2.1 After-Tax Profit

After-tax (AT) profit is what is left when a company deducts expenses and taxes from income. Income is made up of sales (the cash brought in from selling a company's products) and other miscellaneous items.

$$Profit_{AT} = Sales + Other\ Income - Expenses - Taxes \qquad (1.2)$$

Two main expenses exist:

- *Product costs.* This is what it costs to make a company's products. It is comprised of production costs and general expenses. Production costs include raw material costs, packaging material costs, manufacturing costs, warehousing costs, and product delivery costs. General expenses includes costs for research and development, marketing, sales, and corporate administration. Chapter 4 covers these in detail.
- *Interest expense.* Most companies borrow money when they do not have enough cash to fund all the investments they wish to make. These companies carry short-term and long-term loans requiring the payment of interest.

A company pays several kinds of taxes:

- *United States income taxes.* As is the case with personal taxes, companies pay taxes on their ordinary income (sales minus expenses and deductions) and on their capital gains (from the sale of assets). These taxes are payable to the federal government and in some cases to state and municipal governments.
- *Foreign taxes.* Companies doing business outside the U.S. must pay taxes to the governments where they do business. Tax laws vary from country to country. When a company pays foreign taxes, it gets a tax credit against its U.S. taxes.
- *Property taxes.* One accounts for these in Manufacturing Costs, not in the tax category.

In fiscal year 2005, the average tax rate was slightly more than 33% for companies hiring chemical engineers. Table 1.1 shows some industry and company data.

1.8.2.2 Depreciation/Amortization

Depreciation and amortization are tax-advantaged write-off methods dealing respectively with capital investments and the acquisition-associated goodwill. Both are included in expenses but neither are actual expenditures. Because of this, they are not cash flows. They must be added back to AT profit to determine cash flows. The subtracting then adding back of depreciation and amortization has the net effect of reducing taxes paid. Chapter 2 discusses depreciation and its affect on cash flow in more detail. We will not discuss goodwill further.

TABLE 1.1
Chemical Industry Tax Rates[2]

Industry	Company		2005 Tax Rate (%)
Basic Chemicals	Dow		28
	DuPont		28
	FMC		41
	Georgia Gulf		36
	Lyondell Chemical		34
	Olin Corp.		39
		Average	34
Diversified Chemicals	Air Products		28
	Eastman Chemicals		30
	Imperial Chemical		32
	Monsanto Co.		30
	PPG Industries		32
	3M Co.		30
		Average	30
Food Processing	Archer Daniels Midland		33
	Con Agra Foods, Inc.		38
	Del Monte Foods		38
	General Mills		35
	Kellogg Co.		33
	Kraft Foods		31
		Average	35
Household Products	Clorox Co.		34
	Colgate-Palmolive		33
	Kimberly-Clark		23
	Procter & Gamble		30
	Scotts Miracle-Gro		38
		Average	32
Paper/Forest Products	Georgia-Pacific		35
	International Paper		33
	Weyerhaeuser		35
		Average	35
Petroleum, Integrated	BP p.l.c.		35
	Chevron Corp.		37
	Conoco Phillips		45
	Exxon Mobil		39
	Royal Dutch Petroleum		45
		Average	40

TABLE 1.1
Chemical Industry Tax Rates[2] (continued)

Industry	Company		2005 Tax Rate (%)
Pharmaceuticals	Bristol-Myers Squibb		26
	GlaxoSmith Kline		29
	Eli Lilly & Co.		22
	Merck & Co.		28
	Pfizer, Inc.		23
	Schering-Plough		33
		Average	27
Toiletries/Cosmetics	Alberto-Culver		35
	Avon Products		31
	Estee Lauder		37
		Average	33
		Overall Average	33

1.8.2.3 Working Capital Changes

Changes in working capital will result in a cash flow either out of or into a company. Whereas working capital includes cash on hand and taxes payable, it is essentially made up of the following:

- *Inventories* — the raw materials, packaging materials, work in process and finished product owned by or in the control of a company.
- *Accounts receivable* — the money owed to a company for product sold but not yet paid for.
- *Accounts payable* — the money owed by a company for materials or services received but not yet paid for.

Decreases in accounts receivable or increases in accounts payable create cash flows into a company.

1.8.3 INVESTING ACTIVITIES

Investing cash flows are the monies a company invests in itself via capital expenditures and acquisitions plus the gains and losses from the ownership of stocks and bonds.

$$\text{Cash flow}_{investing\ activities} = \text{Asset sales} - \text{Capital investments} - \text{Acquisitions} + \text{Investing gains and losses}$$

1.8.3.1 Asset Sales

From time to time, a company may choose to sell an asset that it previously purchased (e.g., land, a building, a machine, a brand). The proceeds from the sale of the asset result in a cash inflow to the company.

1.8.3.2 Capital Investments

Capital investments can be for facilities (including land purchases) to produce new or improved products, to increase production capacity, to reduce production cost, to replace worn out facilities, or to provide plant environmental or safety upgrades. To qualify as a capital investment, the assets must have more than a one-year operational life; otherwise, they would be considered an expense.

1.8.3.3 Acquisitions

A company may decide to buy another company, a brand name, or a proprietary process. Cash paid for these acquisitions is accounted for here. (Acquisitions paid for with stock are outside the scope of this book.)

1.8.3.4 Investing Gains and Losses

Some companies have invested extensively in the equity market. Gains and losses from these activities are a part of Investing Activities. A gain results in a cash flow into a company.

1.8.4 FINANCING ACTIVITIES

Financing cash flows are associated with how a company uses cash to pay down debt or to borrow more cash (to finance operating and investing), to pay dividends to its shareholders, or to buy back some of its stock.

$$\text{Cash flow}_{\textit{financing activities}} = \text{Dividends to shareholders} - \text{Treasury stock purchases} + \text{Change in debt}$$

1.8.4.1 Dividends to Shareholders

When a company has excess earnings, it may choose to pay dividends to its shareholders or it may choose to reinvest all these earnings back in the company. When a company decides to pay a dividend, this is a cash outflow.

1.8.4.2 Treasury Stock Purchase

Sometimes a company decides to buy some of its own stock. It might do this to reduce the number of shares available in the marketplace or to provide the stock needed to award stock options. If it does, there is a cash outflow.

1.8.4.3 Change in Debt

The cash flow from operating activities is the source of company-produced money. It provides the basic source of funds for investing activities and for dividends and stock purchases. If that cash flow is inadequate, then a company will have to borrow money to finance its cash shortfall. (Another source of funds is the cash held by a company. This will be covered in the next section.) If a company borrows more cash, this is a cash inflow. Conversely, paying off debt is a cash outflow.

1.8.5 YEAR-END CASH BALANCE

The end of year cash balance held by a company is:

$$Cash\ balance_{year\ end} = Cash\ balance_{start\ of\ year} + Cash\ flow_{total}$$
$$+ Exchange\ rate\ changes \qquad (1.3)$$

1.8.5.1 Exchange Rate Changes

Companies doing business outside the U.S. must exchange monies earned in foreign countries into U.S. dollars. If a unit of foreign currency becomes worth more in dollars over the course of a year, this will increase the year-end balance. The converse is also true.

1.8.6 AN EXAMPLE ILLUSTRATES

To illustrate these cash flows, we will use data from Procter & Gamble's 2005 Annual Report.[3] All the dollar amounts are in millions.

1.8.6.1 Operating Activities Cash Flow

First, we will calculate after-tax profits using Equation 1.2:

Sales	$56,741
Other income	346
Total income	**57,087**

And determining expenses:

Production costs	$27,804
General expense	18,010
Product costs	45,814
Interest expense	834
Total expenses	**46,648**

Using this data plus the taxes paid ($3,182), the after-tax profits are:

Income	$57,087
Expenses	−46,648
Taxes	−3,182
AT Profit	**7,257**

We then add back deprecation/amortization charges and the change in working capital:

AT Profit	$7,257
Depreciation/amortization	1,884
Change in working capital	−706
Miscellaneous/minor items	287
Cash flow from Operating Activities	**8,722**

1.8.6.2 Investing Activities Cash Flow

The cash flow from investing activities is:

Proceeds from asset sales	$517
Capital investments	−2,181
Acquisitions	−572
Miscellaneous/minor items	−100
Cash flow from Investing Activities	**−2,336**

1.8.6.3 Financing Activities Cash Flow

The cash flow from financing activities is:

Change in debt (+ means more debt)	$3,111
Dividends to shareholders	−2,731
Treasury stock purchases	−5,026
Miscellaneous/minor items	−478
Cash flow from Financing Activities	**−4,168**

1.8.6.4 Total Cash Flow

Procter & Gamble's total cash flow for fiscal year 2001 was $947 million. Using Equation 1.1:

$$\text{Cash flow}_{total} = \text{Cash flow}_{operating\ activities} + \text{Cash flow}_{investing\ activities}$$
$$+ \text{Cash flow}_{financing\ activities}$$

$$= \$8,722 + (-\$2,336) + (-\$4,168) = \$2,218$$

1.8.6.5 Year-End Cash Balance

Using Equation 1.3:

Cash balance, start of year	$4,232
Total cash flow	2,218
Exchange rate changes	−61
Cash balance, year end	**6,389**

1.9 SUMMARY

Engineers play a major role in profit generation by developing products, processes, and plant designs. When done well, this yields competitively priced products and projects that meet their company's ROI criteria. To create economically viable projects, R&D and plant design engineers must decide whether:

- Is it better to spend more capital and have lower production costs?
- Is it better to spend less capital and have higher production costs?

These two questions are the crux of economic design work, both in R&D and plant design.

1.10 PROBLEMS AND EXERCISES

(Note: You can access Annual Reports and Form 10-K online.)

1. Describe your company's project authorization procedure. (If you are still a student, describe the process for a company that employed you as a co-op, intern, or summer engineer.)
2. Using the Eastman Chemicals, Pfizer, and Colgate Palmolive Annual Reports for 2004, fill in the values for all the variables in the AT profit equation.
3. What is your company's ROI criteria? If it has more than one, what are they? (If you are still a student, describe the process for a company that employed you as a co-op, intern, or summer engineer.)
4. What kinds of changes in a plant might you consider when exploring how to: (a) reduce utility usage, (b) reduce operating labor cost, (c) increase product yields, and (d) increase plant operating efficiency? Define plant operating efficiency as:

$$\frac{\text{In - spec production / wk}}{\text{theovetical production / wk}}$$

5. What kinds of things could an R&D engineer do to ensure the product s/he is working on can be competitively priced?
6. If it saved a dollar of production expenses, how could a company spend that to increase sales?
7. Using data from 2004 Annual Report or Form 10-K, compare the costs of sales, capital investment, and after-tax profit as a percentage of Sales, for the six basic chemicals companies.
8. For one of the following companies: Avon Products, Monsanto, Clorox Co., or General Mills:
   a. List the Operating Activities Cash Flow, the Investing Activities Cash Flow, and the Financing Activities Cash Flow for 2004, 2003, and 2002.
   b. What happened to inventory levels in the period 2004, 2003, and 2002?
   c. From 2002 to 2004, did the amount of dividends paid to shareholders change? By what percentage?
   d. Did debt increase or decrease from 2002 to 2004? How much?

REFERENCES

1. Bajkowski, K., Look at the corporate cash flow statement, *AAII Journal*, June 1999, 26–29.
2. *Value Line Investment Survey, Ratings and Reports*, New York: Value Line Publishing.
3. *Procter & Gamble 2005 Annual Report*, 41, 45.

2 Time Value of Money

Every day, the purchasing power or value of money changes. It changes because of inflation or because it is invested and earns money. Inflation, which devalues money, takes place when the supply of money increases faster than the availability of goods. An investment can be any number of things — a savings account that periodically pays interest, ownership of a stock that appreciates in value and pays periodic dividends, a company's investment of capital in facilities that will make a new product, increasing sales and after-tax profits.

The constant change in money's value complicates economic analyses. When analyzing options, one must compare the cash flows occurring over a period of several years. Because of changing values, cash flows from different years cannot be compared directly to one another. This chapter presents the mathematics and methods that allow one to compare these cash flows, specifically: cash flow diagrams; present worth, future worth, and annuity calculations; inflation and inflation indices; and before- and after-tax considerations.

The material can also be useful managing your personal finances.

2.1 CASH FLOW DIAGRAMS

Cash flow diagrams show how much money comes into and out of an account and when each cash flow occurs. These helpful diagrams have three parts:

- A time line divided into time periods such as years, months, or days.
- Arrows into the time line representing a cash flow (a deposit) into the account.
- Arrows out of the time line representing a cash flow (a withdrawal) out of the account.

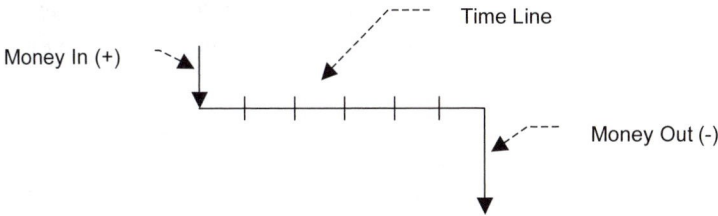

Example 1: Cash Flow Diagram. Draw a cash flow diagram for the following data:

Date	Deposits ($K)	Withdrawals ($K)
7/1/02	2000	—
7/1/06	1500	—
7/1/08	—	500
7/1/10	—	1750

Because all the cash flows occur on July 1 of various years, we will set up the time line on an annual basis. Note the cash flow arrows are drawn to scale. This is not necessary, but one may find it helpful when visualizing a problem. Also note that the cash flow diagram does not show balances in an account. It only shows cash flows into and out of the account.

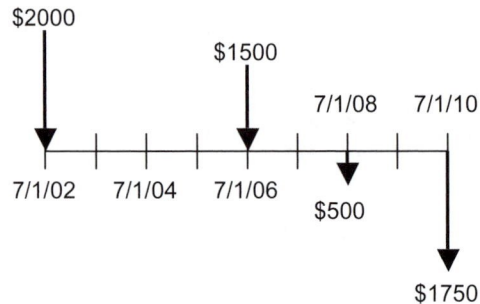

2.2 INTEREST

This is the money earned from an investment. There are two kinds — simple and compound. Almost all situations involve compound interest.

2.2.1 SIMPLE INTEREST

Simple interest means interest is paid each period on the value of the original investment. Assume you borrow $10K for 10 years at 6% per year of simple interest. The annual interest payment would be 0.06 ($10K = $600, and you would repay the principle at the end of 10 years. The cash flow diagram (not to scale) is:

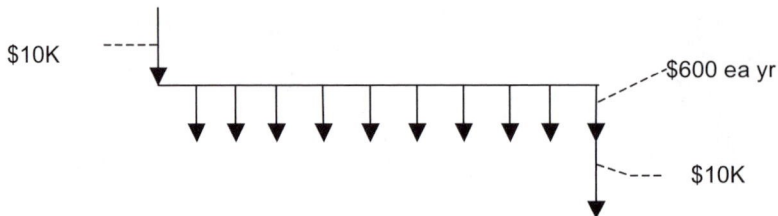

2.2.2 COMPOUND INTEREST

Compound interest is defined by an account where the interest paid is left in the account, and the next period's interest is paid on the principle plus accumulated interest. Assume you deposit $10K for 10 years at 6% interest, compounded annually. Interest is paid each year and left in your account. In the tenth year, you would then withdraw the principle ($10K) and all the accumulated and compounded interest. The cash flow diagram (not to scale) is:

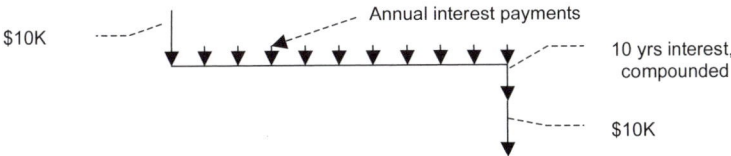

2.2.2.1 Compounding Other Than Annually

Most economic design calculations and analyses use annual compounding. There are some instances (e.g. savings accounts, money market accounts, consumer loans, and some mutual fund dividends) where interest is paid and compounded more often than once per year. In these cases, two interest rates will be quoted — the nominal or *annual percentage rate* (APR) and the effective interest rate. For more details, see the Additional Topics section of this chapter.

2.3 PRESENT WORTH, FUTURE WORTH, AND ANNUITIES

2.3.1 PRESENT AND FUTURE WORTH

Before continuing, we will define four terms:

> *Interest rate* (i): This can be compounded at any interval such as annually, semi-annually, monthly, daily, or so on.
> *Present worth* (P): This is the amount of money in an account at the present time.
> *Future worth* (F): This is the amount of money in an account at some time in the future.
> *Compounding periods* (n): This is the number of years, months, days, and so on that compounding occurs.

Assume you deposit $10K for 10 years at 6% interest, compounded annually. In the tenth year, you would then withdraw the principle ($10K) and all accumulated interest.

- *Year one*: You deposit $10K at the beginning of the year. At the end of the year you have earned interest of:

$$P\ i = \$10,000\ (0.06) = \$600$$

The interest stays in the account. The total in the account is now

$$F = P + Pi = P(1 + i) = \$10,600$$

- *Year two*: At the end of this year, you have earned interest of:

$$[P(1 + i)]i = \$10,600\ (0.06) = \$636$$

The account total is:

$$F = P(1 + i) + [P(1 + i)]i$$

Factoring out $P(1 + i)$ gives

$$P(1 + i)(1 + i) = \$11,236$$

- *Year three*: At the end of this year, you pay interest of:

$$[P(1 + i)(1 + i)]i = \$11,236\ (0.06) = \$674.16$$

$$F = [P(1 + i)(1 + i)] + [P(1 + i)(1 + i)]i$$

Factoring out $[P(1 + i)(1 + i)]$ gives

$$P(1 + i)(1 + i)(1 + i) = \$11,910.16$$

This pattern would continue until year 10. Inspection shows this is a progression; thus, the equation relating present worth (P) and future worth (F) is:

$$F = P(1 + i)^n \qquad (2.1)$$

Example 2: Future worth. If you invest $2500 at 10% interest compounded annually for 5 years, how much money will be in the account at the end of 5 years?

$$P = \$2500$$

$$i = 10\%\ \text{per year}$$

$$n = 5\ \text{years}$$

$$F = P(1 + i)^n = 2500(1 + 0.1)^5 = \$4026$$

Example 3: Present worth. You wish to have $30K in 10 years. If you can invest your money at 9.5% interest per year compounded, how much money do you have to invest today?

Rearranging Equation 2.1:

$$P = F/(1 + i)^n = 30{,}000/(1 + 0.095)^{10} = \$12{,}105$$

Equation 2.1 can also be rearranged to solve for i:

$$i = (F/P)^{1/n} - 1 \tag{2.2}$$

Example 4: Interest rate. You have $3500 to invest. What return (interest rate) must your money earn for you to double your money in 8 years?

$$i = (F/P)^{1/n} - 1 = (7000/3500)^{1/8} - 1 = 9.05$$

2.3.2 ANNUITIES

An *annuity* is a series of uniform payments or withdrawals taking place at equal time intervals. We will call this uniform payment A. Assume you buy a 10-year, 7% annuity for $10K. This annuity will pay you $1424 per year for 10 years. At the end of 10 years, it has no value. The cash flow diagram would be:

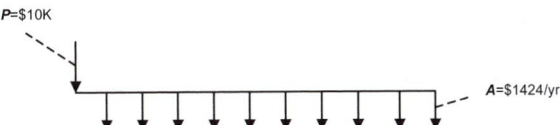

The formula for annuities is:

$$A = P\left[\frac{i(1+i)^n}{(1+i)^n - 1}\right] \tag{2.3}$$

The derivation for this may be found in texts covering Time Value in more detail, such as *Principles of Engineering Economy, 7th Edition* by Grant, Ireson, and Leavenworth.

Example 5: Annuity/present worth. You plan to buy a home and to take out a $150K, 30-year mortgage. Your interest rate will be 7.5%, compounded annually. What will your annual payment be?

Using Equation 2.3:

$$A = 150000 \left[\frac{0.075(1+0.075)^{30}}{(1+0.075)^{30} - 1} \right] = \$12701$$

Example 6: Annuity/present worth. You are considering a cost-reduction project in your department. Company policy states savings projects must return at 15% per year. You have crudely estimated the after-tax savings will be $50K per year. Your company also uses 10 years as the life of a typical cost reduction project. How much capital can you afford to invest in this project?[*]

This is just another form of an annuity problem, where:

$A = \$50,000$ per year

$i = 15\%$

$n = 10$ years

So, we rearrange Equation 2.3 to solve for *P*, the affordable amount of capital:

$$P = A \left[\frac{(1+i)^n - 1}{i(1+i)^n} \right] = 50000 \left[\frac{(1+0.15^{10} - 1}{0.15(1+0.15)^{10}} \right] = \$251000$$

2.3.3 MAKING CALCULATIONS

2.3.3.1 Formulas and Cash Flow Diagrams

By rearranging Equation 2.1 and Equation 2.3, we can develop six different equations for *P*, *F*, and *A*. These equations and their cash flow diagrams are shown in Table 2.1. Note in the two cases dealing with *A* and *F*, the last payment is made in the same period when *F* is withdrawn. This is the only situation where two events take place in the same time period.

Also note the compound interest factors such as (*F*/*P*, *i*%, *n*) appear for the first time. These will be explained later in the chapter.

2.3.3.2 Determining *n*, Counting Periods

So far, the problems have all had a specified *n*. This seldom happens in real life; rather, dates for cash flows are known, leaving one to determine the value of *n*. Because the annuity situation (two cash flows occurring in the same time period) complicates counting periods, use the following guidelines:

---

[*] This will only be an approximate estimate because it doesn't take depreciation into account. Depreciation will be covered later.

TABLE 2.1
Time Value Equations and Cash Flow Diagrams

Present & Future Worth

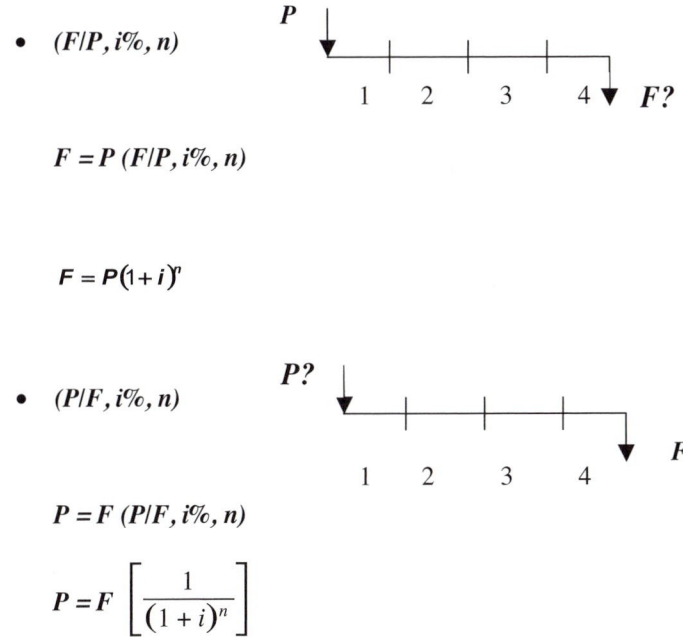

- $(F/P, i\%, n)$

$$F = P \ (F/P, i\%, n)$$

$$F = P(1+i)^n$$

- $(P/F, i\%, n)$

$$P = F \ (P/F, i\%, n)$$

$$P = F \left[\frac{1}{(1+i)^n} \right]$$

Annuities

- $(P/A, i\%, n)$

$$P = A \ (P/A, i\%, n)$$

$$P = A \left[\frac{(1+i)^n - 1}{i(1+i)^n} \right]$$

TABLE 2.1
Time Value Equations and Cash Flow Diagrams (continued)

- $(A/P, i\%, n)$

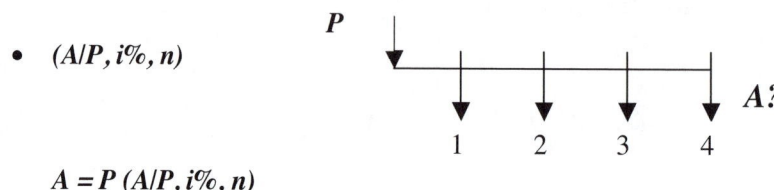

$$A = P\ (A/P, i\%, n)$$

$$A = P \left[\frac{i(1+i)^n}{(1+i)^n - 1} \right]$$

- $(A/F, i\%, n)$

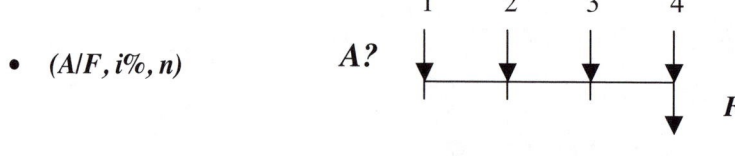

$$A = F\ (F/A, i\%, n)$$

$$A = F \left[\frac{i}{(1+i)^n - 1} \right]$$

- $(F/A, i\%, n)$

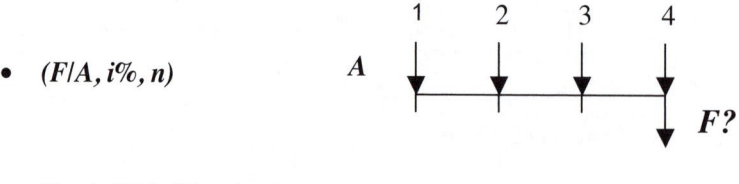

$$F = A\ (F/A, i\%, n)$$

- Payments are at the end of the period.
- For P and F calculations: The number of periods (n) equals the number of periods a sum of money earns interest.
- For annuities: The number of periods equals the number of payments into or out of the annuity. Referring to Table 2.1, note that in A and F calculations there are two cash flows in the last period, an A plus an F.

Example 7: n, *present and future worth.* On 5/1/02, you buy $15K of stock, which you assume will appreciate at 15% per year. You want to calculate how much will the stock be worth on 5/1/07. What is n for this calculation?

First, draw the cash flow diagram:

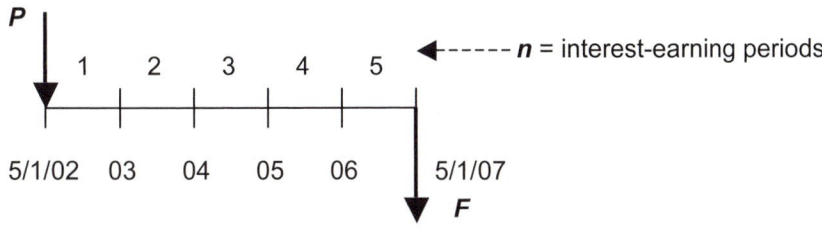

Find the number of periods in which the $15K will grow (earn interest). The first period will be the year from 5/1/02 to 5/1/03, marked "1" on the cash flow diagram. Continue counting this way through the last year — 5/1/06 to 5/1/07, which is year 5. Thus, $n = 5$.

Example 8: n, *annuitiy.* On 10/1/08, you want to have $25K saved so you can pay cash for a new car. Starting on 10/1/02, you plan to start making annual payments into an account where you expect to earn 8.5% per year after tax. You want to calculate how much will you have to deposit each year. What is n for this calculation?

First, draw the cash flow diagram:

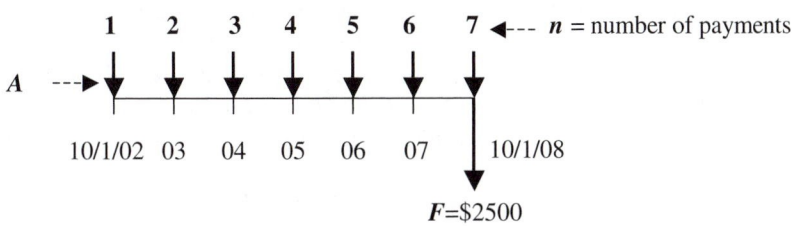

Because this is an annuity and you will make 7 payments, $n = 7$.

Example 9: n, *mixed present and future worth plus annuity.* On 4/1/01, you started making annual investments into an account that you expect will earn 10%

per year after tax. You plan to make payments until 4/1/05. You will then leave the account alone until 4/1/09. You want to calculate how much money will be in the account at that time? What is n for this problem.

Again, start by drawing the cash flow diagram:

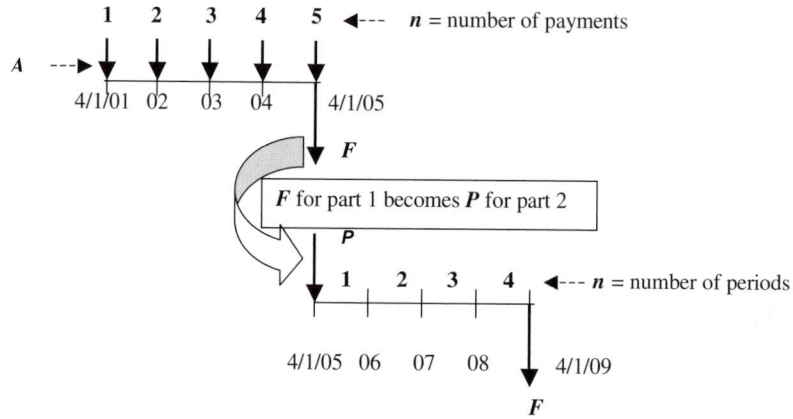

For part 1, the annuity, $n = 5$ (the number of payments); and for part 2, the present/future worth problem, $n = 4$ (the number of time periods the money in the account earns interest). The total equals 9.

2.3.3.3 Compound Interest Tables

Solving the actual time value equations is tedious at best. There are simpler ways, however — the easiest is to use a financial calculator, and the next best is to use the compound interest tables found in Appendix III. We will not address financial calculators here.

The equations in Table 2.1 can all be rearranged to create a series of factors such as F/P and P/F. The factor F/P equals $(1 + i)^n$ and so on. Using these factors, we can create a series of tables that allow for easy solution of the six time value equations. Table 2.2 shows how these tables, the compound interest tables, are built. (Appendix II contains tables for interest rates from 0.25% to 50%.)

If you know P and wish to find F, you would simply multiply the F/P factor by P. (Note the Ps cancel albgebraically, leaving F.) The complete convention for writing a factor is $(F/P, i\%, n)$. Thus the factor, its interest rate, and the number of compounding periods are all specified.

Example 10: Future worth/compound interest tables. You have $10K in an account earning 10% per year. What will be in the account in 5 years?

Because you know P and want to find F, use the F/P factor. Because i is 10%, use the 10% table. To find the correct factor, go down the F/P column to where it crosses the $n = 5$ row. The factor is 1.611.

TABLE 2.2
Compound Interest Table for 10% Interest

n	F/P (F/P, i%, n)	P/F (P/F, i%, n)	A/F (A/F, i%, n)	A/P (A/P, i%, n)	F/A (F/A, i%, n)	P/A (P/A, i%, n)	n
1	1.100	0.909	1.000	1.100	1.000	0.909	1
2	1.210	0.826	0.476	0.576	2.100	1.736	2
3	1.331	0.751	0.302	0.402	3.310	2.487	3
4	1.464	0.683	0.215	0.315	4.641	3.170	4
5	1.611	0.621	0.164	0.264	6.105	3.791	5
6	1.772	0.5645	0.1296	0.230	7.716	4.355	6
7	1.949	0.5132	0.1054	0.205	9.487	4.868	7
8	2.144	0.4665	0.0874	0.187	11.436	5.335	8
9	2.358	0.4241	0.0736	0.174	13.579	5.759	9
10	2.594	0.3855	0.0627	0.163	15.937	6.145	10
11	2.853	0.3505	0.0540	0.154	18.531	6.495	11
12	3.138	0.3186	0.0468	0.147	21.384	6.814	12
13	3.452	0.2897	0.0408	0.141	24.523	7.103	13
14	3.797	0.2633	0.0357	0.136	27.975	7.367	14
15	4.177	0.2394	0.0315	0.131	31.772	7.606	15

$$F = P(F/P, 10\%, 5) = \$10{,}000(1.611) = \$16{,}110$$

Example 11: Annuity/compound interest tables. On 10/1/08, you want to have $25K saved so you can pay cash for a new car. Starting on 10/1/02, you plan to start making annual payments into an account where you expect to earn 10% per year after tax. How much will you have to deposit each year?

$$n = 7 \text{ (from Example 8)}$$

$$F = \$25{,}000$$

$$i = 10\% \text{ per year}$$

$$A = F(A/F, 10\%, 7) = \$25{,}000(0.1054) = \$2635 \text{ per year}$$

Example 12a: Interest rate/compound interest tables. You are assessing whether an energy reduction project for your department is economically feasible. You have roughly estimated the required capital at $220K and the after-tax savings at $60K per year. Your company requires savings projects return more than 15% per year over an average project life of 10 years. Will your project meet the rate-of-return guideline?

To answer this question, you need to calculate *A/P* and then find that value in the compound interest tables in Appendix III.

$n = 10$

$A = \$60,000$ per year (after tax)

$P = \$220,000$

$(A/P, ?\%, 10) = \$60,000/\$220,000 = 0.2727$

From the tables:

$(A/P, 20\%, 10) = 0.239$

$(A/P, 25\%, 10) = 0.280$

Thus, the rate of return is somewhere between 20% and 25% (closer to 25%), and the project exceeds your company's guideline.

Example 12b: Interest rate/compound interest tables. You have \$3500 to invest. What return (interest rate) must your money earn for you to double it in 8 years? Use the compound interest tables to solve this problem.

Calculate P/F: $P/F = 3500/7000 = 0.5$

$n = 8$

In the tables, find where $(P/F, ?\%, 8) = 0.5$. From the tables:

$(P/F, 9\%, 8) = 0.5019$

$(P/F, 10\%, 8) = 0.4665$

Thus, i is between 9% and 10%. Interpolating, $i = 9.054$. (Note this is a repeat of Example 4, where the calculated i was 9.05%.)

2.4 BEFORE- AND AFTER-TAX CONSIDERATIONS

Recall from Chapter 1 that companies, as well as individuals, must pay income tax on their earnings. Because after-tax dollars are what remains in a company's account, it makes sense to use after-tax earnings or savings when making economic comparisons. The following equation relates after-tax (AT) and before-tax (BT) earnings:

$$AT\ Earnings = BT\ Earnings\ (1 - Tax\ Rate) \qquad (2.4)$$

The tax rate varies and is dependent upon the amount of taxable income. In Chapter 1, we mentioned the average tax rate (federal, state and local) for chemical industry companies was 33% in fiscal year 2005.

Example 13: AT earnings. You are working on a cost-reduction project for which the expected BT annual savings are $100K per year. If your company's effective tax rate is 33%, what is the annual AT income increase?

Use Equation 2.4:

$$AT\ Earnings = \$100K \text{ per year } (1-0.33) = \$67K \text{ per year}$$

2.4.1 DEPRECIATION

We know the value of a car decreases as it gets older because it gradually wears out and because it becomes obsolete. The same is true for the value of plants and equipment. For plants, obsolescence occurs because the product made has been superceded by something better, because the equipment has worn out or because the process is less efficient than more modern processes.

To provide tax relief for this decline in plant value, the Internal Revenue Service (IRS) in the United States permits companies to deduct a portion of the value of their capital assets from revenues each year. This decreases taxes by reducing taxable income. The depreciation charge is a book/accounting transaction; no expenditure of money is involved.

Since 1934, there have been six major changes in tax laws. Enacted in 1986, the present law specifies the use of either the straight-line or the "Modified Accelerated Cost Recovery System" methods for all equipment installed in 1987 and later. Neither method allows the use of salvage value, which is the estimated value of an asset at disposal.

2.4.1.1 Straight-Line

Th straight-line method is the simplest. It allows for a uniform amount to be deducted from revenues each year:

$$Annual\ deprecation\ writeoff = \frac{Capital\ investment}{Equipment\ life} \tag{2.5}$$

2.4.1.2 Modified Accelerated Cost Recovery System (MACRS)

The Modified Accelerated Cost Recover System (MACRS) is a combination of the declining balance and the straight-line methods of calculating depreciation. Table 2.3 shows the MACRS recovery periods, or equipment life, allowed for several different classes of equipment. Table 2.4 shows the annual depreciation write-offs allowed. Note the depreciation allowances are shifted forward as compared to the straight-line method. The table uses the "half-year" convention. This assumes the capital is installed at mid-year, so the first and last year depreciation percentages are for a half year.

Example 14: Depreciation. For a capital investment of $250K for vegetable oil product equipment, what is the depreciation write-off for Year 3?

TABLE 2.3
MACRS Recovery Periods

Type of Equipment	Recovery Period (years)
Cement manufacturing	15
Chemical and allied product manufacturing	5
Food and beverage manufacturing	3
Glass products manufacturing	7
Industrial steam and electric generation or distribution systems	15
Petroleum refining	10
Pulp and paper manufacturing	7
Rubber product manufacturing	7
Vegetable oil and vegetable oil product manufacturing	10

Source: Internal Revenue Service, How to Depreciate Property, *Publication 946*, Washington, DC, 1999.

TABLE 2.4
MACRS Annual Depreciation Percentages

	Annual Depreciation Percentage by Recovery Period				
Year	3 yrs	5 yrs	7 yrs	10 yrs	15 yrs
1	33.33	20.00	14.29	10.00	5.00
2	44.45	32.00	24.49	18.00	9.50
3	14.81	19.20	17.49	14.40	8.55
4	7.41	11.52	12.49	11.52	7.70
5		11.52	8.93	9.22	6.93
6		5.76	8.92	7.37	6.23
7			8.93	6.55	5.90
8			4.46	6.55	5.90
9				6.56	5.91
10				6.55	5.90
11				3.28	5.91
12					5.90
13					5.91
14					5.90
15					5.91
16					2.95

Source: Internal Revenue Service, How to Depreciate Property, *Publicaion 946*, Washington, DC, 1999.

For the straight-line method, assume a useful life of 10 years and use Equation 2.5. (Recall the write-off is the same every year.)

$$\text{Depreciation write-off}_{Year\ 3} = \$250K/10 = \$25K$$

For MACRS, use Table 2.3 and Table 2.4. Table 2.3 specifies a 10-year recovery period for vegetable oil product manufacturing. Going to Table 2.4 and using the column for a 10-year period, you find the depreciation percentage for Year 3 is 14.4%:

$$Depreciation\ write\text{-}off_{Year\ 3} = \$250K\ (0.144) = \$36K$$

2.4.2 AFTER-TAX CASH FLOWS

Usually, companies account for the depreciation write-off in production costs, where it looks like another expense or a negative cash flow. However, because no expenditures are involved it is not a cash flow; it is simply a deduction from expenses to reduce taxes. This requires special treatment when calculating AT cash flows. Because depreciation was subtracted form revenues for tax calculation purposes and because it is not a cash flow, it must be added back to the quantity: *Revenues – Taxes*.

$$AT\ Cash\ flow = (Revenues - Expenses)\ (1 - Tax\ rate) + Depreciation\ write\text{-}off$$

$$= (BT\ Earnings)(1 - Tax\ rate) + Depreciation\ write\text{-}off \tag{2.6}$$

Note that this equation is simply Equation 2.4 plus the depreciation write-off.

Example 15: AT cash flow. You are working on a proposal for an energy reduction project. You have estimated it will require a capital investment of \$225K and will return a net annual savings (BT) of \$105K per year.[*] Your company uses straight-line depreciation for savings project calculations. Assume a 10-year equipment life and a 35% tax rate. Use Equation 2.5 to find the annual depreciation write-off and Equation 2.6 to calculate the AT cash flow.

$$Annual\ depreciation\ write\text{-}off = \$225K/10 = \$22.5K\ per\ year$$

Net annual savings are equivalent to BT earnings. Thus:

$$AT\ Cash\ flow = \$105K\ per\ year\ (1 - 0.35) + \$22.5K = \$90.8K\ per\ year$$

2.5 INFLATION AND INDICES

Inflation is always with us. To keep track of its effects, different organizations have created and publish cost indices for a variety of items, such as construction costs, energy costs, the cost of consumer goods, the cost of different industrial products, farm costs, labor costs, and so on. The relationship between costs and the indices is:

$$\frac{Cost_{at\ time\ 2}}{Cost_{at\ time\ 1}} = \frac{Index_{at\ time\ 2}}{Index_{at\ time\ 1}} \tag{2.7}$$

---

[*] Net savings are the savings in energy costs minus the increase in costs related to the new capital investment. These increased costs are for maintenance, insurance and taxes, operating supplies, overhead and miscellaneous, and depreciation.

Two indices are of most interest to us: one for the cost of process-plant equipment and construction, and the other for the cost of chemical-type products.

2.5.1 PROCESS PLANT EQUIPMENT AND CONSTRUCTION

There are several indices tracking plant equipment and construction, such as the Chemical Engineering Plant Cost Index (CEPI), the Marshall & Swift Equipment Cost Index, and the Engineering News-Record Construction Cost Index.

Based upon chemical plant construction, the CEPI is probably the best index for inflating chemical-type equipment and plant costs. It includes the costs to design, purchase, and install chemical plant equipment and is weighted as follows:

- 61% for equipment, machinery, and supports
- 22% for construction labor
- 7% for buildings
- 10% for engineering and supervision

Chemical Engineering updates the index monthly. Appendix II lists the yearly average CEPIs from 1956 to 2005. The period 1957 to 1959 is defined as an index of 100.

2.5.2 CHEMICAL PRODUCT COSTS

On a monthly basis, the Department of Labor publishes a wide variety of indices of the selling prices of different products. The Producer Price Index (PPI) for Chemical & Allied Products is most likely the best for indexing production costs for chemical-type products. PPIs for 1984 to 2005 are also in Appendix II. The Department of Labor has defined December 1984 as an index of 100. Some authors have proposed using the Department of Labor index or the Consumer Price Index to index production costs. I believe it is more appropriate to use the PPI.

When you examine the increase in the CEPI and PPI over time, you find their inflation rates (Table 2.5) are different.

Thus when making inflation adjustments using Equation 2.7, it is wise to use an index that represents the category of costs you are dealing with.

Example 16: Indices/PPI. If the manufacturing costs for one of your company's chemical plants was $2.00 per unit in 2000, what would you expect them to be in 2004?

Because you are dealing with chemical product costs, use the PPI to adjust costs. Rearranging and substituting into Equation 2.7:

$$Cost_{2004} = Cost_{2000}(PPI_{2004}/PPI_{2000}) = \$2.00/unit \ (172.8/156.7) =$$
$2.21/unit

Example 17: Indices/CEPI. If a distillation tower cost $50K in 1994, what would a similar tower cost in 2002?

Use the CEPI because the distillation tower is a piece of equipment in a chemical-type plant. Again using Equation 2.7:

TABLE 2.5
Ten-Year Inflation Rates (%)

	1980–1980	1985–1995	1990–2000
CEPI	3.2	1.6	1.0
PPI*	no data	3.6	2.6

* Chemical and Allied Products

$$Cost_{2002} = Cost_{1994}(CEPI_{2002}/CEPI_{1994}) = \$50K(395.6/368.1) = \$53.7K$$

2.6 SUMMARY

This chapter has covered the basics of the time value of money. With this information, one can make present worth, future worth, and annuity calculations on either a BT or AT basis and can adjust costs for the effect of inflation. That will allow one to compare investment options and to make personal financial projections. It also provides the foundation for the economic comparison of design options. The methods for comparing design options will be covered in Chapter 5.

2.7 PROBLEMS AND EXERCISES

1. Compare the following investments. Assume that your initial investment is \$10K and that the rate of return will be the same each year. How much money will be in each account after 30 years? Make the calculations on an AT basis.
   a. A 30 year CD earning 5% per year BT (3.5% AT).
   b. A corporate bond fund having a yield of 7% per year BT (4.9% AT). Assume AT interest is reinvested and there is no capital appreciation over the 30 years.
   c. An equity mutual fund that appreciates at 8% per year and pays 2% per year dividends (1.4% AT). Assume AT dividends are reinvested. Capital gains tax of 15% will have to be paid on the difference between the original investment of \$10K and the appreciated value of the fund.
2. You want to have \$30K saved by 8/1/10 so you can make a down payment on a house. If you invest the same amount of money each year starting on 8/1/05, how much will you have to invest each year? Assume you will earn 8% per year AT. Draw the cash flow diagram.
3. If a chemical plant cost \$18M in mid-1994, what would you expect it to cost at the end of 2004?
4. What was the rate of chemical product price inflation from 1985 to 1994?
5. You have just purchased a 30-year, 6% per year annuity for \$250K. What will it pay you each year? (Payments will begin in one year. At the end of 30 years, there will be nothing left in the account.)

6. In 2004, your grandparents gave you an annuity that will pay $1500 per year for 20 years. You will get the first payment on 7/1/09. You have offered to sell the annuity on 7/1/05. Assuming the interest rate today is 5% per year, for how much would you be able to sell the annuity? Include a cash flow diagram as part of your solution.

7. In July 2005 when you start your new job, you plan to begin making annual payments of $3000 into an IRA. Assuming your account earns 10% per year, how much will be in your account when you are 55? (Assume you will be 22 when you begin contributing to the IRA.)

8. What was the annual rate of chemical plant inflation from 1990 to 2003?

9. You invest $6000 on 1/1/06 and $8000 on 1/1/09. Your investment earns 9.5% per year (AT). How much will be in your account on 1/1/35?

10. If you wish to have $50K in your brokerage account on 1/1/10, how much will you have to invest on 7/1/05? Assume your investment will grow at 7% per year, AT.

11. If you invest $5K and it grows to $12,050 in 13 years, what was the annual growth rate (interest rate)? Use the compund interest tables to solve.

12. You wish to have $2.5M (in today's dollars) available when you retire in 35 years. Assuming this money will be in tax-sheltered accounts such as IRAs or 401Ks and will grow at 10% per year, how much money do you have to put in these accounts each year?

13. Develop compound interest factors (to five decimal places) for:
    a. $(P/F, 13.8\%, 10)$
    b. $(A/P, 7.35\%, 9.5)$

2.8 ADDITIONAL TOPIC

2.8.1 COMPOUNDING OTHER THAN ANNUALLY

So far, the discussions and problems have focused on annual compounding. It is important to understand that all the formulas in Table 2.1 and the compound interest tables (Appendix III) can apply to any compounding frequency — annual, quarterly, monthly, and so on. In all cases, i and n must be based on the same compounding period. For example, if the compounding period were monthly, then i would be expressed in percent per month and n would be the number of months in which compounding occurs.

Converting annual interest rates to some other compounding frequency is easy; simply divide the annual rate by the number of compounding periods in a year. To keep terms straight, we slightly modify our terms:

i_a = The annual compounding frequency = the annual percentage rate (APR)
i_c = The interest rate at which compounding is done. In the case of annual compounding, $i_c = i_a$.
m = The number of compounding periods per year

Then,

$$i_c = \frac{i_a}{m} \qquad\qquad (\text{AT2.1})$$

Example AT-1. What is the compounding interest rate when you are quoted a rate of 6% per year, compounded monthly?
Using Equation AT 2.1:

$$i_a = 6\% \text{ per year}$$

$$m = 12 \text{ months/year}$$

$$i_c = \frac{i_a}{m} = \frac{6\%/\text{yr}}{12\text{mo}/\text{yr}} = 0.5\%/\text{mo}$$

Whenever compounding is done at a frequency greater than once per year, the effective annualized interest rate will be greater than i_a. We will call this the *effective interest rate*, or i_e; it is the interest rate per year after compounding:

$$i_e = (1 + i_c)^m - 1 \qquad\qquad (\text{AT2.2})$$

Example AT-2. What is the effective interest rate when the APR is 6% per annum compounded monthly and daily?
Monthly compounding:
From Example AT-1, $i_c = 0.5\%$ and $m = 12$ months/year.
Using Equation AT2.1:

$$i_e = (1 + 0.5)^{12} - 1 = 0.0617 \text{ or } 6.17\%/\text{yr}$$

Daily compounding:
Using Equation AT2.1 to calculate i_c:

$$i_c = \frac{i_a}{m} = \frac{5.5\%/\text{yr}}{12\text{mo}/\text{yr}} = 0.458\%/\text{mo}$$

Using Equation AT2.2:

$$i_e = (1 + 0.00016438)^{365} - 1 = 0.0618 \text{ or } 6.18\%/\text{yr}$$

There are two techniques one can use to solve present worth, future worth, and annuity problems when compounding is other than yearly. Referring to the equations in Table 2.1, note the term $(1 + i)^n$ is found in all the equations. The first technique uses i_c and m. In this case, the term $(1 + i)^n$ becomes $(1 + i_c)^{mn}$. Note that the compounding interest rate (i_c) and the compounding frequency (mn) have the same basis. Thus, the future worth equation becomes:

$$F = P(1+i_c)^{mn}$$

The second technique would be to calculate i_e, which has an annual compounding frequency, and use it in your calculations. The future worth equation then becomes:

$$F = P(1 + i_e)^n$$

Example AT-3. You are considering buying a house. You will need to have a $150K mortgage. For a 30-year fixed-rate mortgage, what will the monthly payments be when the interest rate is 5.5% per year, compounded monthly?

$$P = \$150K$$

$$m = 12 \text{ months/year}$$

$$i_a = 5.5\% \text{ per year}$$

$$i_c = \frac{i_a}{m} = \frac{5.5\%/\text{yr}}{12 \text{ mo}/\text{yr}} = 0.458\%/\text{mo}$$

The number of compounding periods is $mn = (12 \text{ months/year})(30 \text{ years}) = 360$ months. Solving for A, using the second annuity equation in Table 2.1:

$$A = P\left[\frac{i(1+i)^n}{(1+i)^n - 1}\right] = P\left[\frac{i_c(1+i_c)^{mn}}{(1+i_c)^{mn} - 1}\right] = \$150K\left[\frac{0.00458(1+0.00458)^{360}}{(1+0.00458)^{360} - 1}\right] = \$851.31/\text{mo}$$

Example AT-4. You have just bought furniture for your new home. The price, including tax, was $8500. No down payment was required. You have 3 years to pay for the furniture, and your monthly payments will be $282.32. You know your loan is compounded monthly. Find the following:

• The number of compounding periods
• The compounding interest rate
• The APR
• The effective interest rate

The number of compounding periods = $mn = (12 \text{ months/year})(3 \text{ years}) = 36$ months.

The compounding interest rate can be found by calculating $P/A = \$8500/\$282.32 = 30.108$ and finding the compound interest table where $P/A = 30.108$ and the number of compounding periods is 36. This occurs at an i of 1%; thus, $i_c = 1\%$ per month.

The APR is the "per annum" rate, or i_a. Rearranging Equation AT2.1:

$$APR = i_a = mi_c = (12 \text{ months/year})(1\% \text{ per month}) = 12\% \text{ per year}$$

The effective interest rate is found using Equation AT2.2:

$$r_e = (1 + 0.01)^{12} \; 1 = 0.127 \text{ or } 12.7\% \text{ per year}$$

2.9 PROBLEMS AND EXERCISES

1. Joe's Used Car Lot will sell you a car for $5K. No down payment is needed. Your low monthly payments will be only $145.00 for 42 months. What APR and effective interest rates will you pay?
2. You have been quoted an APR of 8.37% per year, compounded monthly, for a loan. What is the effective interest rate?
3. What is the compounding interest rate when the APR is 8.7% per year, compounded quarterly?
4. You have just purchased appliances for your house. The total cost was $4834.75, which includes taxes. After a down payment, you still owe $4200. The dealer will finance for this $2\frac{1}{2}$ years at a rate of 8.5% per annum, compounded monthly. What will your monthly payments be?
5. You have purchased a $5K, 5-year certificate of deposit. It pays 3.1% per year, compounded daily. When the certificate matures in 5 years, how much money will you receive?

3 Estimating Investments

This chapter discusses how to estimate the investments in a project — capital, startup expense, and working capital. Because this book focuses on the earlier stages of a project, a time when few design details are known, it will present only order-of-magnitude (OOM) and study methods. These types of estimates are rather inaccurate.[*] In spite of that, they are satisfactory in the early part of a design when the engineers are making broad decisions and evaluating options. A portion of this chapter includes an update of an article written by T.R. Brown in 2000 and published in *Hydrocarbon Processing*.[1]

3.1 CAPITAL COSTS DEFINED

Capital costs include a broad range of cost categories. Whereas companies group cost components in their own way, the total list of components will be the same. It is important to understand these to ensure a complete estimate. The main categories used in this book are: Equipment, Yard/Site Work, Buildings, Equipment Related, Engineering, Construction Overhead, and Contingency. These are detailed below.[2-4]

Estimate Categories	Subcategories	Typical Equipment
Equipment	Processing	Centrifuges
		Columns
		Compressors
		Cyclones/dust filters
		Dryers
		Evaporators
		Filters
		Furnaces
		Heat exchangers
		Pumps
		Reactors
		Tanks/pressure vessels
	Packaging	Cappers
		Case packers
		Conveyors
		Fillers
		Labelers
		Palletizers
		Shrink wrappers
		Uncasers

---

[*] In his book *Process Engineering Economics*, Couper defines the accuracy for OOM estimates as –30% to +50% and for study estimates as 15% to +30%.

Estimate Categories	Subcategories	Typical Equipment
	Utilities	Boilers
		Compressors
		Cooling towers
		Electric generators
		Refrigeration systems
		Substations
		Water systems
	Environmental	Cyclones/dust filters
		Effluent filters
		Fume containment systems
		Gas scrubbers
		Incinerators
		Precipitators
		Sewage treatment systems
		Spill containment systems
		Waste compactors
Yard/Site Work	Construction labor	
	Fire protection equipment	
	Grading	
	Landscaping	
	Parking	
	Railroad tracks	
	Roads	
	Security systems	
	Sewers	
	Site preparation	
	Yard lighting	
Buildings	Construction labor	
	Air conditioning	
	Control rooms	
	Employee facilities (lockers, cafeteria, restrooms, etc.)	
	Laboratories	
	Lighting	
	Maintenance facilities/shop equipment	
	Office buildings/furnishings	
	Process & packaging buildings	
	Telephone systems	
	Warehouses/loading equipment	
Equipment related	Construction labor	
	Electrical	
	Equipment installation	
	Insulation	
	Foundations/supports	
	Piping/chutes/ducts	
	Safety devices	
Instrumentation/controls		
Engineering	Company engineering effort	
	Engineering contractor costs, including overhead & profit	

Estimate Categories	Subcategories	Typical Equipment
Construction Overhead	Benefits	
	Contractor profit	
	Construction planning/field engineering	
	Equipment/tool rental	
	Field supervision	
	Temporary facilities	
Contingency	An allowance to account for	
	uncertainties and unknowns	

3.2 ESTIMATING CAPITAL

Because the design is not well-defined when doing feasibility studies or option analyses, estimating methods are quite simple, and of necessity are based on minimal scope definition.

3.2.1 INFLATION ADJUSTMENTS

The data on which an estimate is based is often quoted at a date different from the date of the estimate. Let's say you want to know the cost of a heat exchanger in today's dollars but your data is based on 2004 costs. When this occurs, costs are adjusted for inflation using the Chemical Engineering Plant Cost Index (CEPI).

Recall that the relationship between costs and the indices is given by Equation 2.7:

$$\frac{Cost_{at\ time\ 2}}{Cost_{at\ time\ 1}} = \frac{Index_{at\ time\ 2}}{Index_{at\ time\ 1}}$$

A good guideline is the one proposed by Vatavuk: Don't escalate costs over a period of more than five years.[5] Where this is a useful rule, there may be times it is fitting to escalate over a longer period. In those cases, accuracy will most likely suffer.

Example 1. If the cost of a plant were $30M in 1996, what would the cost of that same plant be in 2001?

The CEPI was 381.7 in 1996 and 394.3 in 2001. Rearranging Equation 2.7:

$$Cost_{2001} = Cost_{1996} \left(\frac{CEPI_{2001}}{CEPI_{1996}} \right)$$

$$= \$30M \left(\frac{394.3}{381.7} \right)$$

$$= \$31.0M$$

3.2.2 ORDER-OF-MAGNITUDE ESTIMATES (OOM)

These are the simplest and least accurate of all the estimates. They are most often based upon a slightly developed design concept. At times, only the concept or an idea exists. For example, an engineer might be asked the question, "What would be the cost of a 150,000 ton per year cyclohexane plant on the West Coast that's like the one in Baton Rouge?" In this case, no engineering has been done — only the question about building another plant exists. There are two common OOM methods.

3.2.2.1 Ratioing by Capacity

This is one of the most useful relationships in cost engineering. Normally, when the cost of different capacity items having the same design features is plotted on log-log paper versus capacity, the line is straight. This leads to:

$$\frac{\text{Cost}_{\text{size 2}}}{\text{Cost}_{\text{size 1}}} = \left(\frac{\text{Capacity}_{\text{size 2}}}{\text{Capacity}_{\text{size 1}}}\right)^{n} \tag{3.1}$$

where n is the size exponent.

This equation is valid for both equipment purchase and plant and process costs. Size exponents are a function of the type of plant, process, or equipment. Garrett, in his book *Chemical Engineering Economics*, and Guthrie, in *Process Plant Estimating, Evaluation, and Control*, include log-log plots for several hundred types of chemical plants.[6,7] The cost data for these charts is now 20 to 45 years old, so they would have to be used with a great degree of caution; but the size exponents will be of continuing value. Remer and Chai also published over 600 size exponents for a variety of chemical plants.[8] The average exponent for these plants was 0.67. This value compares well to previously calculated averages (Guthrie and Chilton).[9,10] In 1970, Chase proposed a method to calculate plant and process size exponents from the exponents of the different equipment classes used in the plant.[11] When a plant size exponent is unknown, one should use the average, 0.67. For equipment, use 0.6.

Example 2. If a 100-ton per year plant cost $32.9M, what would a 150-ton per year plant cost?

Because the size exponent is not known, use the average, 0.67. Rearranging Equation 3.1:

$$\text{Cost}_{150 \text{ TPY}} = \text{Cost}_{100 \text{ TPY}} \left(\frac{150 \text{ TPY}}{100 \text{ TPY}}\right)^{0.67}$$

$$= \$32.9\text{M} \left(\frac{150}{100}\right)^{0.67}$$

$$= \$43.2\text{M}$$

Example 3. You have recent quotes for two fixed tube sheet heat exchangers.

Area	Cost
500 ft²	$9,450
2000 ft²	$20,000

Using this data, estimate the cost of a 900-ft^2 fixed tube sheet heat exchanger.

There are two ways to answer this question. First, you can calculate the size exponent for your type of heat exchanger and solve for the new price using Equation 3.1. Second, you can plot the costs you know as a straight line on log-log paper. The cost of the 900 ft^2 heat exchanger will be on that line.

Size exponent method. Rearrange Equation 3.1 to solve for the exponent:

$$\frac{Cost_{size\ 2}}{Cost_{size\ 1}} = \left(\frac{Capacity_{size\ 2}}{Capacity_{size\ 1}}\right)^n$$

Take logarithms of both sides and rearrange:

$$\log\left(\frac{Cost_{size\ 2}}{Cost_{size\ 1}}\right) = n \log\left(\frac{Capacity_{size\ 2}}{Capacity_{size\ 1}}\right)$$

$$n = \left[\frac{\log\left(\dfrac{Cost_{size\ 2}}{Cost_{size\ 1}}\right)}{\log\dfrac{Capacity_{size\ 2}}{Capacity_{size\ 1}}}\right]$$

$$n = \left[\frac{\log\left(\dfrac{20000}{9450}\right)}{\log\left(\dfrac{2000}{500}\right)}\right] = 0.54$$

Rearranging Equation 3.1 and using this exponent:

$$Cost_{900} = 9450\left(\frac{900}{500}\right)^{0.54} = \$13K$$

Figure 3.1 shows the graphical solution for the problem.

3.2.2.2 Scaling by Capital/Unit of Capacity or by Capital/Unit of Sales

These are often used but are flawed methods. They are flawed because the ratios, capital/unit of capacity or sales, are not constants. Rather, the capital cost in each ratio is a variable, dependent on the capacity of the plant. To use these ratios accurately, one would have to know what the base capacity or sales volume was and then develop a ratio of the capital using a size exponent.

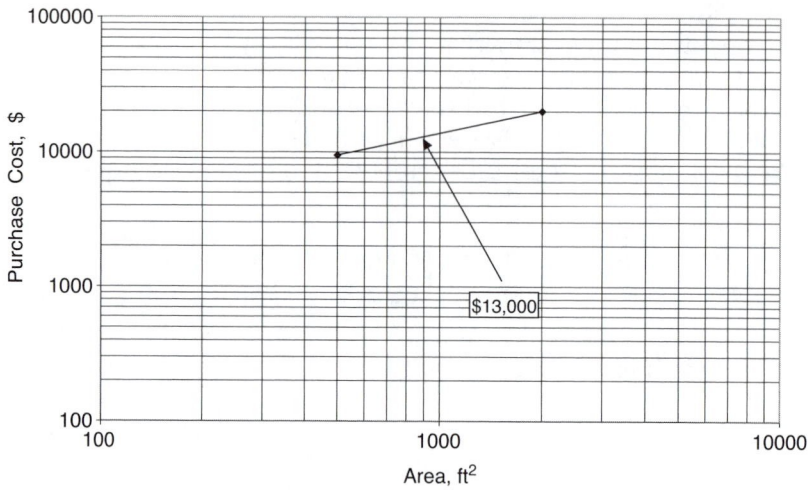

FIGURE 3.1 Fixed tube sheet exchanger costs.

3.2.2.3 Other

Viola proposes a method based on the number of major operating steps in a process.[12] To use this method, one must create a graph relating cost and the complexity of a plant. The graph plots costs and complexities of plants or processes that are fairly similar. Viola's method adjusts the costs for pressure and materials of construction.

Several other methods are described in the 7th edition of *Perry's Chemical Engineering Handbook*.[13] As well, Allen and Page compare some of these methods and propose an improved system.[14] These procedures are based on a number of limiting assumptions. Whereas these methods may be of value in a pre-design situation, I believe a better approach would be to do a small amount of design and to use the "study estimate" technique described in the next section.

3.2.3 STUDY ESTIMATES

Although not much design information is known when doing a study estimate, it is quite a bit more than exists when doing an OOM estimate. At a minimum, one should have block flow diagrams, first draft material/energy balances, plus an environmental risk review. These are used to develop rough sizes and specifications for the major process, packaging, utility, and environmental equipment.

The study estimate method is based upon using these sizes and specifications to estimate equipment purchase prices. The purchase costs are then multiplied by several factors to account for all the other components in the estimate.

3.2.3.1 The Lang Method

H.J. Lang did the original work on this method of estimating in the 1940s. He published three articles in 1947 and 1948.[15] His studies included estimates of 14 processing plants built mostly of carbon steel.

His method involves summing the equipment purchase costs and multiplying it by a factor:

$$Capital\ cost = \Sigma(Equipment\ purchase\ cost) * Lang\ factor$$

Lang's purchase costs included those for raw material handling and storage equipment, processing equipment, finished product handling and storage equipment, and instruments.

He developed factors for three types of plants: solids processing, solids and fluids processing, and fluids processing. I have updated Lang's original factors for use with study estimates, making four changes:

- Included the cost of services (utilities) equipment in the equipment purchase cost. This will improve accuracy because services costs are so variable; therefore, services costs were removed from the factors. (In Lang's original data, they were 9.9% of the cost of a plant.)
- Included the cost of environmental equipment in the equipment purchase cost. Today, environmental systems are often a large part of the cost of a plant. In the 1940s, almost nothing was spent in this area. This does not change the factors.
- Because instrument scope is not known for OOM and study estimates, one cannot estimate the purchase cost of instruments. How to account for the cost of instrumentation is covered later.
- Because the cost of buildings will differ quite a bit from project to project, I have used data from Peters and Timmerhaus to modify the Lang factors for different buildings situations.[16] Note that for each type of plant, there are three levels of building scope.

TABLE 3.1
Lang Factors

Type of Plant	Original Factors	Modified Study Estimate Factors		
		New Plant/ New Site	New Unit, Existing Site	Expansion, Existing Site
Solids processing	3.10	3.2	2.7	2.6
Solids and fluids processing	3.63	3.5	3.3	3.1
Fluids processing	4.74	4.5	4.2	4.1

TABLE 3.2
Hand Factors (CS Equipment)

Equipment Type	Factor
Fractionating columns	4
Pressure vessels/tanks	4
Heat exchangers	3.5*
Fired heaters	2
Pumps	4
Compressors	2.5
Instruments	4
Miscellaneous equipment	2.5

* The Hand factor for plate and frame heat exchangers with SS plates is also 3.5.

Source: Brown, T.R., Capital Cost Estimating, *Hydrocarbon Processing*, October 2000, 93–100. With permission.

If the plant is built of materials other than carbon steel, the factors also have to be adjusted for this change. This will be addressed shortly.

3.2.3.2 The Hand Method

In 1958, W.E. Hand published an article in the *Petroleum Refiner* in which he modified Lang's method.[17] His method involves multiplying the purchase cost of each piece of equipment by its own factor and summing these to arrive at the total capital cost. He developed factors for eight types of carbon steel equipment.

As did Lang, Hand includes instrumentation as a type of equipment, so instrumentation costs are not included in his other factors.

3.2.3.3 Adjusting Lang and Hand Factors for Materials Other Than Carbon Steel

Lang and Hand's factors were developed for carbon steel equipment. When other materials are used, the factors have to be adjusted using a *materials factor* (F_m). Clerk published the basis for the materials factor in 1963.[18] The adjustment is needed because the yard/site, buildings, engineering, construction overhead, and some of the equipment-related costs are independent of metallurgy. Only piping/chutes/ducts, some instruments, some safety devices, and some contingency costs increase when metallurgy is upgraded.

The materials factor (F_m) is determined by using the chart in Figure 3.2. First, calculate the material cost ratio (the cost of the alloy divided by the cost of carbon steel). Then go vertically from this value to the curve and read F_m on the y-axis.

The material cost ratio and F_m are fairly easy to develop for a single piece of equipment, as would be done when using the Hand factor. However, when one uses

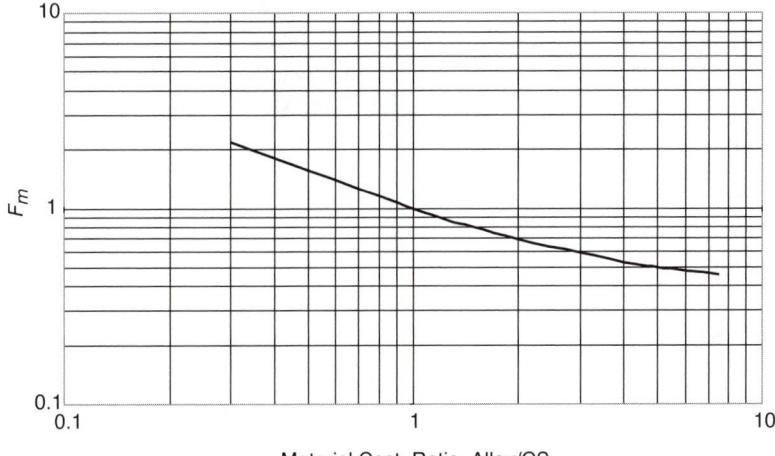

FIGURE 3.2 F_m Material adjustment factor.

the Lang method, estimating the material cost ratio for an entire plant is more complicated. This is illustrated in Section 3.2.5.1.

Example 4. The Hand factor for heat exchangers is 3.5. What would the adjusted Hand factor be for a U-tube heat exchanger with titanium tubes and a CS shell? Assume a material cost ratio is 2.6.

Enter Figure 3.2 with a material cost ratio of 2.6. Go vertically to the curve and find $F_m = 0.63$ on the y-axis. The adjusted Hand Factor is then:

$$(Hand\ factor)\ F_m = 3.5\ (0.63) = 2.2$$

3.2.3.4   Other Factors

To complete the estimate, one must factor in instrumentation costs, building costs (only for Hand factor estimates), and adjustments for non-U.S. construction.

3.2.3.4.1   Instrumentation Factors

Instrumentation factors are a function of the amount of instrumentation in a process or plant. The table below gives factors for three levels of control. These are based upon data from Aries and Newton, Garrett, Guthrie, Jelen and Black, Liptak, plus Rodriguez and Coronel adjusted for recent (1997–1998) industrial experience and tempered by my judgment.[19–25]

TABLE 3.3
F_i Values

Local controls	1.15
A typical chemical or processing plant	1.35
Extensive controls, central control, computerization	1.55

3.2.3.4.2 Building Factors (Only for Use with Hand Factors)

Hand factors include almost no building costs, so for any Hand factor estimate money must be added to correctly account for building costs. This is done using a *building factor* (F_b). The following building factors were developed from data in Peters and Timmerhaus.[26]

TABLE 3.4
F_b Values

Type of Plant	New Plant/ New Site	New Unit at an Existing Site	Expansion at an Existing Site
Solids processing	1.68	1.25	1.15
Solids and fluids processing	1.47	1.29	1.07
Fluids processing	1.45	1.11	1.06

3.2.3.4.3 Place Factors

There are times when one wishes to know the cost of a process or plant outside the U.S., and no one who understands costs and construction in the other country is around. When that is the case, one can develop the plant cost assuming it is in the U.S. and then adjust this cost with a *place factor*. By definition, the place factor for the U.S. is 1.0. In 1996, McConville published factors for many different countries.[27] Some of these are shown in Table 3.5. Because economic conditions in countries relative to the U.S. are always changing, place factors will not be reliable for more than one or two years.

TABLE 3.5
1996 F_p Values

Country	Factor	Country	Factor
Brazil	0.9	Japan	1.15
Canada	1.16	Malaysia	0.9
China	0.97	Mexico	0.93
Czech Republic	1.15	Saudi Arabia	1.3
France	0.96	South Korea	0.93
Germany	1.05	United Kingdom	1.14

3.2.3.5 The Methods in Final Form

When the instrument (F_i), building (F_b), and place (F_p) factors are added to the Lang and Hand factor methods, the equations become:

The Lang equation:

$$Capital\ cost = \Sigma(Equipment\ purchase\ cost) \times Lang\ factor \times F_m \times F_i \times F_p \qquad (3.2)$$

The Hand equation:

$$Capital\ cost = \Sigma[Equipment\ purchase\ cost \times (Hand\ factor \times F_m)] \times F_i \times F_b \times F_p \quad (3.3)$$

Figure 3.3 shows how the costs for Equation 3.3 relate to each other.

3.2.3.6 Creating One's Own Factors

Over the years, a number of people have added to Hand's work. Most of the factors, but not all, are similar to his — say ±20% from his original. One may want to create their own factors for any number of reasons — to ensure they are based on current data, to have factors for specific types of Hand's "miscellaneous equipment," to include instrumentation in the factor, and so on. Below are two examples of heat exchanger factors.

Heat Exchanger Factor

Heat exchanger cost	$10,000
Piping, foundation/supports, insulation, safety devices	6,100
Installation, direct labor	5,900
Freight, insurance and taxes	1,300
Construction overhead	4,100
Contractor engineering expense	2,400
Contingency	4,400
Contractor fee	900
Total module capital	35,100

Factor = 35,100/10,000 ~ 3.5

Source data: Adapted from Ulrich[28] and Guthrie,[29] but without instrumentation.

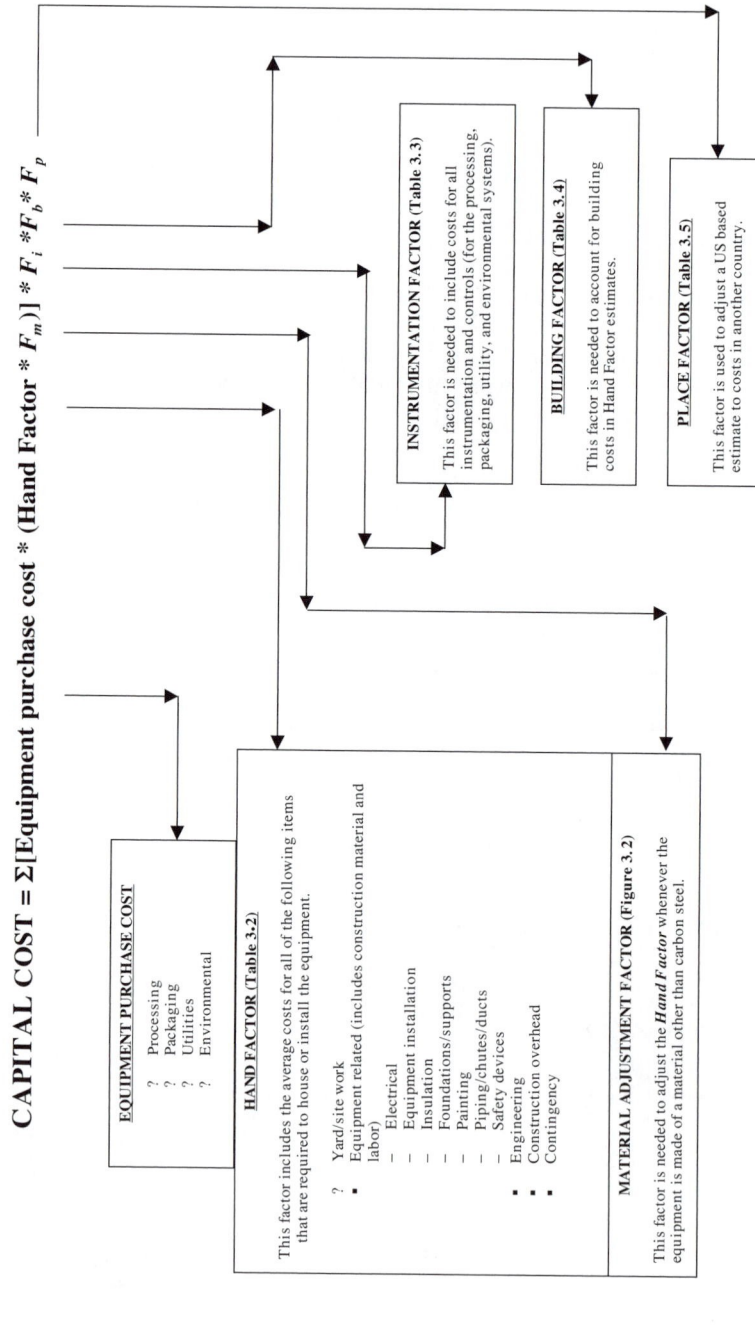

FIGURE 3.3 Hand factors and estimate structure.

Heat Exchanger Factor

U-tube exchanger, 400 ft^2, carbon steel, 150 psig	$2100
Installation	500
Supports	200
Piping and safety device, 100 ft of 2-in schedule 40, carbon steel pipe	1300
Construction field office and fee	1600
Engineering, corporate and contractor	1100
Contingency	1000
Total	7800

Factor = 7800/2100 ~ 3.7

Source: Brown[30]

For reference, the Hand factor for heat exchangers is 3.5.

3.2.3.7 Other Factor Methods

There are at least two other factor methods designed for use with study estimates — Garrett's and Wroth's.[31,32] Brown compared the Garrett factor method with the Hand factor method and found the results to be within 3.5% of each other.[33] Couper compared the Wroth factor method to the Hand factor method and found it produced a cost 25% higher.[34] A closer look at Wroth's factors shows the difference is caused by larger factors for a few types of equipment — heat exchangers (4.8 versus 3.5), pumps and motors (7.7 versus 4), and miscellaneous (4 versus 2.5). Whereas these larger factors may have worked for Wroth's company, I question their general validity. Additionally, they and the Garrett method add considerable complexity to the study estimating process. Thus, I see little reason to use other than Hand or Lang factors for study estimates. (You will find a comparison of Lang and Hand factor results later in this chapter.)

3.2.4 EQUIPMENT PURCHASE COST ESTIMATING

At the heart of both methods is estimating equipment purchase costs. All the sources of price data — budget quotes, company or personal data, purchased data and software, and published data — have different levels of accuracy because economic conditions are constantly changing. For example, when the construction industry is booming, fabrication shops run closer to capacity and costs are higher. The converse is true when there is little construction activity. On a smaller scale, when an individual shop is close to its capacity, its costs will be higher regardless of the situation in the industry.

3.2.4.1   Budget Quotes

The most accurate method of pricing equipment for a study estimate is to get a budget quotation from a vendor. By necessity, the quote will be based upon rough equipment specifications. It is important to ask for a budget quote, not a firm price bid, or the vendor will of necessity ask for more detail to ensure an accurate quote. Even when asked for a budget figure, they may balk out of fear they will be held to a quote that is less precise than a firm price bid. Asking two or three vendors for a quote improves accuracy. As it is based on the present economic situation and on your company's unique requirements, this method is the most exact.

Getting budget quotes is time consuming, so it should be used only for the more costly pieces of equipment. It's okay to use less precise methods for the less expensive items.

3.2.4.2   Company or Personal Data

The next most accurate method is to use data you or your company have collected. Although it is usually not current data, this method does reflect the unique equipment requirements of your company. When using this kind of data, make sure your operating temperatures and pressures and metallurgy are the same as in your data base. If not, you will need to make adjustments to the price. The data in Appendix IV contains data that will help when making these types of adjustments. Because this data is almost always historical, it will have to be time indexed using Equation 2.7.

You may also have to size ratio the data using Equation 3.1 so that it will match the capacity of your equipment. See Appendix IV for size exponent data.

3.2.4.3   Purchased Data and Software

The accuracy is this type of data is determined by the source of the base data. Some purchased data and software is based upon vender quotations and upon actual prices and bids obtained by engineering contractors and companies that own processing plants. As a result, the quality of this data is equivalent to company and personal data. Some purchased data and software, however, is based upon published data, and its accuracy is equal to published data. Some are a blend of purchased data and of vender quotations and actual prices or bids. Because one would not know which data came from what source, it would be wise to treat it all as published data.

Before using purchased data or software, check on the source of the base data.

3.2.4.4   Published Data

Published data is by far the least accurate. It is never economically current and does not take into account a company's unique requirements. Because conditions were different when different databases were created, they seldom agree with each other.

However, there are times when this kind of data should be used — when no other data is available, when speed is of the essence, or when accuracy is not critical (e.g., for minor, less expensive equipment). This data will always have to be time indexed, may have to be size ratioed, and may have to be adjusted for materials, pressure, and other factors.

There is another valuable use for published data — adjusting historical price data. To illustrate, turn to the Heat Exchanger page in Appendix IV. Note the factors in the bottom half of the page. These are used to adjust purchase prices when the exchanger in question is different from the basis of the graph. For shell and tube units, the basis is a carbon steel U-tube rated at 150 psig. If the exchanger being priced was a fixed tube sheet, having a carbon steel shell and stainless steel tubes, and was rated at 400 psig, one would adjust the graph price by multiplying it by 1.05 (fixed tube sheet), 1.16 (400 psig), and 1.7 (SS tubes/CS shell). Of course, one can use this data in Appendix IV to adjust any purchase price data.

Example 5. You need to estimate the purchase price of a 2500 ft^2 floating head heat exchanger rated at 150 psig and having a CS shell and Monel tubes. One month ago, you got a bid of $30K for a 2500 ft^2 fixed tube sheet unit rated at 600 psig and having a CS shell and tubes. What is the price of the new exchanger?

You must adjust the quotation for the type of exchanger (floating head versus fixed tube), for the pressure rating (150 versus 600 psig), and for materials (CS/Monel versus all CS).

$$\$_{new\ unit} = \$_{quote} * \left(\frac{floating\ head\ factor}{fixed\ tube\ sheet\ factor} \right) * \left(\frac{150\ psig\ factor}{600\ psig\ factor} \right) * \left(\frac{CS\ /\ Monel\ factor}{CS\ /\ CS\ factor} \right)$$

$$= \$30K * \left(\frac{1.3}{1.05} \right) * \left(\frac{1}{1.26} \right) * \left(\frac{3.3}{1} \right) = \$97.3K$$

Several comprehensive chemical equipment cost databases are:

- Appendix IV in this book.
- Ulrich, G.D., *Chemical Engineering Process Design and Economics: A Practical Guide, 2nd Edition*, Process Publishing, 2004.
- Peters, M.S. and Timmerhaus, K.D., *Plant Design and Economics for Chemical Engineers, 5th Edition*, McGraw-Hill, 2003.
- Chauvel, A. et al., *Manual of Process Economic Evaluation*, Editions TECHNIP, 2003.
- Walas, S.M., *Chemical Process Equipment, Selection and Design*, Butterworth-Heinemann, 1990.
- Garrett, D.E., *Chemical Engineering Economics*, Van Nostrand Reinhold, 1989, Appendix 1.
- Guthrie, K.M., *Process Plant Estimating, Evaluation, and Control*, Craftsman Book Co., 1974.

3.2.5 An Example Illustrates the Methods

Estimate the capital cost, in mid-2000 dollars, of an upgrade to a fatty acid separation process using the equipment list and prices in Table 3.6. Prices are for the year 2000. The process will be automated as is typical for a chemical plant and will be in the U.S.

TABLE 3.6
Fatty Acid Separation Process Data

Equipment	Preliminary Specs	Material Ratio: Alloy/CS	Purchase Cost ($K)
Vacuum dryer heater	U-tube, 200 ft², 304 SS tubes/304 SS shell, 150 psig maximum allowable pressure	3.5	$26.1
Vacuum dryer	Diameter: 2 ft, height: 6 ft, 304 SS, full vacuum/150 psig maximum allowable pressure	1.7	1.9
Vacuum dryer steam ejector	2 stage, 50 mmHg absolute pressure, 100 lb/hr equivalent air flow, no condenser, 304 SS	2.0	9.9
Vacuum dryer pump	ANSI, 240 gpm, Head: 130 psi, 304 SS	2.0	12.8
Still heater	U-tube, 450 ft², 304 SS tubes/304 SS shell, 800 psig maximum allowable pressure	3.5	50.7
Fatty acid still	Diameter: 10 ft, height: 35 ft, heating coil and internals included, 304SS, full vacuum/800 psig maximum allowable pressure	1.7	55
Still bottoms pump	ANSI, 100 gpm, Head: 50 psi, 304 SS	2.0	4
Overheads condenser	U-tube, 570 ft², 304 SS tubes/304 SS shell, 150 psig maximum allowable pressure	3.5	54
Overheads surge tank	Diameter: 2 ft, height: 4 ft, 304 SS, 50 psig maximum allowable pressure	1.7	1.6
Overheads pump	ANSI, 150 gpm, Head: 120 psi, 304 SS	2.0	6
Still steam ejector	3 stages: ejectors and condensers, 15 mmHg absolute pressure, 20 lb/hr air load, 304SS	2.0	124.3
		Total	346.3

3.2.5.1 The Lang Method

Use Equation 3.2:

$$Capital\ cost = \Sigma(Equipment\ purchase\ cost) * Lang\ factor * F_m * F_i * F_p$$

Find $\Sigma(Equipment\ purchase\ cost)$ from the equipment list/cost table. This is equal to $346.3K. Use a Lang Factor of 4.1 because the plant is a fluid processing plant being added to an existing site.

Find the factors F_m, F_i, and F_p.

F_m: Find the weighted alloy/CS ratio average for the 11 material ratios:

Average ratio = ($26.1K + $50.7K + $54K)(3.5) + ($1.9K + $55K + $1.6K)(1.7)
              + ($9.9K + $12.8K + $4K + $6K + $124.3K)(2)/$346.3K
              = 2.5

From Figure 3.2, F_m = 0.63.

- F_i: Because the process is fully automated, use 1.35.
- F_p: Because the process will be in the U.S., use 1.0.

Calculate the capital cost:

$$ \$ = (\$346.3K)(4.1)(0.63)(1.35)(1) = \$1.208M $$

3.2.5.2 The Hand Method

Use Equation 3.3:

Capital cost = Σ[*Equipment purchase cost* * (*Hand factor* * F_m)] * F_i * F_b * F_p

Find Σ [*Equipment purchase cost* * (*Hand factor* * F_m)]. Use the equipment list/cost in Table 3.7, the listing of Hand factors in Table 3.2, and Figure 3.2.

- F_i: Because the process is fully automated, use 1.35.
- F_b: Because the process is an upgrade (to an existing plant), use 1.06.
- F_p: Because the process will be in the U.S., use 1.0.

Calculate the capital cost:

$$ \$ = (\$726.4K)(1.35)(1.06)\ (1) = \$1.039M $$

3.2.5.3 Observations/Conclusions

This example shows:

- The Lang method produced a result 16% higher than the Hand method. This is within the accuracy range for this type of estimate.
- The five major pieces of equipment (those whose purchase cost is more than $25K) account for about 88% of the capital cost. This follows the conclusions of Pareto, an Italian engineer, who found that a relatively small amount of effort (equipment purchase costs, in this case) will produce the majority of the results.

From this, one can conclude that the Lang method yields a higher estimated cost. A review of the two factors confirms this should be the case. Early in a project

TABLE 3.7
Fatty Acid Separation Process

Type Equipment	Purchase Cost ($K)	Hand Factor	Alloy/CS Ratio	F_m	K^*
Vacuum dryer heater (heat exchanger)	26.1	3.5	3.5	0.55	50.2
Vacuum dryer (pressure vessel)	1.9	4	1.7	0.77	5.9
Vacuum dryer steam ejector (steam ejector)	9.9	2.5	2.0	0.69	17.1
Vacuum dryer pump	12.8	4	2.0	0.69	35.3
Still heater (heat exchanger)	50.7	3.5	3.5	0.55	97.6
Fatty acid still (pressure vessel)	55	4	1.7	0.77	169.4
Still bottoms pump	4	4	2.0	0.69	11
Overheads condenser (heat exchanger)	54	3.5	3.5	0.55	104
Overheads surge tank (pressure vessel)	1.6	4	1.7	0.77	4.9
Overheads pump	6	4	2.0	0.69	16.6
Still steam ejector (steam ejector)	124.3	2.5	2.0	0.69	214.4
				Total	726.4

* [Equipment purchase cost * (Hand factor * F_m)]

when one has only sized the major equipment, the Lang method would be the preferred because it is more conservative. One would also be wise to add an allowance for minor equipment. The allowance will depend upon how uncertain one feels about the quality of their scope definition. It might be as small as 10% and as high as 50% of the purchase price of the major equipment.

3.2.6 How to Use the Factors with Spare and Used Equipment

Often one will use spare or used equipment in a design. The cost of that equipment is much less than the cost of a similar piece of new equipment. Because the Lang and Hand factors are based on full-price equipment, the question arises, how does one get a complete estimate when using spare or used equipment? The answer is quite simple. One must first estimate the cost of the piece of spare or used equipment as if it were new. The as-if-new cost is multiplied by the Lang or Hand factor (adjusted if needed by an F_m) plus any appropriate other factors — F_i, F_b, or F_p. The as-if-new cost minus the cost to the project of the spare or used equipment is then subtracted from that value. This way, the full installation, instrumentation, building, and place factor costs are included along with the actual cost of the spare or used equipment.

Example 6. As part of a plant upgrade project, you plan to use a spare 1000 ft² U-tube heat exchanger rated at 150 psig and built with a carbon steel shell and stainless steel tubes. The plant will charge you $2000 for the exchanger. Assume the instrumentation will be typical for a processing plant and that your plant is in the U.S. What is the capital cost of this exchanger?

First, you would determine the as-if-new price of the exchanger. Using your company's database, you find this would be $20K. Next, you would find the appropriate Hand factor and adjust it because you have stainless steel tubes. The Hand factor for heat exchangers (Table 3.2) is 3.5.

Find F_m: Find the alloy/CS ratio. Use the factors on the Heat Exchanger Price Graph in Appendix IV. You find this to be 1.7. From Figure 3.2, $F_m = 0.8$.

Find the other factors — F_i, F_b, F_p:

$F_i = 1.35$ (Table 3.3 for typical controls)
$F_b = 1.06$ (Table 3.4 for expansion of fluid processing at an existing site)
$F_p = 1.0$ (By definition, construction in the U.S. = 1.0)

Find the Capital cost:

$$ \$ = (\$20K)(3.5)(0.8)(1.35)(1.06)(1) - (\$20K - 2K) = \$62.1K $$

3.3 ESTIMATING THE OTHER INVESTMENTS

3.3.1 WORKING CAPITAL

As explained in Chapter 1, working capital is mostly made up of inventories, accounts receivable, and accounts payable. Inventories generally amount to 3 to 6 months of materials on hand, comprised of:

- Raw materials and packaging materials: 1 to 2 months of supply on hand
- In-process materials: 1 to 2 months of production on hand
- Finished product: 1 to 2 months of sales on hand

Roughly, each month of inventory can be valued by multiplying an average month's production by the production costs.

A simplifying assumption for accounts receivable and accounts payable is that they are equal. With these assumptions, working capital is roughly equal to the value of 3 to 6 months of production.

3.3.2 STARTUP EXPENSES

Companies define and account for startup costs differently. Because it is all-inclusive, I prefer expenses above normal due to the startup of a new or modified process. Because we are talking about expenses, costs to solve equipment problems during startup (these are capital) are excluded. Included are:

- Salaries, travel expenses (if any), and benefits for operators and managers hired early

- Training operators and managers to operate the new process
- Costs above what is expected once the process is started up and operating well:
  - Material and product losses due to operating problems during startup
  - Reprocessing or disposal of off-quality materials made during startup
  - Added labor and supervision due to lower operating efficiencies
  - Higher utility costs due to lower operating efficiencies

Startup expenses range from 5% to 15% of capital.

3.3.3 SUPPLIER ADVANCES AND ROYALTIES

No rules-of-thumb or factors exist for estimating supplier advances or royalties. If either are a part of one's project, whoever is contracting with the suppliers or patent/license holders should estimate these costs.

3.4 SUMMARY

To make OOM and study estimates, an engineer has to:

- Be able to adjust costs to different time periods. Use Equation 2.7 and one of the construction cost indices, preferably the CEPI for chemical processes, to adjust costs.
- Be able to ratio costs for different capacities. Use Equation 3.1 and a size exponent list to ratio costs. When one does not know the size exponent, use 0.67 for plants and 0.60 for equipment.
- Be able to estimate equipment purchase costs using budget quotes, company and personal data, purchased data/software, or published data.
- Be able to convert equipment purchase costs into capital costs using the Lang method (Equation 3.2) or the Hand factor method (Equation 3.3).

3.5 PROBLEMS/EXERCISES

1. Using the equipment cost graphs in Appendix IV, find the purchase cost (at a CEPI of 460) of the following:
   a. A U-tube HEX with 304 SS shell and tubes, a pressure rating of 600 psig, and 16 ft tubes. The HEX is a three-shell unit with 200 ft^2 in each shell.
   b. A centrifugal air compressor having a capacity of 900 scfm and a discharge pressure of 150 psig.
   c. A system comprised of a tank, agitators, and a pump. The tank is made of CS and has a capacity of 70K gal. The tank will have three side-mounted, CS 20 HP propeller agitators with stuffing boxes. The pump is a cast iron rotary positive displacement unit rated at 200 gpm and 100 ft discharge pressure.
   d. A 304 SS, 4000 gal vertical pressure vessel having a design rating of 150 psig.

    e. A 316 SS stripping column. The column is 6 ft in diameter, is packed with 3-in SS Pall rings, and rated at 50 psig. The packed height is 80 ft and there is 5 ft for gas distribution at the bottom and another 5 ft for gas disengagement at the top.

    f. A 100K lb/hr package boiler that generates 450 psig steam.

    g. A CS, 50 psig, agitated reactor having a capacity of 8K gal. The agitator is a 25 HP two-bladed turbine with a seal.

2. What materials factor would you use to adjust the Hand factor for a SS centrifugal ANSI pump?

3. You are estimating the cost of a $4M fluid process using Hand factors. The process, which will be centrally controlled, is a new unit in an existing site.

    a. What building factor would you use?

    b. What instrumentation factor would you use?

4. Estimate the capital cost (in 2007 dollars) for the batch hydrogenation process described by the flowsheet in Figure 3.4 and equipment list in Table 3.8. Instrumentation is to be typical of a processing plant. The process will be built in an existing plant in Germany. Assume the CEPI for 2007 is 500.

3.6 ADDITIONAL TOPIC

3.6.1 TYPES OF ESTIMATES

The main part of this chapter reviewed the type of estimating used in the earliest stages of a project — order-of-magnitude and study estimates. These are ideally suited for early work and for studies because they are designed to estimate the full project cost when there are just a few design details. Estimating continues throughout the life of a project. These later estimates require much more design detail and take a considerable amount of time to complete. As an example, a detailed estimate done during the design phase of a $200M project might cost $200K and could well contain 300 to 400 pages of estimating (not design) detail. Whereas a typical engineer can make order-of-magnitude and study estimates, detailed estimates are most often done by engineers specializing in cost engineering. Table AT3.1 shows the different types of estimates, their accuracy range, the purpose of each, the general type of design information available, and in what project phase they are used. The table lists six project phases. These and their purposes are:

Project Phase	Purpose of the Phase
Process Development	To define the key process steps and operating conditions, to develop raw material and packaging material specifications, and to develop process design data
Feasibility	To decide whether or not a proposal is economically feasible
Conceptual Engineering	To develop the major features of the design for the selected feasible plan
Definition Engineering	To complete the design in enough detail so that construction drawings and instructions can be made
Design Engineering	To completely engineer the details and to create construction drawings and instructions so the plant can be built
Construction	To build the plant per construction drawings and instructions

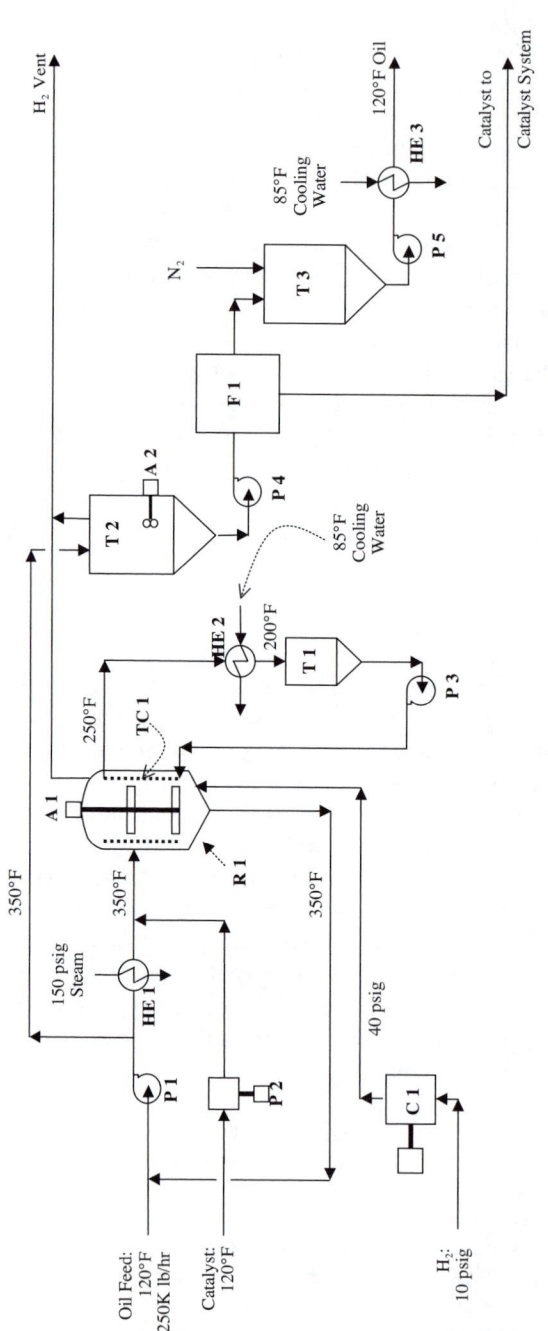

FIGURE 3.4 Problem 4, batch oil hydrogenation.

TABLE 3.8
Problem 4, Hydrogenation Equipment List

Equipment Number/Description	Comments
P 1: Oil feed pump; ANSI centrifugal, cast steel, 270 gpm @ 350 ft of head	This is an existing pump. The cost to replace it (CEPI = 400) is $9.8K.
HE 1: Reactor heater: 2800 ft^2 U-tube, CS/CS (shell/tubes), 150 psig MAP	
P 2: Catalyst pump: diaphragm, CS, 1 gpm @ 150 psig	A similar pump, but rated at 1.5 gpm @ 150 psig, was purchased in 2000 for $5300.
C 1: Hydrogen compressor: reciprocating, 260 scfm, 40 psig discharge pressure, 50 HP	
R 1:Reactor: 12K gal, CS, 150 psig MAP	A similar reactor, but 10K gal, SS, and 50 psig pressure rating, was purchased in 1999 for $58.5K.
A 1: Reactor agitator, Two-bladed turbine, CS, 150 rpm, 40 HP with mechanical seal	
TC 1: Reactor heating and cooling coil: 250 ft^2, CS	Use a price of $14/ft^2 @ CEPI of 500
T 2: Filter feed tank: 12K gal, CS, 150 psig PV	
A 2: Filter feed tank agitator: Side-entering propeller with mechanical seal, CS, 5 HP	
P 4: Filter feed pump: ANSI centrifugal, cast steel, 270 gpm @ 350 ft of head	
F 1: Filter, 40 ft^2, candle, CS	In 2002, a similar candle filter, but having 60 ft^2, was bought for the Memphis plant for $12.2K.
T 3: Filtrate tank: 2000 gal, CS, atmospheric	
P 5: Filtrate pump: ANSI centrifugal, cast steel,70 gpm @ 350 ft of head	
HE 3: Oil cooler: 660 ft^2 U-tube, CS/CS, 150 psig	
T 1: Water surge tank: 100 gal, CS, 150 psig PV	This is an existing tank. Its replacement cost (CEPI = 400) is $1800.
P 3: Recirculating water pump: ANSI centrifugal, cast steel, 70 gpm @ 350 ft of head	
HE 2: Water cooler: 50 ft^2 U-tube, CS/CS 150 psig	

TABLE AT3.1
Estimate Types and Project Phases

Estimate Type	Order of Magnitude	Study	Preliminary	Definitive	Detailed
Estimate accuracy[35]	30% to +50%	25% to +30%	20% to + 25%	10% to +20%	5% to +10%
Estimate purpose	Option analysis Preliminary feasibility analysis	Option analysis Feasibility analysis Engineering funding	Full start project funding	Project cost control	Construction cost control Final funding
Available design details	Type of plant Plant capacity General location	A. Block flow diagram — first cut material and energy balances	A. First draft P&IDs B. First draft process description C. Major equipment and motor list D. Control strategy and 1st cut logic diagrams E. Study models and preliminary equipment layouts F. Preliminary building and utility requirements G. Preliminary site layout	A. P&IDs > 95% complete B. Design bases > 95% complete C. Equipment and motor specifications > 95% complete D. Logic diagrams > 95% complete E. Construction models and equipment layouts > 95% complete F. Final building and utility requirements	100% engineering completion on all items
		H. Preliminary HSE risk assessment	H. Preliminary HSE mitigation plan	G. Final site layout H. Final HSE mitigation plan	
Used in these project phases	Process development	All when doing option analyses Process development Feasibility	Conceptual Engineering Definition Engineering	Design	Construction

REFERENCES

1. Brown, T.R., Capital Cost Estimating, *Hydrocarbon Processing*, October 2000, 93–100.
2. Guthrie, K.M., Capital cost estimating, *Chemical Engineering*, March 24, 1969, 114–142.
3. Garrett, D.E., *Chemical Engineering Economics*, New York: Van Nostrand Reinhold, 1989, 27–33, 38–40.
4. Humphreys, K.K. and Wellman, P., *Basic Cost Engineering*, New York: Marcel Dekker, 1987.
5. Vatavuk, W.M., Updating the *CE* cost index, *Chemical Engineering*, January 2002, 62–70.
6. Garrett, D.E., *Chemical Engineering Economics*, New York: Van Nostrand Reinhold, 1989, 309–353.
7. Guthrie, K.M., *Process Plant Estimating, Evaluation, and Control*, Solana Beach, CA: Craftsman Book Co., 1974.
8. Remer, D.S. and Chai, L.H., Estimate costs of scaled-up process plants, *Chemical Engineering*, April 1990, 138–175.
9. Guthrie, K.M., *Process Plant Estimating, Evaluation, and Control*, Solana Beach, CA: Craftsman Book Co., 1974.
10. Chilton, C.H., Six-tenths factor, *Chemical Engineering*, April 1950, 112–114.
11. Chase, J.D., Plant cost vs. capacity: new way to use exponents, *Chemical Engineering*, April 6, 1970, 113–118.
12. Viola, J.L., Estimate capital costs via a new shortcut method, *Chemical Engineering*, April 6, 1981, 80–86.
13. Perry, R.H. and Green, D.W. (Eds.), *Perry's Chemical Engineering Handbook*, New York: McGraw-Hill, 1997, 9-64 to 9-67.
14. Allen, D.H. and Page, R.C., Revised technique for predesign cost estimating, *Chemical Engineering*, March 3, 1975, 142–150.
15. Lang, H.J., Engineering approach to preliminary cost estimates, *Chemical Engineering*, September 1947, 130–133; Cost relationships in preliminary cost estimation, *Chemical Engineering*, October 1947, 117–121; Simplified approach to preliminary cost estimates, *Chemical Engineering*, June 1948, 112–113.
16. Peters, M.S. and Timmerhaus, K.D., *Plant Design and Economics for Chemical Engineers, 4th Edition*, New York, McGraw-Hill, 1991, 175.
17. Hand, W.E., From flow sheet to cost estimate, *Petroleum Refiner*, September 1958, 331–334.
18. Clerk, J., Multiplying factors give installed costs of process equipment, *Chemical Engineering*, February 1963, 182, 184.
19. Aries, R.S. and Newton, R.D., *Chemical Engineering Cost Estimation*, New York, McGraw-Hill, 1955.
20. Garrett, D.E., *Chemical Engineering Economics*, New York: Van Nostrand Reinhold, 1989.
21. Guthrie, K.M., *Process Plant Estimating, Evaluation, and Control*, Solana Beach, CA: Craftsman Book Co., 1974.
22. Jelen, F.C. and Black, J.H., *Cost and Optimization Engineering*, New York: McGraw-Hill, 1983.
23. Liptak, B.G., Costs of process instruments, *Chemical Engineering*, September 7, 1970, 60.

24. Rodriguez, A.L.M. and Coronel, F.J.M., A better way to price plants, *Chemical Engineering*, March 1992, 124–127.

25. Grimm, F.B. and Shafer, S.L., Unpublished correspondence, 1998.

26. Peters, M.S. and Timmerhaus, K.D., *Plant Design and Economics for Chemical Engineers, 4th Edition*, New York, McGraw-Hill, 1991, 175.

27. McConville, J.G., *The 1996 International Construction Costs and Reference Data Yearbook*, New York: John Wiley & Sons, 1996.

28. Ulrich, G.D., *A Guide to Chemical Engineering Process Design and Economics*, New York: John Wiley & Sons, 1984, 272–275.

29. Guthrie, K.M., *Process Plant Estimating, Evaluation, and Control*, Solana Beach, CA: Craftsman Book Co., 1974.

30. Brown, T.R., Unpublished report, 1970.

31. Garrett, D.E., *Chemical Engineering Economics*, New York: Van Nostrand Reinhold, 1989.

32. Wroth, W.F., No. 42: Factors in cost estimating, *Chemical Engineering*, October 17, 1960, 204.

33. Brown, T.R., Capital cost estimating, *Hydrocarbon Processing*, October 2000, 99–100.

34. Couper, J.R., *Process Engineering Economics*, Marcel Dekker, Inc., New York, 2003, 83–85.

35. Couper, J.R., *Process Engineering Economics*, Marcel Dekker, Inc., New York, 2003, 58.

4 Estimating Production Cost

In addition to capital cost estimating, an engineer must be able to estimate production costs to create economically viable designs. This chapter will review the methods for creating study-grade production cost estimates. These have a couple of similarities to order of magnitude and study-grade capital estimates:

- They are used to determine economic feasibility and to analyze options.
- They have a wide range of accuracy. My observation is the accuracy range is about ±50%. The main reason the range is so broad is that plant design details, and thus production cost details, are sketchy.

Part of this chapter includes an update of an article written by T.R. Brown in 2000 and published in *Chemical Engineering*.[1]

4.1 PRODUCTION COSTS DEFINED

Companies categorize or group their production costs depending on their structure and organization. The usual ways to do this are by controllable activity and by fixed and variable costs. Some texts use the terms "direct" and "indirect" costs as synonymous with fixed and variable costs. This book uses *controllable activity*.* If one's company has set up its cost structure differently, they can adjust the categories.

4.1.1 COST CATEGORIES[2-4]

Categories	Subcategories	Components
Raw materials	In-freight	
	Product ingredients	Ideal material usage
		Losses
		Overpack of product
	Catalyst and solvents	Ideal usage
		Losses
	By-product credits	
Packaging materials	In-freight	

---

* Controllable activity accounting identifies costs of specific work activities. Each activity has someone assigned the responsibility for it. The activities are defined so that the responsible person has the power to influence its cost.

Categories	Subcategories	Components
	Product packaging (i.e., drums, bags, pails, plastic bottles, fiber cartons, corrugated containers)	Ideal usage Losses
Manufacturing	Operating labor (wages)	Process Packaging Warehousing Material supply
	Utilities	Steam Water: processing, chilling, and cooling Electrical power Refrigeration Fuel Compressed air Inert gases (not a part of the product) Sewage and waste disposal Losses, both thermal and material
	Employee benefits	Medical/dental coverage Pension/retirement plans Vacation Unemployment and disability insurance Social Security and Medicare
	Supervision (wages and benefits)	Plant manager plus Production, Staff, and Maintenance managers
	Laboratory	Labor (wages and benefits) Supplies
	Maintenance	Labor (wages and benefits) Replacement and maintenance parts Contract maintenance
	Insurance and taxes	Property and other non-income taxes Fire, theft, property damage, and liability insurance Permits
	Operating (consumable) supplies	Lubricants Filter aids Custodial supplies
	Plant overhead	Clerical staff (accounting, human resources, quality control — wages and benefits) Medical department Recruiting Plant security Groundskeeping
	Depreciation	Depreciation as allowed by the IRS
	Contract manufacturing	Includes the costs of any contract manufacturing
Product delivery	Shipping costs	Truck Rail Air

4.2 FIXED AND VARIABLE COSTS

One can also classify production costs as fixed or variable. *Fixed costs* are those that do not change when production volume in a plant changes. Alternatively, *variable costs* vary with volume. Production costs usually range from 60% to 80% variable, with 70% being a good average to use when no other data is available. Table 4.1 classifies the cost categories as fixed or variable. If a category is partially fixed and partially variable, both are indicated.

TABLE 4.1
Fixed/Variable Classifications for Production Costs

	Fixed	Variable
Raw materials		X
Packaging materials		X
Manufacturing		
Operating labor	X	X
Utilities	X	X
Employee benefits	X	X
Supervision	X	X
Laboratory	X	X
Maintenance	X	X
Insurance and taxes	X	
Operating supplies		X
Plant overhead	X	
Depreciation	X	
Product delivery		X

Source: Brown, T.R. Estimating product costs, *Chemical Engineering*, August 2000, 86–89. With permission.

4.3 ESTIMATING METHODS

4.3.1 GENERAL

4.3.1.1 Inflation Adjustments

Similar to capital estimates, the data for product costs will often be quoted for a time different from that for your estimate. Use Equation 2.7 and the producer price index (PPI) for Chemical and Allied Products to adjust these costs. The PPIs for 1984 to 2005 are found in Appendix II.

4.3.1.2 Plant Production vs. Design Rate

Plants operate at less than 100% of the design rate. When estimating production costs, one must take this into account. The efficiency losses come from scheduled maintenance, sales fluctuations (causing excess inventory and production slow-downs), breakdowns, process changeovers, power/feedstock outages, and equipment cleaning/washouts. Garrett states, "… the current percent of design capacity operation … is only about 80%."[5] When making study estimates, a good simplifying assumption is to divide design rate costs by the percent of design capacity operation. If you do not have data from your company on its percent design rate performance, use 80%. For example, if the production cost is $525/T at design rate and if the plant is expected to operate at 80% of design rate, the actual production cost would be

$$(\$/T)_{actual} = \frac{\$525/T}{0.80} = \$656/T.$$

4.3.2 Flexing Existing Costs for Production Volume Changes

This method is useful when:

1. One is estimating production cost for a plant or process that is similar to an existing facility. In this case, one could use data from the existing facility and adjust it to the volume level of the new unit.
2. One is estimating production costs in an existing plant at a new volume level.

Two examples will illustrate how to adjust production costs for volume changes.

Example 1: If a plant is budgeted to produce 10M cases of product for $100M and if its budget is 70% variable, what should its budget be when making 12M cases?

Variable costs = 0.7 ($100M) = $70M; *Fixed costs* = 0.3 (100M) = $30M

Adjust the variable costs:

New variable costs = 70M (12M cases/10M cases) = $84M

New budget = *New variable costs* + *Fixed costs* = $84M + $30M = $114M

New unit cost = *New budget/New production* = $114M/12M cases = $9.50/case

Example 2: (This is a rephrasing of Example 1.) If a plant is budgeted to produce 10M cases of product for $10.00/case and if its budget is 70% variable, what should its budget be when making 12M cases?

Variable costs = 0.7 ($10.00/case) = $7.00/case; *Fixed costs* = 0.3 (10.00) = $3.00/case

Because the variable costs are stated in $/case, the new variable costs are also $7.00/case. Find the new fixed costs in $/case:

New fixed costs = [($3.00/case) (10M cases)]/12M cases = $2.50/case

New unit cost = *Variable costs* + *Fixed costs* = $7.00/case + $2.50/case = $9.50/case

4.3.3 STUDY ESTIMATES

For study estimates, the estimating of product costs is a combination of direct estimating and factoring. Table 4.2 shows which costs are estimated in detail and which are factored.

TABLE 4.2
Estimating Methods

	Detailed Estimate	Factored Estimate
Production Costs		
Raw materials	X	
Packaging materials	X	
Manufacturing		
Operating labor	X	
Utilities	X	
Employee benefits		X
Supervision		X
Laboratory		X
Maintenance		X
Insurance and taxes		X
Operating supplies		X
Plant overhead		X
Depreciation		X
Product delivery	X	

Source: Brown, T.R. Estimating product costs, *Chemical Engineering*, August 2000, 86–89. With permission.

4.3.3.1 Detailed Estimate Subcategories

4.3.3.1.1 Raw/Packaging Materials and Utility Costs

Three ways exist to estimate materials and utility costs — from material and energy balances, from company data, or from public data for the same process. When using material balance data, one must add 2% to 3% to the ideal raw/packaging material usage to account for plant losses and overpacking. For thermal utilities, add 5% to 10% for thermal and material losses. Electrical losses are negligible; ignore them.

Current company price data, supplier budget quotations, or published data are the sources for raw material purchase prices. Two good public sources are the *Chemical Market Reporter* for chemicals and *The Wall Street Journal* for commodities such as grains and feeds (corn, oats), foods (coffee, flour, orange juice concentrate), fats and oils (soybean oil, corn oil), or petroleum products (#2 fuel oil, natural gas).

For packaging material costs, use current company price data or supplier budget quotations. *Perry's Chemical Engineering Handbook* lists 1996 prices for some packaging materials (See Table 4.3). I am not aware of routinely updated public sources for cost data.

TABLE 4.3
1996 Packaging Material Costs[6]

	Price ($/unit)
Drums	
55 gal, Steel, open head	70/drum
55 gal, Steel, tight head	50/drum
55 gal, Fiber (for dry materials)	21/drum
Bags	
2 ft^3 multiwall paper, sewn open-mouth	0.50/bag
2 ft^3 multiwall paper, sewn valve	0.65/bag
1.33 ft^3 paper with polyethylene liner, sewn valve	0.60/bag
Polyethylene pouch, $8^3/_4$ in $\times 16^3/_4$ in, 115 fluid oz capacity	0.13/pouch
Paper bag, sugar type, 6 in $\times 2^3/_4$ in $\times 16^3/_4$ in, 115 fluid oz capacity	0.11/bag
Folding box (white), $9^1/_2$ in $\times 4^1/_2$ in $\times 15$ in, 0.37 ft^3	0.47/box
Corrugated carton, 24 in $\times$ 16 in $\times$ 6 in, 275 lb test	1.25/box

For individual utilities, use company price data or quotes from the utility companies for electric power, fuel oil, natural gas, sewage treatment, and landfill. For items such as cooling water, refrigeration, and steam, company cost data is probably the most accurate. Not much high quality public data is found. One can find what there is in textbooks on plant design or in magazines like *Chemical Engineering*. Because utility costs are highly variable from one location to another, use caution when using data from other locations or from public sources. Table 4.4 shows typical costs from 1997.

Example 3: Calculate the utility costs in $/lb of product for a process having the following energy balance. Use the 1997 costs from Table 4.4.

Final product	100 lb
Steam	20 lb
Cooling water	60 gal

TABLE 4.4
1997 Utility Costs[7]

	Price ($/unit)
Steam (175 psig)	4.00/1000 lb
Electricity	0.04/kWh
Cooling water (85°F)	0.05/1000 gal
Process water	0.05/1000 gal
Refrigeration (−30°F)	1.95/ton-day; 6.77/MBtu
Natural gas	2.60/MBtu
Biological sewage treatment	0.05–0.20/lb of organics
Landfill	0.06/lb (dry)

Because these are utilities where thermal losses will be included, you must add 5% to 10% for material and thermal losses. Use 7.5%.

Steam: $(20 \text{ lb}_{steam}/100 \text{ lb}_{product})$ $(\$4.00/1000 \text{ lb}_{steam})(1.075) = \$0.00086/\text{lb}_{product}$

Water: $(60 \text{ gal}/100 \text{ lb}_{product})(\$0.05/1000 \text{ gal}) (1.075) = \$0.00003/\text{lb}_{product}$

Utility cost $= \$0.00086/\text{lb}_{product} + \$0.00003/\text{lb}_{product} = \$0.00089/\text{lb}_{product}$

4.3.3.1.2 *Operating Labor*

To determine labor costs, one must estimate the number of operators (crew size) and the wage rate. The preferred way to determine labor needs is to get a manufacturing estimate. However, because there is not enough time or because manufacturing has not staffed a project, that is often not practical. If it is not, scaling of data from a similar operation will work well. For this, use the Wessell ratio, Equation 4.1. When no like operations are available, one can use the Ulrich method for processes. I am not aware of similar methods for packaging, warehousing, or material supply operations. For these, one must get some type of estimate from manufacturing or scale from a similar operation.

Wage rates vary considerably depending on the skills and responsibility of the operator and on the region of the country. To estimate rates, work with manufacturing or use the rate from a similar plant in the area. As a last resort, both *Chemical Week* and *Chemical and Engineering News* publish national average rates.

4.3.3.1.2.1 *The Wessell Ratio*

In 1952, H.E. Wessell published an article in which he correlated operating hours to production.[8] Using operating hours per ton data published by Chilton in the February 1951 issue of *Chemical Engineering*, Wessell developed three curves relating operating hours, processing steps, and plant capacity to each other. The three curves were for highly manual batch operations, highly automated operations, and a blend of batch and automated operations. The correlation only applies to processing operations. Translating the curves into equation form gives:

$$\frac{\text{operating hours}}{\text{Ton}} = b\left(\frac{\text{process steps}}{\text{TPD}^{.78}}\right)$$

This method is difficult to use — actually, impractical — because no guidelines exist for defining the number of processing steps. The real value of Wessell is the rearrangement of the equation into a ratio form:

$$\frac{\left(\text{hr/unit of production}\right)_2}{\left(\text{hr/unit of production}\right)_1} = \left(\frac{\text{Capacity}_1}{\text{Capacity}_2}\right)^n \qquad (4.1)$$

On log-log paper, this is a straight line having a negative slope. The average n is 0.78.

Example 4: Estimate the crew size of a proposed 300 TPD plant that will operate 24 hrs/day, 7 days/wk and 50 wks/yr (8400 hrs/yr). It will be similar to two other plants in your company. These plants also operate 8400 hrs/yr.

	Plant A	Plant B
Capacity (TPD)	150	240
Crew size, people/shift (8 hrs/shift)	7	8

Find operating hours/T for the existing plants:

$$\left(\frac{\text{hrs}}{T}\right)_A = \frac{\left(7\,\dfrac{\text{people}}{\text{shift}}\right)\left(8\,\dfrac{\text{hrs}}{\text{person}}\right)\left(3\,\dfrac{\text{shifts}}{\text{day}}\right)}{150\,\dfrac{T}{\text{day}}} = 1.12$$

$$\left(\frac{\text{hrs}}{T}\right)_B = \frac{\left(8\,\dfrac{\text{people}}{\text{shift}}\right)\left(8\,\dfrac{\text{hrs}}{\text{person}}\right)\left(3\,\dfrac{\text{shifts}}{\text{day}}\right)}{240\,\dfrac{T}{\text{day}}} = 0.80$$

Plot these on log-log paper and extend the line to 300 TPD to find the hrs/T for the proposed plant. This is shown in Figure 4.1.

From the chart, the hrs/T = 0.68. Thus, the crew size is:

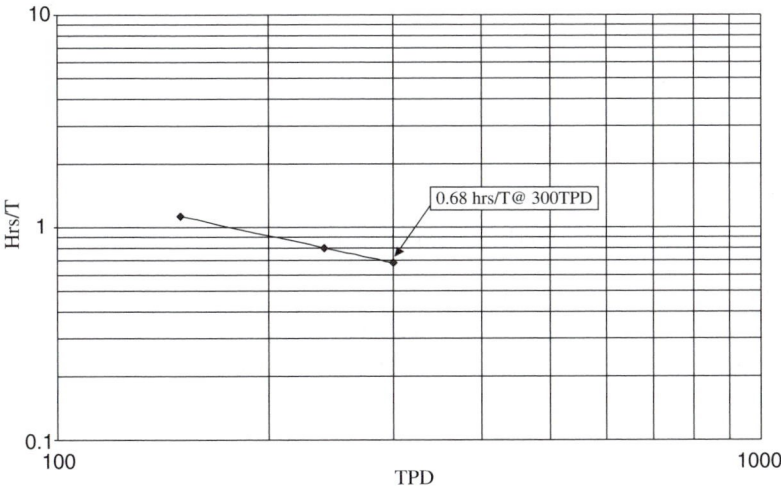

FIGURE 4.1 Example 3, the Wessell Ratio.

$$\frac{\text{people}}{\text{shift}} = \frac{\left(0.68\,\dfrac{\text{hr}}{\text{T}}\right)\left(300\,\dfrac{\text{T}}{\text{day}}\right)}{\left(8\,\dfrac{\text{hr}}{\text{person}}\right)\left(3\,\dfrac{\text{shifts}}{\text{day}}\right)} = 8.5$$

Looking back at this example, note that a number of conversion factors are found on both sides of the equation. Thus:

$$\frac{\text{hrs}}{\text{unit}} \propto \frac{\text{hrs}}{\text{T}} \propto \frac{\text{people}\big/\text{shift}}{\text{unit}} \propto \frac{\text{crew size}\big/\text{shift}}{\text{unit}}$$

Similarly:

$$\frac{\text{hrs}}{\text{unit}} \propto \frac{\text{people}}{\text{T}} \propto \frac{\text{people}}{\text{lb}} \propto \frac{\text{people}}{\text{case}}$$

In addition, one can use two commonly available pieces of data with the Wessel method: unit cost ($/unit) and wage rate ($/hr):

$$\frac{\$}{\text{unit}} = \left(\frac{\text{hrs}}{\text{unit}}\right)(\text{wage rate})$$

Rearranging:

$$\frac{\text{hrs}}{\text{unit}} = \frac{\$/\text{unit}}{\text{wage rate}}$$

4.3.3.1.2.2   The Ulrich Method

This method is straightforward and easy to use. It requires that one know what equipment is in the process and use Table 4.5 to estimate the crew size. Published in Ulrich's book, *A Guide to Chemical Engineering Process Design and Economics*, this technique assumes the process is well automated.[9]

Example 5: Find the crew size for a process that has the following equipment:

Equipment Type	Number of Units
Reactors	1
Heat exchangers	7
Centrifugal separators and filters	2
Pressure vessels	4
Towers, including auxiliary pumps and exchangers	1
Storage vessels	2
Cooling towers	1
Mechanical refrigeration units	1

Using Table 4.5, the operators per shift are:

Reactors: 1 @ 0.25 =	0.25
Heat exchangers: 7 @ 0.05 =	0.35
Centrifugal separators and filters: 2 @ 0.06 =	0.12
Pressure vessels: 4 @ 0 =	0
Towers, including auxiliary pumps and exchangers: 1 @ 0.2 =	0.20
Storage vessels: 2 @ 0 =	0
Cooling towers: 1 @ 0.5 =	0.50
Mechanical refrigeration units: 1 @ 0.25 =	0.25
Total:	1.67

4.3.3.1.3 Product Delivery

Product delivery costs are difficult to estimate because it's hard to define shipping destinations simply enough for a study estimate and to estimate the per-mile shipping rates. A typical company will ship its product to thousands of locations. This in itself becomes a rate estimation and calculation nightmare. When doing

TABLE 4.5
Estimating Direct Labor*

Generic Equipment Type	Operators/Unit/Shift
Auxiliary facilities	
Air plants	0.5
Boilers	0.5
Cooling towers	0.5
Water demineralizers	0.25
Electric generating plants:	
Stationary	1.4
Portable	0.25
Electric substations	0
Incinerators	0.9
Mechanical refrigeration units	0.25
Wastewater treatment plants	0.9
Water treatment plants	0.9
Blowers and compressors	0.05–0.09
Conveyors	0.1
Crushers, mills, grinders	0.25–0.5
Drives and power recovery units	0
Evaporators	0.15
Fans	0.03
Furnaces	0.25
Gas-solids contacting equipment	0.05–0.15
Heat exchangers	0.05
Mixers	0.13
Process vessels	
Pressure vessels	0
Towers, including auxiliary pumps and exchangers	0.1-0.25
Drums	0
Pumps	0
Reactors	0.25
Separators	
Clarifiers and thickeners0.1	
Centrifugal separators and filters	0.03–0.1
Cyclones	0
Bag filters	0.1
Electrostatic precipitators	0.1
Rotary and belt filters	0.05
Plate and frame, shell, and leaf filters	0.5
Expression equipment	0.1
Screens	0.03
Size-enlargement equipment	0.05–0.15
Storage vessels	0
Vaporizers	0.03

* Adjusted for labor productivity to 2003.

study estimates, the shipping destinations usually have not been defined, so one must make a few simplifying assumptions. Either:

- Identify the 10 to 15 largest customers and assume all the product is shipped to these locations. Because 70% to 85% of a company's product is usually shipped to their largest 10 to 15 customers, this is a good approximation. If several locations are near each other, one can calculate what amounts to a "center of mass" for the nearby locations, then estimate mileage based on shipping from your plant to these centers of customer mass.
- Working with Sales, split the country into 5 to 10 shipping regions, then roughly estimate the amount of product that will be shipped to each region. Lastly, approximate the center of customer mass for each region and estimate mileage based on shipping to the centers of mass.

For more details, see the section in Chapter 10 entitled "The Economics of Plant Siting."

Backhauling and the existence of "transportation corridors" make "per mile" shipping costs difficult to estimate.* It is best to get a rough estimate from the experts — one's traffic department. Next best is to ask for budget quotes from trucking companies and railroads. Because the backhauling and transportation corridor issues are complex and not easily dealt with by someone unfamiliar with transportation, these quotes will be questionable. Table 4.6 dimensions some of the different costs on a per-mile basis.

TABLE 4.6
Shipping Costs[10]

Shipping Method	Price ($/mile, mid-1997)
Rail — 20,000 gal tank cars	2.05
Tank Truck — 45K lb	2.75
Semi-trailer (dry van) — 45K lb*	1.35

* When shipping drums, the drums and pallets weigh 4K lb. This leaves 41K lb for product.

Source: Brown, T.R. Estimating product costs, *Chemical Engineering*, August 2000, 86–89. With permission.

* *Backhauling* amounts to arranging for a truck or rail car to carry a load back from your shipping destination. If that is not done, your company will pay mileage for a round trip even though a load is being hauled only one direction.

4.3.3.2 Factored Subcategories[11]

- *Employee benefits.* These range from 22% to 45% of the cost of operating labor. For study estimates, use your company's percentage.
- *Supervision.* The range for these costs is 10% to 30% of the cost of operating labor.
- *Laboratory.* These costs range from 10% to 20% of operating labor.
- *Maintenance.* For new plants, maintenance costs range from 2% to 10% of the cost of the plant. The average is 6%. For older plants, these costs range from 5% to 31% of the depreciated cost of the plant, the average being 12%.
- *Insurance and Taxes.* These costs run from 3% to 5% of the capital investment in the plant or process.
- *Operating Supplies.* These range from $1/2$% to 3% of the capital investment.
- *Plant Overhead.* Plant overhead ranges from 1% to 5% of the capital investment.
- *Depreciation.* Of the two accepted methods for calculating depreciation, use the straight-line method because of its simplicity for study estimates. The annual straight depreciation write off is given by Equation 2.5:

$$Annual\ deprecation\ writeoff = \frac{Capital\ investment}{Project\ life}$$

4.4 SUMMARY

Study grade product cost estimates are a combination of detailed and factored estimating. Table 4.7 and Table 4.8 summarize the methods.

4.4.1 An Example Illustrates

You are doing the conceptual design of a hydrogenation process for a vegetable oil blend containing 70% soybean oil, 20% cottonseed oil, and 19% sunflower oil. Your next step is to estimate the production cost, in 2004 dollars, for the process that will be installed in your existing Newark, NJ plant.

Plant design basis and cost:

Capacity	250M lb/yr
Operating hours per year	6000 (24 hrs/day, 5 days/wk, 50 wk/yr)
Flowsheet	See Figure 3.4 (Chapter 3, Problem 4)
Equipment list	See Table 3.8 (Chapter 3, Problem 4)
Capital Cost	$700K
Economic life	10 yrs

Feed oil shipping:

Soybean oil will be shipped from Decatur, IL

Cottonseed oil will be shipped from Memphis, TN

Sunflower oil will be shipped from Winnipeg, Manitoba

All feed oils will be shipped to the Newark plant in 150K lb tank cars

The Newark plant will pay the in-freight from the shipping points to the plant

Product usage and shipping:

The Newark plant will use 100M lb/yr of the hydrogenated oil product

The Kansas City, MO plant will use 80M lb/yr

The Atlanta, GA plant will use 60M lb/yr

The hydrogenated oil for Kansas City and Atlanta will be shipped from Newark in 150K lb tank cars

Material and energy balances:

Ideal material usage:

Oil (feed basis)	100 lb
Catalyst 0.05 lb	(Catalyst can be reused 4 times)
H_2 (20% excess)	25.6 scf (@60°F and 1 atm)
Product	100 lb

Ideal utility usage:

Oil (feed basis)	100 lb
150 psig steam	15.7 lb
85°F cooling water	60 gal
Electricity	0.65 kWh

Prices (in 2004 dollars.):

Prices in 2004 dollars:

Feed oils:

Soybean	$0.28/lb (FOB Decatur, IL)
Cottonseed	$0.35/lb (FOB Memphis, TN)
Sunflower	$0.32/lb (FOB Winnipeg, Manitoba)
Hydrogen	$0.70/1000 scf (FOB Newark)
Catalyst	$35.00/lb (FOB Newark)
Wage rates	$25/hr
Labor hrs to unload a tank car	1.5hrs
Labor hrs to load a tank car	0.75 hrs

Utilities:

150 psig Steam	$6.54/1000 lb
Cooling Water	$0.072/1000 gal
Electricity	$0.055/kWh

Rail Shipping:

150,000 lb tank car	$2.80/mile

TABLE 4.7
Study Estimates — Detailed Items

		Cost Information Sources
Raw and packaging materials	To determine usage: • Develop from the material balances and add losses • Use actual data, including losses, from the same process in another plant	• Budget quotes • Company price data • Public data: - *Chemical Market Reporter* - *The Wall Street Journal*
Utilities	To determine usage: • Develop from the energy balances and add losses • Use actual data, including losses, from the same process in another plant	• For purchased items: - Budget quotes - Company price data • For company-generated items, use company cost data • Public data
Operating labor	To determine crew size: • Manufacturing estimate • Scaling from data for a similar process using the Wessell ratio • Ulrich method	• Manufacturing cost data • Data from a similar operation • Public data: - *Chemical Week* - *Chemical and Engineering News*
Product delivery	Determine mileage from your shipping point to the destinations: • Locate the destinations for the 10 to 15 largest customers, finding the center of customer mass for nearby locations • Split the country into 5 to 10 regions and find the center of customer mass for each	• CompanyTraffic department • Budget quotes from shippers

TABLE 4.8
Study Estimate Factors

	Factor Basis	**Range of Costs (%)**	**Factor for Study Estimates (%)**
Manufacturing			
Employee benefits	% of operating labor	22–45	40
Supervision	% of operating labor	10–30	20
Laboratory	% of operating labor	10–20	10
Maintenance	% of capital	2–10	6
Insurance and taxes	% of capital	3–5	3
Operating supplies	% of capital	$1/_2$–3	1
Plant overhead	% of capital	1–5	1
Depreciation	Equation 2.5	—	—

Solution:

1. Calculate raw material costs:

Oil: 0.7 lb soybean oil @ \$0.28/lb = 0.196/lb$_{oil}$
     0.2 lb cottonseed oil @ \$0.35/lb = 0.070/lb$_{oil}$
     0.1 lb sunflower oil @ \$0.32/lb = 0.032/lb$_{oil}$
     Total: \$0.296/lb$_{oil}$

$$\text{Catalyst}: \quad \left(\frac{0.05\,\text{lb}_{catalyst}}{100\,\text{lb}_{oil}}\right)\left(\frac{1}{4\ \text{uses}}\right) = \$0.0044 \,/\,\text{lb}_{oil}$$

$$\text{Hydrogen}: \quad \left(\frac{25.6\ \text{scf}}{100\,\text{lb}_{oil}}\right)\left(\frac{\$0.70}{1000\ \text{scf}}\right) = \$0.0002 \,/\,\text{lb}_{oil}$$

Total = 0.298 + 0.0044 + 0.0002 = \$0.3026
  Losses @ 2.5% = 0.0076
  Raw materials = \$0.3102/lb$_{oil}$

2. Packaging materials — none required

3. Manufacturing costs:
  Operating labor:
   Process: Use the Ulrich method

   Heat exchangers, 4 @ 0.05 operators/unit/shift = 0.20 operator/shift
   Reactor, 1 @ 0.25 operators/unit/shift   = 0.25 operator/shift
   Compressor, 1 @ 0.07 operators/unit/shift = 0.07 operator/shift
   Filter, 1 @ 0.5 operators/unit/shift   = 0.50 operator/shift
   Total: = 1.02 operator/shift
                (Use 1.0 operator/shift)

$$\$\,/\,\text{lb}_{oil} = \frac{1\ \text{operator} * 6000\text{hr}\,/\,\text{yr} * \$\,\&\,25\,/\,\text{hr}}{250\text{M lb}_{oil}} = \$0.0006\,/\,\text{lb}_{oil}$$

  Unloading and loading:

  Unloading @ 1.5 person-hrs/tank car and loading @ 0.75 person-hrs/tank
   car

  Find the number of tank cars to be unloaded per year:

$$TC / yr = \left(\frac{lb_{oil} \text{unloaded} / yr}{150K \ lb / TC} \right) = \frac{250M \ lb / yr}{150K \ lb / TC} = 1667 \ TC / yr$$

Find the number of tank cars to be loaded per year:

$$TC / yr = \frac{\text{Shipments}}{150K \ lb / TC} = \frac{(80M + 60M)}{150K} = 933 \ TC / yr$$

Find the hrs/yr and cost of unloading and loading:

$$hr_{unloading} / yr = (1667 \ TC / yr)(1.5 \ hr / TC) = 2500 \ hr / yr$$

$$hr_{loading} / yr = (933 \ TC / yr)(0.75 \ hr / TC) = 700 \ hr / yr$$

$$\text{Total hrs} = 2500 + 700 = 3200 \ hr / yr$$

$$\$ / lb_{oil} = \frac{3200 \ hr / yr * \$25 / hr}{250M \ lb / yr} = \$0.00032 / lb_{oil}$$

Total = 0.0006 + 0.0003 = $0.0009/lb_{oil}$

Labor related costs: Use the study estimate factors from Table 4.8 — employee benefits @ 40% of operating labor, supervision @ 20%, and laboratory @ 10% or 70% in total.

$$\$/lb = 0.7 * \$0.0009/lb = \$0.0006/lb$$

Utilities (including material and thermal losses for steam and cooling water):

$$\text{Steam} : \$ / lb_{oil} = \left(\frac{15.7 lb_{steam}}{100 lb_{oil}} \right) \left(\frac{\$6.54}{1000 lb_{steam}} \right)(1.075) = \$0.0011 / lb_{oil}$$

$$\text{Cooling water} : \$ / lb_{oil} = \left(\frac{60 \ gal_{water}}{100 lb_{oil}} \right) \left(\frac{\$0.072}{1000 gal_{water}} \right)(1.075) = \$0.00005 / lb_{oil}$$

$$\text{Electrical} : \$ / lb_{oil} = \left(\frac{0.65 kWh}{100 lb_{oil}} \right) \left(\frac{\$0.055}{kWh} \right) = \$0.00036 / lb_{oil}$$

Total utilities $ / lb_{oil}$ = $0.0011 + 0.00005 + 0.00036 = $0.0015 / lb_{oil}$

Capital related: Maintenance @ 6% of capital, Insurance and taxes @ 3%, Operating supplies @ 1%, and Plant overhead @ 1% or 11% in total.

$$\$ / lb = \frac{0.11 * \$700K}{250M \ lb_{oil} / yr} = \$0.0003 / lb_{oil}$$

Depreciation:

$$\$/yr = Capital/Life = \$700K/10 \ yrs = \$70K/yr$$

$$\$ / lb = \frac{\$700 / yr}{250M \ lb / yr_{oil}} = \$0.0003 / lb_{oil}$$

Manufacturing Summary:

Operating labor	$\$0.0009/lb_{oil}$
Labor related	$0.0006/lb_{oil}$
Utilities	$0.0015/lb_{oil}$
Capital related	$0.0003/lb_{oil}$
Depreciation	$0.0003/lb_{oil}$
Total:	$\$0.0036/lb_{oil}$

4. Shipping Costs, In-Freight and Product Delivery:
   Mileage data from MapQuest®; Newark to Atlanta, 1185 mi; to Memphis, 1088 mi; to Kansas City, 864 mi; to Decatur, 870 mi; and to Winnipeg, 1645 mi.
   Find shipping costs @ $2.80/mi:

	M-lb/yr	Number TCs/yr	Miles	$K/yr
In-Freight				
Decatur to Newark[a]	175	1167	870	2843
Memphis to Newark	50	334	1088	1017
Winnipeg to Newark	25	167	1645	769
			Total In-Freight:	4629
Product Delivery				
Newark to Atlanta	60	400	1185	1327
Newark to Kansas City	80	534	864	1292
			Total Product Delivery:	2619

[a] Example calculation: number of TCs from Decatur to Newark:

$$\left[NumberTCs / yr = \frac{lb / yr \ shipped}{lb / TC} = \frac{0.70 * 250M \ lb / yr = 175M \ lb / yr}{150K \ lb / yr} \right]$$

In-freight = ($4629M/yr)/(250M-lb/yr) = $0.0185/lb$_{oil}$
Product delivery = ($2619M/yr)/(250M-lb/yr) = $0.0105/lb$_{oil}$

5. Production Cost Summary:

Raw Materials:		
Materials		$0.3102/lb$_{oil}$
In-freight		0.0185/lb$_{oil}$
	Subtotal	0.3287/lb$_{oil}$
Manufacturing:		
Operating labor		0.0009/lb$_{oil}$
Employee benefits, supervision, laboratory		0.0006/lb$_{oil}$
Utilities		0.0015/lb$_{oil}$
Maintenance, Insurance and Taxes, Operating		
Supplies, Plant Overhead		0.0003/lb$_{oil}$
Depreciation		0.0003/lb$_{oil}$
	Subtotal	0.0036/lb$_{oil}$
Product Delivery:		0.0105/lb$_{oil}$
Total Production Cost:		$0.343/lb$_{oil}$

4.5 PROBLEMS AND EXERCISES

1. If a plant produces 100 TPY of product for $250/$T$ and its production cost is 70% variable, what would the cost/T be when producing 125 TPY?
2. If a plant produced 1000K cases per year of product for $12,000K/yr and its production cost is 75% variable, what would the cost per case be when producing 1250K cases per year?
3. Estimate the operating crew size for a process having the following equipment: two compressors, four heat exchangers, nine pumps, one reactor, one plate and frame filter, and ten storage tanks.
4. Estimate the crew size for a proposed 300 TPD plant that will operate 24 hrs/day, 7 days/wk, and 50 wks/yr. It will be similar to another plant in your company that also operates 8400 hr/yr. This plant produces 150 TPD and has a crew of 30.
5. What is the operating crew size/shift for a 100 TPY process having one reactor, one plate and frame filter, seven heat exchangers, four pressure vessels, and nine centrifugal pumps? If this process runs for 24 hrs/day, 7 days/wk, and 50 wks/yr, and if the wage rate is $25/hr, what is the operating labor cost in $/$T$?

6. Complete the following production cost estimate in $/lb. Assume the plant will produce 500 TPY and was built for $1.5M. Use an economic life of 10 years.

	$K/yr
Raw materials	1500
Packaging materials	30
Manufacturing cost	—
Operating labor	360
Employee benefits	
Supervision	
Laboratory	
Maintenance	
Utilities	140
Depreciation	
Insurance and taxes	
Plant overhead	
Product delivery	350
Production cost	

7. You are studying the feasibility of building a processing plant on the West Coast. The plant, which will have a capacity of 25M units/yr, is similar to another larger plant in the Midwest. Here are comparative data for the two plants:

	Midwest	West Coast
Capacity	70M units/yr	25M units/yr
Capital cost (in today's dollars)	$210M	$105M
Wage rate	$15/hr	$18/hr
Raw materials	$8.00/unit	$8.70/unit
Packaging materials	$3.20/unit	$3.35/unit
Average shipping distance to customers	90 mi	70 mi

The following are the actual production costs for the Midwest plant.

	$/unit	
Raw materials	8.00	
Packaging materials	3.20	
Manufacturing		
Operating labor	1.46	
Employee benefits	0.60	These costs are
Supervision	0.29	70.5% of operating labor
Laboratory	0.14	
Maintenance	0.12	
Utilities	0.33	
Depreciation	0.30	
Insurance and taxes	0.09	
Operating supplies	0.02	
Plant overhead	0.05	
Total manufacturing	3.40	
Product delivery	1.40	
Production cost	16.00	

Depreciation is based upon a 10-year economic life. What is your estimate of the production costs for the proposed West Coast plant?

8. Find the present prices for the following:

Acetaldehyde	Flour	Tallow
Ammonium sulfate	#2 heating oil	Toluene
Benzene	Sodium hydroxide	Xylene
Hydrochloric acid	Soybean oil	

9. You are working to estimate production costs for a plant that is under construction. Your estimate of $640/T$ is based upon the plant's design capacity of 200 TPD. For the first two to three years of operation, you expect the plant to run at about 75% of its design rate. What would you expect the production costs to be in that period?

4.6 ADDITIONAL TOPIC

4.6.1 PRODUCT COST AND GENERAL EXPENSE

While production costs are the costs engineers can affect, they are only a portion of the cost to make, sell, and market a product. The total cost of a product is the *product cost*. It is the sum of production costs and general expense.

General expenses are categorized as follows:

Cost Categories	Subcategories
Research and Development	Wages and benefits (managers, technicians, secretaries)
	Raw materials and packaging materials (for pilot plant production)
	Pilot plant operating costs
	Laboratory costs
	Product research/testing
	Miscellaneous (travel, building space costs, recruiting)
Marketing	Wages and benefits (managers and secretaries)
	Advertising expense (TV, radio, magazines)
	Market research
	Miscellaneous (travel, building space costs, recruiting)
Sales	Wages, benefits, and commissions (managers and secretaries)
	Sales allowances/promotions
	Miscellaneous (travel, building space costs, recruiting)
Administrative	All costs associated with corporate functions such as: general management, finance, accounting, legal, public relations, information systems, data centers, health and safety, environmental quality, and so on

General expense is often expressed as a percent of sales. Table AT4.1 shows 2004 data for companies employing process engineers.[*] The general expense percentage is greatly dependent upon the kind of products a company makes. If a company makes and sells in mature product markets, its general expense percentage will be less than a company in a dynamic and changing market.[†] Note the difference between the integrated petroleum industry — a mature market — and the pharmaceutical industry, a very dynamic market. General expense is only 3.5% of sales for petroleum and 50.3% for pharmaceuticals.

The best way to estimate general expense is to use one's own companies "percent of sales." If for some reason that is not available, use the average for whatever industry in which one's company is included.

---

[*] The data is taken from either the 2004 annual reports or Forms 10-K for each of the companies.

[†] A *mature market* is one where there are few differences between products, where there is little product innovation, and where customers make their purchase decisions mainly based on price.

TABLE AT4.1
General Expense as a Percent of Sales

Industry	Company	Annual Report Year	General Expense, ($ or £)	Sales	General Expense (% of Sales)
Basic Chemicals	Dow	2004	2.54	40.2	6%
	DuPont	2004	4.47	27.3	16%
	FMC	2004	0.35	2.1	17%
	Georgia Gulf	2004	0.06	2.2	3%
	Lyondell Chemical	2004	0.39	6.0	7%
	Olin Corp	2004	0.01	2.0	1%
				Average:	8%
Diversified Chemicals	Air Products	2004	1.16	8.1	14%
	Eastman Chemical	2004	0.60	6.6	9%
	Imperial Chemical*	2004	0.78	3.5	22%
	Monsanto Co.	2004	1.66	5.5	30%
	PPG Industries	2004	2.02	9.5	21%
	3M Co.	2004	5.48	20.0	27%
				Average:	21%
Food Processing	Archer Daniels Midland	2004	1.40	36.2	4%
	Con Agra Foods	2005	1.84	14.6	13%
	Del Monte Foods	2005	0.48	3.2	15%
	General Mills	2005	2.42	11.2	22%
	Kellogg Co.	2004	2.63	9.6	27%
	Kraft Foods	2004	6.66	32.2	21%
				Average:	17%
Household Products	Clorox Co.	2005	1.07	4.4	25%
	Colgate-Palmolive	2004	3.63	10.6	34%
	Kimberly-Clark	2004	2.51	15.1	17%
	Procter & Gamble	2005	18.00	56.7	32%
	Scotts Miracle-Gro	2004	0.53	2.0	26%
				Average:	27%
Paper/Forest Products	Georgia Pacific	2004	2.07	19.7	11%
	International Paper	2004	3.06	25.5	12%
	Weyerhaeuser	2004	1.50	20.2	7%
				Average:	10%
Petroleum, Integrated	British Petroleum	2004	15.63	294.8	5%
	Chevron	2004	5.25	150.9	4%
	Conoco Phillips	2004	2.83	135.1	2%
	ExxonMobil	2004	14.95	291.3	5%
	Royal Dutch Petroleum	2004	17.23	337.5	5%
				Average:	4%
Pharmaceuticals	Bristol-Myers Squibb	2004	8.99	19.4	46%
	GlaxoSmith Kline*	2004	9.90	20.4	49%
	Eli Lilly & Co.	2004	6.98	13.9	50%
	Merck & Co.	2004	11.36	23.9	47%
	Pfizer	2004	25.66	52.5	49%

TABLE AT4.1
General Expense as a Percent of Sales (continued)

Industry	Company	Annual Report Year	General Expense ($ or £)	Sales	General Expense (% of Sales)
	Schering-Plough	2004	5.42	8.3	66%
			Average:		51%
Toiletries	Alberto-Culver	2004	1.33	3.3	41%
	Avon Products	2004	3.61	7.7	47%
	Estee Lauder	2004	3.65	5.8	63%
			Average:		50%

* Data quoted in £.

REFERENCES

1. Brown, T.R., Estimating product costs, *Chemical Engineering*, August 2000, 86–89.
2. Garrett, D.E., *Chemical Engineering Economics*, New York, Van Nostrand Reinhold, 1989, 44–62.
3. Ulrich, G.D., *A Guide to Chemical Engineering Process Design and Economics*, New York: John Wiley & Sons, 1984, 325–327, 334–338.
4. Peters, M.S. and Timmerhaus, K.D., *Plant Design and Economics for Chemical Engineers*, New York: McGraw-Hill, 1991, 194–207.
5. Garrett, D.E, *Chemical Engineering Economics*, New York, Van Nostrand Reinhold, 1989, 45.
6. Perry, R.H. and Green, D.W., *Perry's Chemical Engineering Handbook, 7th Edition*, New York: McGraw-Hill, 1997, 21–37 to 2139.
7. Seider, W.D., Seader, J.D., and Lewin, D.R., *Chemical Process Synthesis, Design and Evaluation*, New York: John Wiley & Sons, 1999.
8. Wessell, H.E., New graph correlates operating labor data for chemical process, *Chemical Engineering*, July 1952, 209–210
9. Ulrich, G.D., *A Guide to Chemical Engineering Process Design and Economics*, New York: John Wiley & Sons, 1984, 328–330.
10. Neumeister, J.M., Unpublished correspondence, April 8, 1998, and August 19, 1998.
11. Garrett, D.E., *Chemical Engineering Economics*, New York, Van Nostrand Reinhold, 1989, 52–55.

5 Economic Evaluation Methods

Chapter 2 provided the basis for doing economic option comparisons. This chapter covers the more common methods for the economic evaluation of projects and options. They are:

- *Net present value (NPV)*. This is the sum of the present values for each cash flow in a cash flow series.
- *Equivalent annual cost (AC)*. This is the sum of the annuity values for a cash flow series.
- *Return on investment (ROI)*. This is the rate of interest that an investment (a project) returns. It is equal to the i where the NPV for a cash flow series equals 0.
- *Breakeven volume*. This estimates how much volume must be generated for a proposal to break even financially. It can be used for proposals intended to develop added volume.

The chapter also covers two routinely used risk appraisal tools: sensitivity analysis and decision trees.

5.1 EVALUATION METHODS

5.1.1 NET PRESENT VALUE (NPV)

Net present value (NPV) is simply the sum of the present worths for an AT cash flow series. To calculate the NPV, one must select an economic life for the option/project, an n, and a discount rate, i, for the calculations. (When making NPV calculations, i is generally called the *discount rate*, the rate at which future cash flows are discounted to convert them to present values.) Economic life is equal to the time period between startup and obsolescence of the process or product, or between startup and when the equipment wears out. In some companies, the Financial Department may specify a standard economic life based upon company experience with project obsolescence. The discount rate is often equal to a company's minimum return on investment (or hurdle rate) for project funding.

Example 1 (NPV/simple cash flow): Find the NPV for the following cash flows. Use a 15% discount rate and a five-year economic life. Assume no depreciation is involved.

Year	AT Cash Flow
0	–220K
1	40K
2	80K
3	–30K
4	80K
5	100K

To convert these cash flows/future values to present values, use the P/F factors from the 15% Compound Interest Table. It is helpful to create the following table:

Year	n	AT Cash Flow ($K)	(P/F, 15%, n)	P ($K)
0	0	–220	1.000	–220.0
1	1	40	0.870	34.8
2	2	80	0.756	60.5
3	3	–30	0.658	–19.7
4	4	80	0.572	45.8
5	5	100	0.497	49.7

NPV = The sum of the Ps = – $48.9K

Example 2 (NPV/single project): You are working on a project that will improve labor productivity in your plant. Total BT labor and labor-related savings are expected to be $925K per year. These savings will begin one year after the making the capital investment. The project requires a capital investment of $1560K. You have estimated the added costs due to the capital investment will be $102K per year plus depreciation.* You expect your project to have a life of 10 years. Your company's hurdle rate is 20% for cost reduction projects, its tax rate is 35%, and it uses the straight-line method for depreciation. What is the NPV for your project?

The labor savings minus the added costs and depreciation occur every year from Year 1 (one year after making the capital investment) to Year 10 (the end of the project's life). Notice that this meets the definition of an annuity.

Find the annual depreciation writeoff using Equation 2.5:

Depreciation = $1560K/10 = $156K per year

The annual BT earnings increase is:

BT cash flow = Labor savings – Added costs – Depreciation

= $925K – 102K – 156K = $667K per year

---

* These increased costs are for maintenance, insurance and taxes, operating supplies, and overhead and miscellaneous.

Drawing the cash flow diagram:

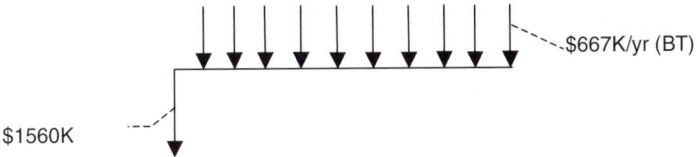

Find the AT cash flow using Equation 2.6:

$$AT\ Cash\ Flow = (BT\ Earnings)(1 - Tax\ rate) + Depreciation\ write\text{-}off$$

$$= \$667K/yr\ (1 - 0.35) + 156K/yr = \$590K\ per\ year$$

Find the NPV:

$$P_{capital} = -\$1560K$$

(This is not discounted because it is spent in Year 0. Also note that this is negative because it is a cash flow out.)

$$P_{AT\ cash\ flow} = A(P/A,\ 20\%,\ 10) = (\$590K)(4.192) = \$2473K$$

$$NPV = P_{capital} + P_{AT\ cash\ flow} = -\$1560K + 2473K = \$913K$$

5.1.2 ANNUAL COST (AC)

Annual cost (AC) is the sum of the equivalent annuities for each AT cash flow in a cash flow series. As with NPV, one must select an *n* and an *i* for the calculations. This is done in the same way as for NPV calculations. AC can be computed in two different ways: converting each cash flow into its equivalent annuity, or finding the NPV for the cash flow series and then converting that into an AC (an annuity). When calculating an equivalent annuity, one must first convert the cash flows to a present worth in Year 0.

Example 3 (AC/simple cash flow): What is the AC for the cash flows in Example 1? Again, use a 15% discount rate, a five-year economic life, and assume no depreciation.

Year	AT Cash Flow
0	−220K
1	40K
2	80K
3	−30K
4	80K
5	100K

Because the second method of AC calculation is the easiest to use in this situation, we will use it.

From Example 1, the NPV was –$49K. Converting this to an annuity:

$$A = P(A/P, 15\%, 5) = (-\$49K)(0.298) = -\$14.6K \text{ per year}$$

Example 4 (AC/single project): What is the AC for Example 2?

For this problem, the first method is easier. Because the annual AT cash flow calculated in Example 2 is already an annuity, it is the AC for the annual cash flows. What is left to do is convert the capital investment into an annuity.

$$AC_{AT \text{ cash flow}} = \$590K \text{ per year (from Example 2)}$$

$$AC_{Capital} = P(A/P, 20\%, 10) = (-\$1560K)(0.239) = \$-373K \text{ per year}$$

$$AC_{total} = AC_{AT \text{ cash flow}} + AC_{Capital} = \$590K/yr + (-\$373K/yr) = \$217K \text{ per year}$$

Checking this answer by converting the NPV from Example 2 gives:

$$AC_{total} = NPV(A/P, 20\%, 10) = (\$913K)(0.239) = \$218K \text{ per year}$$

Considering rounding errors, this is the same answer as above.

5.1.3 WHEN IS YEAR 1 FOR A TYPICAL PROJECT?

Cash flows during a project are not as simple as the examples used so far. Typically, the investments (capital, working capital, startup expenses, and introductory marketing) take place over a several year period. Increased revenues often begin as many as several years after the investments start. A good guideline is to designate Year 1 as the year revenues begin. That means some investments will happen in Year 0, Year 1, Year 2, and so on. Cash flows might look like those shown in Table 5.1.

Example 5 (NPV/complete venture): What is the NPV for the cash flows in Table 5.1? Use a 35% tax rate, a 15% discount rate, straight-line depreciation, and a 10-year project life. Assume there is no salvage value.

To solve the example, we will complete a table, Table 5.2, similar to the one in Example 1.

We must convert the capital investments in Year–2 and Year–1 into future values (in Year 0) using *F/P* factors. For Year–2:

$$F = P(F/P, 15\%, 2) = -7(1.323) = -9.3$$

Working capital must be returned at the end of the project because the economic assumption is that the project is obsolete and would be closed down. When that occurs, the working capital — inventories, accounts payable, and accounts receivable — would be eliminated. Note that working capital is returned at its original value. While other ways exist of handling the return value, this way is simple and conservative.

TABLE 5.1
Example 5: Project Cash Flows ($M)

Year	Capital	Working Capital	One-Time Startup Expense	Introductory Marketing Expense	Product Costs	Revenues
-2	-7	—	—	—	—	—
-1	-25	—	—	—	—	—
0	-8	—	-2.5	—	—	—
1	—	-30	-2.5	-8	-125	150
2	—	—	—	-6	-145	175
3	—	—	—	—	-160	195
4	—	—	—	—	-160	195
5	—	—	—	—	-160	195
6	—	—	—	—	-160	195
7	—	—	—	—	-160	195
8	—	—	—	—	-160	195
9	—	—	—	—	-160	195
10	—	30	—	—	-160	195

Next, we will calculate the BT cash flow. This is the sum of the revenues — expenses for a given year. For Year 1:

$$BT\ cash\ flow = (-2.5) + (-8) + (-125) + 150 = 14.5$$

Now we can calculate depreciation (Equation 2.5) and find the AT cash flow (Equation 2.6) for Year 1:

$$Annual\ depreciation\ write\text{-}off = (7 + 25 + 8)/10 = 4$$

$$AT\ cash\ flow = 14.5\ (1 - 0.35) + 4 = 13.4$$

Depreciation charges begin in Year 1, so for Year 0 the AT cash flow will be:

$$AT\ cash\ flow = 2.5\ (1 - 0.35) = -1.6$$

We then calculate $P_{AT\ cash\ flow}$ for Year 1:

$$P_{AT\ cash\ flow} = (13.4)(P/F, 15\%, 1) = (13.4)(0.870) = 11.7$$

The rest of the table is filled out in a similar manner, and $P_{capital}$ and $P_{AT\ cash\ flow}$ are found by summing all the individual Ps for each year. Then:

$$NPV = P_{capital} + P_{AT\ cash\ flow} = -64.8 + 115.8 = \$51M$$

TABLE 5.2
Example 5: Cash Flows and NPV ($M)

Year	F/P, 15%, n	P/F, 15%, n	Capital	Working Capital	P, Capital	One-Time Startup Expense	Introductory Marketing Expenses	Product Costs	Revenues	BT cash Flow	AT Cash Flow	P, AT Cash Flow
−2	1.323	—	−7	—	−9.3	—	—	—	—	—	—	—
−1	1.15	—	−25	—	−28.8	—	—	—	—	—	—	—
0	1	1	−8	—	−8.0	−2.5	—	—	—	−2.5	−1.6	−1.6
1	—	0.870	—	−30	−26.1	−2.5	−8	−125	150	14.5	13.4	11.7
2	—	0.756	—	—	—	—	−6	−145	175	24	19.6	14.8
3	—	0.658	—	—	—	—	—	−160	195	35	26.8	17.6
4	—	0.572	—	—	—	—	—	−160	195	35	26.8	15.3
5	—	0.497	—	—	—	—	—	−160	195	35	26.8	13.3
6	—	0.4323	—	—	—	—	—	−160	195	35	26.8	11.6
7	—	0.3759	—	—	—	—	—	−160	195	35	26.8	10.1
8	—	0.3269	—	—	—	—	—	−160	195	35	26.8	8.8
9	—	0.2843	—	—	—	—	—	−160	195	35	26.8	7.6
10	—	0.2472	—	30	7.4	—	—	−160	195	35	26.8	6.6
					Σ = −64.8						Σ =	115.8

5.1.4 NPV OR AC?

Which method should one use when comparing alternatives? There are two answers found. The first is to use the method specified by the Finance Department in your company. If they have no preference, use the method you find easiest to apply.

Both NPV and AC measure the same thing; they just express it differently. When there is one investment and the annual profit increase is a uniform cash flow, then both NPV and AC require the same number of time-value calculations — two. However, if there are several investments in different years and if the profit increase changes from year to year, then NPV is a simpler method.

Overall, NPV is the easier method to use and is the one I recommend when you have a choice.

5.1.5 RETURN ON INVESTMENT (ROI)

Return on investment (ROI) is the annual return on an investment. It is equal to the discount rate where the NPV of a cash flow series is zero. To calculate it, one must select an *n* as was done for NPV calculations.

For simple cash flows as in Example 2, one can find the ROI by calculating either *P/A* or *A/P* and finding that result in the compound interest tables.

For complex cash flows as in Example 5, one calculates NPV using several different discount rates, *i*, until the *i* where the NPV = 0 is found. A good starting point would be the company's minimum acceptable *i* (often called the *hurdle rate*). Select subsequent *i*'s using these guidelines:

- When the NPV is positive, the ROI is greater than the selected *i*;
- When it is negative, the ROI is less than the selected *i*.

Example 6 (ROI/simple cash flows): Find the ROI for the project described in Example 2.

From Example 2:

$$P = Capital = \$1560K$$

$$A = AT\ cash\ flow = \$590K\ per\ year$$

$$P/A = \$1560K/\$590K = 2.644$$

From the compound interest tables:

$$(P/A,\ 35\%,\ 10) = 2.715$$

$$(P/A,\ 40\%,\ 10) = 2.414$$

Thus, the ROI is between 35% and 40%. Interpolating, we find that the ROI = 36.1%

Example 7 (ROI/complete venture): Find the ROI for the cash flows in Example 5. From Example 5: The project life is 10 years and the NPV at 15% = \$51.1M.

Because the 15% NPV is positive, we will try a higher *i*, 20%. To find the NPV, use Table 5.3, changing the *F/P* and *P/F* factors.

$$NPV\ @\ i = 15\% = -68.3 + 94.6 = \$26.3M$$

Similarly, when *i* = 30%, NPV = \$6.6M, and when *i* = 35%, NPV = \$18M. Plotting this data in Figure 5.1, we find that the ROI is 27.5%.

5.1.6 BREAKEVEN VOLUME

If a proposal is intended to generate added volume, it is important check to figure out how much volume must be produced, so that the added revenues balance the added costs. This volume is called the *breakeven volume*. Usually, in breakeven discussions one sees revenues and costs considered on a before-tax basis. I believe

TABLE 5.3
Example 7: NPV at 20%

Year	F/P, 20%, n	P/F, 20%, n	Capital	Working Capital	P, Capital	AT Cash Flow	P, AT Cash Flow
−2	1.440	—	−7	—	−10.1	—	—
−1	1.200	—	−25	—	−30.0	—	—
0	1	1	−8	—	−8.0	−1.6	−1.6
1	—	0.833	—	-30	−25.0	13.4	11.2
2	—	0.694	—	—	—	19.6	13.6
3	—	0.579	—	—	—	26.8	15.5
4	—	0.482	—	—	—	26.8	12.9
5	—	0.402	—	—	—	26.8	10.8
6	—	0.3349	—	—	—	26.8	9.0
7	—	0.2791	—	—	—	26.8	7.5
8	—	0.2326	—	—	—	26.8	6.2
9	—	0.1938	—	—	—	26.8	5.2
10	—	0.1615	—	30	4.8	26.8	4.3
				=	−68.3		94.6

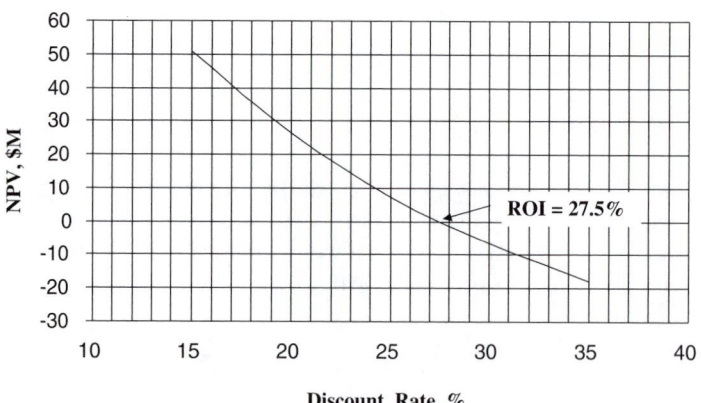

FIGURE 5.1 Example 7, NPV vs. discount rate.

a better method is to base the analysis on an after-tax basis. Consider the equation for AT cash flow, Equation 2.6:

$$AT\ Cash\ Flow = (Revenues - Expenses)(1 - Tax\ rate) + Depreciation$$

At breakeven, the AT cash flow will be zero, so rearranging Equation (2.6) and substituting *Product Costs* for *Expenses* gives an expression for the cash flow at breakeven:

$$Revenues\ (1 - Tax\ rate) + Depreciation = Product\ Cost\ (1 - Tax\ rate)\quad (5.1)$$

Breakeven volume can be illustrated by the graph in Figure 5.2. Here one plots AT product cost cash flow and AT revenues vs. volume. Breakeven volume occurs where the cost and revenue lines cross. The product cost line intersects the *y*-axis at a value equal to the fixed costs, and the revenue line intersects it at the value equal to the depreciation.

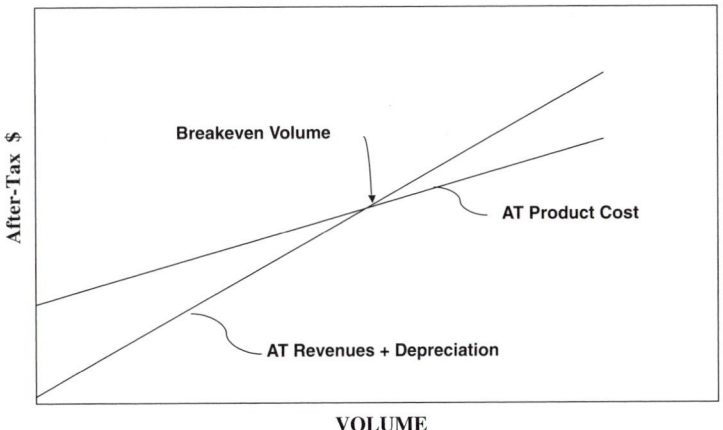

FIGURE 5.2 Breakeven volume.

Example 8: Find the breakeven volume when:

- The Product Cost is $12/case at 1000K cases per year
- The Product Cost is 75% variable
- AT Sales Revenues are $13/case
- The tax rate is 35%
- The annual depreciation write-off is $600K per year

This problem can be solved either algebraically or graphically. Use Equation 5.1 to solve algebraically. Let *x* = *Breakeven Volume*:

$$Revenues\ (1 - Tax\ rate) + Depreciation = Product\ Cost\ (1 - Tax\ rate)$$

$$= (Variable\ Costs + Fixed\ Costs)(1 - Tax\ rate)$$

$$[(\$13/case)(x)](1 - 0.35) + 600K$$
$$= [(0.75)(\$12/case)(x) + 0.25\ (\$12/case)\ (1000K\ cases)](1 - 0.35)$$

$$8.45x + 600K = 5.85x + 1950K$$

$$x = 519K \text{ cases}$$

Solving graphically, we know that at a volume of 0, the sales revenue is 0, so the AT revenue value is equal to the depreciation or $600K, and the AT product cost equals the AT fixed cost, which is $(0.25)(\$12/\text{case})(1000K \text{ cases})(1 - 0.35)$, or $1950K.

To plot the revenues and product costs, we need another point for each curve. At 1000K cases, the AT sales revenues plus depreciation are:

AT Sales Revenue + Depreciation = $(\$13/\text{case})(1000K \text{ cases})(1 - 0.035)$ + 600K = $9050K

At this same volume, we know the product costs are $12/case, or:

AT Product Costs = $(\$12/\text{case})(1000 \text{ cases})(1 - 0.35) = \$7800K$

Both curves are plotted in Figure 5.3; they intersect at a volume of 519K cases.

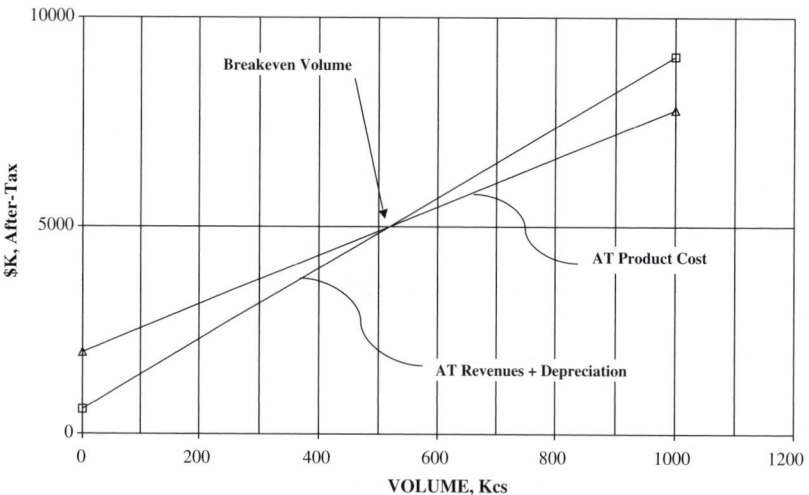

FIGURE 5.3 Example 8, breakeven volume.

If this problem had been solved using BT figures, the breakeven volume would be 750K cases. To anyone's way of thinking, this is a significant difference, again pointing out why AT numbers must be used in economic analyses. This reason is shown below:

Product Cost = Variable Costs + Fixed Costs = Sales Revenue

Let x = *Breakeven Volume*

$(0.75)(\$12/\text{case})(x) + (0.25)(\$12/\text{case})(1000\text{K cases}) = (\$13/\text{case})(x)$

$9x + 3000\text{K} = 13x$

$x = 750\text{K cases per year}$

5.2 OTHER METHODS

Several other methods are worth mentioning — benefit-cost ratio, years-to-payout, and capitalized cost. As the field of economic comparisons has matured, they have largely been replaced with the methods presented in this book.

5.2.1 BENEFIT–COST RATIO

The *benefit–cost ratio* simply divides the benefits by the cost. Most authors do not specify exactly what the benefits and costs are. They could be an annual savings and the investment that yields the savings; they could be NPVs or ACs; and they could be expressed in AT or BT dollars. Upon examination, one sees this method is a rapid ROI estimator. As such, it has the same shortcomings as ROI when comparing related options. The "Additional Topics" section of this chapter shows how to develop benefit/cost or rapid ROI calculation guidelines and discusses the use of AT and BT dollars. If one has an interest in this measure, read the material at the end of this chapter.

5.2.2 YEARS-TO-PAYOUT

Years-to-payout is defined as the year in which the total of the income generated by a project is equal to the total of its investments. Whereas most authors do not specify whether AT or BT dollars should be used when finding the payout year, I suggest using AT dollars. Years-to-payout should not be used as the only measure for alternate comparison because it has two major shortcomings. First, it cannot differentiate between projects having a low ROI from those having a high ROI. Second, it ignores cash flows after the payout year. Even so, years-to-payout used in conjunction with ROI, NPV, or AC provides some added data about a project.

5.2.3 CAPITALIZED COST

The *capitalized cost* is defined as the amount of capital required to replace a piece of equipment in perpetuity. While it is as valid a method for option comparison as NPV or AC, I find it is not as conceptually meaningful as the other two methods. I can easily conceive of what NPV or AC are, but not so for capitalized cost. It is also more difficult to calculate because one must first calculate either the NPV or AC for the project. With one of those in hand, one can then calculate the capitalized cost for the project.

5.3 ANALYSIS OF RISK

Before starting this section, quickly reviewing the section entitled "Project Risks and Risk Analysis" in Chapter 1 would be helpful.

5.3.1 SENSTIVITY ANALYSIS

Sensitivity analysis is a method that helps quantify the level of risk in a project proposal. It provides data to the decision-makers that shows how the economic results would change when key project factors vary from their estimated values. Recall that until after startup, the key factors are nothing but estimates. Actual costs will most likely be different from the estimates. These variations can have either a positive or negative impact on results. For example, the ROI would increase if less capital were spent and would decrease if more were spent.

Economic results are items such as ROI, NPV, AC, or some other factor used by a company. Key project factors include items like capital cost, production cost (or any of its components), sales volume, selling price, marketing expense (introductory or ongoing), startup expense, working capital, life of the project, and so on. The results of a sensitivity analysis are usually plotted in a diagram like the one in Figure 5.4.

By noticing the slope of the different lines, one can tell at a glance which variables have the greatest impact on financial results, The greater the slope of the line, the larger the economic impact. Looking at Figure 5.4, one can see that deviations from the Sales Volume estimate will have the biggest effect on ROI, that Production Cost variations are also significant, that Capital is not particularly worrisome, and that Startup Expense variations are insignificant.

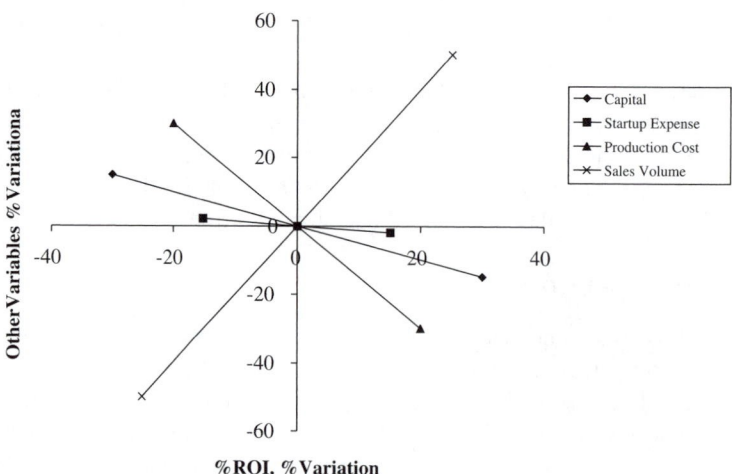

FIGURE 5.4 Sensitivity diagram.

Once one understands how the potential deviations from the estimates can affect economic results, one must decide what, if anything, must be done. Using the data in Figure 5.4 and assuming the economic risk is too high, one might reduce the risk of not meeting the ROI target by:

- Reducing the quoted ROI, assuming the new quote would still be above the company hurdle rate.
- Reducing the sales volume or production cost estimates upon which the ROI is based.
- Improving the accuracy of the sales volume or the production cost estimates. This might well involve doing a sensitivity analysis on the subparts of volume and production costs. For volume, one might analyze the variability of the estimates by sales region to see which have the greatest effect on the total estimate. For production costs, one would look at the different cost components to see which of those most effect costs.

One could then take action to reduce the variability of the most significant subparts. For example, if the raw material cost estimates were causing most of the variability, one might spend time questioning the raw material specifications, working to improve process yields or reduce material losses, doing a more detailed cost inquiry from several material venders, reexamining whether to make or buy the materials in question, and so on.

Example 9: Perform a sensitivity analysis for capital spending deviations from the expected capital investment for the problem in Example 7 in this chapter. For that project, the most probable estimates are:

ROI	27.5%
Project life	10 years
Capital cash flow	$7M, Year–2
	$25M, Year–1
	$8M, Year 0
Capital estimate accuracy	As high as $60M (+50%)
	As low as $32M (–20%)

You must now calculate the ROI and percent variation from the target of 27.5% for the two extremes of capital spending.

Assume the same year by year capital split as for $40M:

Year	For $40M	For $32M	For $60M
– 2	7M	5.6M	10.5M
– 1	25M	20.0M	37.5M
0	8M	6.4M	12.0M

Find the ROI for \$32M of capital. We will first assume a discount rate of 35%. The cash flows are shown in Table 5.4.

TABLE 5.4
Example 9: Cash Flows

Year	F/P, 35%, n	P/F, 35%, n	Capital	Working Capital	P, Capital	AT Cash Flow	P, AT Cash Flow
−2	1.8225	—	−5.6	—	−10.2	—	—
−1	1.35	—	−20.0	—	−27.0	—	—
0	1	1	−6.4	—	−6.4	−1.6	−1.6
1	—	0.7407	—	−30	−22.2	13.4	9.9
2	—	0.5487	—	—	—	19.6	10.8
3	—	0.4064	—	—	—	26.8	10.9
4	—	0.3011	—	—	—	26.8	8.1
5	—	0.2230	—	—	—	26.8	6.0
6	—	0.1652	—	—	—	26.8	4.4
7	—	0.1224	—	—	—	26.8	3.3
8	—	0.0906	—	—	—	26.8	2.4
9	—	0.0671	—	—	—	26.8	1.8
10	—	0.0497	—	30	1.5	26.8	1.3
		subtotals		=	−64.3		57.3
						NPV=	−7.0

Similarly, one calculates the ROI for a number of discount rates until NPV = 0. For this above set of cash flows, this occurs at an i of 31.7%, the ROI. The same process is used to find the ROI for cash flows for \$60M of capital; this ROI is 20.5%. These results are plotted in Figure 5.5. Note that the line is not straight, a situation that often occurs.

5.3.2 DECISION TREES

Decision trees are useful when comparing the possible outcomes of decision choices. In diagram form, they show:

- Each decision being considered, including connected future events
- The projected outcomes for all the future events and an estimate of the probabilities for each event
- The probabilities and outcomes are then used to estimate the probability-weighted outcome for each decision

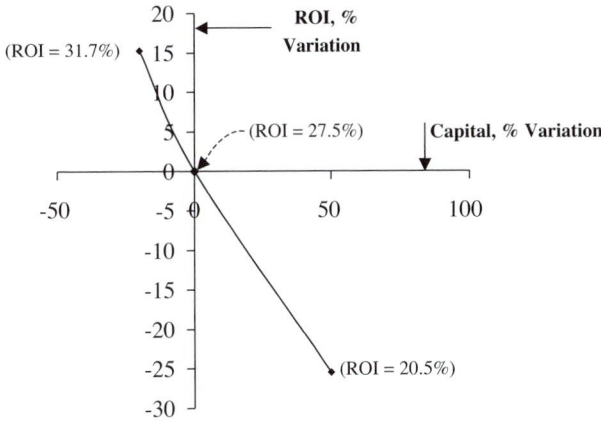

FIGURE 5.5 Example 9, sensitivity diagram.

We will use two diagramming conventions — decisions are shown by a square and probable outcomes by a circle.

Consider the following situation: you have a chance to win $450 if you can correctly guess whether a coin flip will be heads or tails. For the opportunity to win, you must pay $250. If you win, your net gain will be $200 ($450 – $250). Assume you choose heads. (The probability of the flip coming up heads is 50% every time the coin is tossed.) The decision tree is shown in Figure 5.6.

	Probability	Outcome	Weighted Outcome
Heads	0.5	$450	= $225
Tails	0.5	0	= 0
			$225
No choice made	1.0	0	= 0

Flip the coin: -$250

Don't flip

FIGURE 5.6 Decision tree, coin flipping.

For the branch "Flip the coin," the cost is $250 and the probability-weighted outcome is $225. Thus, the expected value of this branch is the cost plus the probability-weighted outcome, or –$25; and the expected value of the "Don't flip" branch is 0 + 0 = 0. Because the "Don't flip" branch has the higher expected value, it is the best economic choice. (For simplicity, taxes were not considered in this example.)

Although more complex, decision trees for project decisions are constructed in a similar manner. Several differences are worth mentioning:

- Investments and outcomes will often occur in different years. When that is the case, NPVs or ACs must be used to compare the expected outcomes.
- AT cash flows should be used for the reasons explained earlier.
- Estimating of the probabilities of the different branches for project decisions is not an exact science. It is most often performed by the people who know the most about the project. They may meet as a group, coming to a consensus opinion, or they may be interviewed by someone who has experience assessing probabilities. In the latter case, the interviewer develops the estimates based on the data from the interviews.

Example 10: As part of a new product feasibility, you are studying how much capacity should be built for making and packing a new product. Because a great deal of uncertainty exists about how much volume the new product will generate, you have decided to do a decision tree analysis to help sort out the best course of action. At present, the estimates for market development (assume it takes two years to reach final market volume) are:

- Full development: 400K units per year. The probability of this happening is about 30%.
- Medium development: 280K units per year. The probability is about 60%.
- The product fails. The probability is about 10%.

Other data:

- The capital cost for the 400K-unit plant is $7.5M, and for the 280K-unit plant, it is $6M. Assume it is spent in Year 0.
- Annual AT Profit for the 400K-unit plant is estimated to be $9/unit per year when making 400K units per year. When running at 280K units, the 400K-unit plant will lose $5/unit per year.
- If a 280K-unit plant is built and the market is 280K units per year, the profit will be $6/unit per year. If the market develops to 400K units per year, a contract manufacturer can be online at the end of Year 2 to produce the added units. That production will generate an AT profit of $5/unit per year.
- If the product fails, the plant will be mothballed and the equipment reused for other purposes. Assume this would happen at the end of Year 2. Assume losses of $2M per year for the first two years (until the plant is shutdown). For the 400K-unit plant, the salvage value is $2.1M and the AT mothballing expenses are $0.5M. For the 280K-unit plant, the salvage value is $1.7M and AT mothballing costs are $0.4M.
- The Marketing and Sales Departments have estimated the product will have about a 10-year life, so you plan to use that as the economic life for your study.

- Use a hurdle (discount) rate of 15%.
- For simplicity, ignore the salvage value in Year 10, startup expenses, and working capital.

First, draw the decision tree (Figure 5.7).

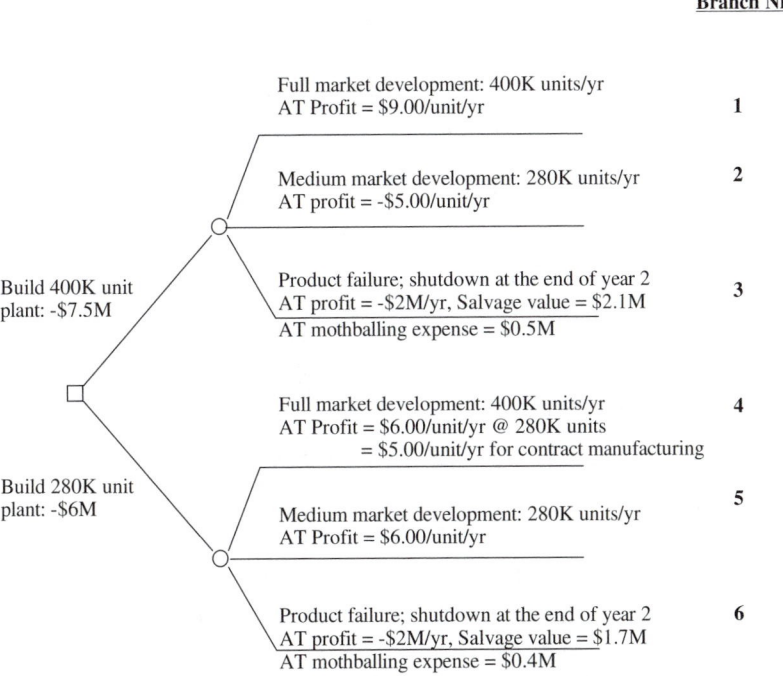

Branch Nr

Full market development: 400K units/yr
AT Profit = $9.00/unit/yr — **1**

Medium market development: 280K units/yr — **2**
AT profit = -$5.00/unit/yr

Product failure; shutdown at the end of year 2 — **3**
AT profit = -$2M/yr, Salvage value = $2.1M
AT mothballing expense = $0.5M

Build 400K unit plant: -$7.5M

Full market development: 400K units/yr — **4**
AT Profit = $6.00/unit/yr @ 280K units
= $5.00/unit/yr for contract manufacturing

Build 280K unit plant: -$6M

Medium market development: 280K units/yr — **5**
AT Profit = $6.00/unit/yr

Product failure; shutdown at the end of year 2 — **6**
AT profit = -$2M/yr, Salvage value = $1.7M
AT mothballing expense = $0.4M

FIGURE 5.7 Example 10, decision tree.

Find the probability-weighted NPV for each of the possible decisions. For the decision to build the larger plant, this means finding the probability-weighted NPVs for branches 1, 2, and 3. Similarly for the smaller plant, calculate branches 4, 5, and 6. For branch 1, the AT profit for Year 1 through Year 10 is:

$$AT\ profit/yr = \$9/unit * 400K\ units/yr = \$3.6M\ per\ year$$

Next, find the present value of the profit in Year 1 and multiply that by the probability for the branch (30%):

$$Probability\text{-}weighted\ P = (\$3.6M)(P/F, 15\%, 1)(0.3) = (3.6)(0.8696)(0.3) = \$0.94M$$

The rest of the calculations are similar. Because a number of the cash flows are not uniform, the NPVs will be easier to calculate using a table similar to the one in Example 9. Table 5.5 is for the 400K-unit per year plant and Table 5.6 is for the 280K-unit plant.

TABLE 5.5
Example 10: Probabilty-Weighted NPV, 400 Unit/Year-Plant

Year	P/F, 15%, n	Initial Capital	P, Capital	Branch 1				Branch 2				Branch 3						Total P, All Branches
				AT Profit	P	Probability	Probability Weighted P	AT Profit	P	Probability	Probability Weighted P	AT Profit	Salvage Value	Mothballing	P	Probability	Probability Weighted P	
0	1	−7.5	−7.5	—	—	0.3	—	—	—	0.6	—	—	—	—	—	0.1	—	−7.5
1	0.8696	—	—	3.6	3.13	0.3	0.94	−2.0	−1.74	0.6	−1.04	−2	—	—	−1.74	0.1	−0.17	−0.28
2	0.7561	—	—	3.6	2.72	0.3	0.82	−2.0	−1.51	0.6	−0.91	−2	2.1	0.5	−0.30	0.1	−0.03	−0.12
3	0.6575	—	—	3.6	2.37	0.3	0.71	−2.0	−1.32	0.6	−0.79	—	—	—	0.00	0.1	0.00	−0.08
4	0.5718	—	—	3.6	2.06	0.3	0.62	−2.0	−1.14	0.6	−0.69	—	—	—	0.00	0.1	0.00	−0.07
5	0.4972	—	—	3.6	1.79	0.3	0.54	−2.0	−0.99	0.6	−0.60	—	—	—	0.00	0.1	0.00	−0.06
6	0.4323	—	—	3.6	1.56	0.3	0.47	−2.0	−0.86	0.6	−0.52	—	—	—	0.00	0.1	0.00	−0.05
7	0.3759	—	—	3.6	1.35	0.3	0.41	−2.0	−0.75	0.6	−0.45	—	—	—	0.00	0.1	0.00	−0.05
8	0.3269	—	—	3.6	1.18	0.3	0.35	−2.0	−0.65	0.6	−0.39	—	—	—	0.00	0.1	0.00	−0.04
9	0.2843	—	—	3.6	1.02	0.3	0.31	−2.0	−0.57	0.6	−0.34	—	—	—	0.00	0.1	0.00	−0.03
10	0.2472	—	—	3.6	0.89	0.3	0.27	−2.0	−0.49	0.6	−0.30	—	—	—	0.00	0.1	0.00	−0.03
																NPV=	−8.31

TABLE 5.6
Example 10: Probability-Weighted NPV, 280 Unit/Year-Plant

Year	P/F, 15%, n	Initial Capital	P, capital	Branch 4				Branch 5				Branch 6						Total P, All Branches
				AT Profit	P	Probability	Weighted P	AT Profit	P	Probability	Weighted P	AT Profit	Salvage Value	Mothballing	P	Probability	Weighted P	
0	1.0000	−6	−6	—	—	0.3	—	—	—	0.6	—	—	—	—	—	0.1	—	−6.00
1	0.8333	—	—	1.68	1.40	0.3	0.42	1.68	1.40	0.6	0.84	−2	—	—	−1.67	0.1	−0.17	1.09
2	0.6944	—	—	1.68	1.17	0.3	0.35	1.68	1.17	0.6	0.70	−2	−1.7	−0.4	−2.85	0.1	−0.28	0.77
3	0.5787	—	—	2.28	1.32	0.3	0.40	1.68	0.97	0.6	0.58	—	—	—	0.00	0.1	0.00	0.98
4	0.4823	—	—	2.28	1.10	0.3	0.33	1.68	0.81	0.6	0.49	—	—	—	0.00	0.1	0.00	0.82
5	0.4019	—	—	2.28	0.92	0.3	0.27	1.68	0.68	0.6	0.41	—	—	—	0.00	0.1	0.00	0.68
6	0.3349	—	—	2.28	0.76	0.3	0.23	1.68	0.56	0.6	0.34	—	—	—	0.00	0.1	0.00	0.57
7	0.2791	—	—	2.28	0.64	0.3	0.19	1.68	0.47	0.6	0.28	—	—	—	0.00	0.1	0.00	0.47
8	0.2326	—	—	2.28	0.53	0.3	0.16	1.68	0.39	0.6	0.23	—	—	—	0.00	0.1	0.00	0.39
9	0.1938	—	—	2.28	0.44	0.3	0.13	1.68	0.33	0.6	0.20	—	—	—	0.00	0.1	0.00	0.33
10	0.1615	—	—	2.28	0.37	0.3	0.11	1.68	0.27	0.6	0.16	—	—	—	0.00	0.1	0.00	0.27
																NPV=		0.37

The calculations show that building the smaller plant is best economically because its NPV is \$0.37M, as compared to –\$8.31M for the larger plant. From this data, one can also see that the probability-weighted ROI for the small plant is just above 15% (the discount rate) because the NPV is slightly positive.

Also note that if one only considered the large plant at full market development (an optimistic assumption), the NPV would be \$10.6M and the ROI around 47%. Had company decision-makers not seen the decision tree and were only given the optimistic estimate, they would have mistakenly authorized the large plant. This points out the value of decision trees whenever a great deal of uncertainty exists in the project assumptions.

5.4 SUMMARY

For NPV, AC, and ROI calculations:

- Use your company's hurdle rate as the discount rate; or if a specific discount rate has been established for a project, use that.
- Establish Year 1 as the year in which revenues begin.
- Use AT cash flows per Equation 2.6.
- Use obsolescence (product, process, equipment) or company financial guidelines to set the economic life for the comparison.
- Options having an ROI greater than the company's hurdle rate would be acceptable.

For finding the breakeven volume, use AT cash flows as defined by Equation 2.6. Solve for the breakeven point using Equation 5.1.

Two useful tools are available for analyzing and understanding the economic risks in a project — sensitivity analysis and decision trees. Sensitivity analysis shows how economic results would vary when the key project factors, such as capital cost, are different from their estimated amounts. Decision trees enable project decision-makers to assess the impact of their decisions by probability-blending all the potential outcomes.

5.5 PROBLEMS AND EXERCISES

1. Find the NPV for a savings project that has cash flows of:
   a. Capital = \$1000K (in Year 0)
   b. Annual energy savings (BT) = \$600K (savings start one year after the capital is invested)
      Use a discount rate of 15%, a project life of 10 years, and a tax rate of 35%.
2. Find the AC for the cash flows in Problem 1.
3. Find the ROI for the cash flows in Problem 1.

4. What is the breakeven volume when:
   a. The production cost is $14.50/case when producing 1000K cases per year
   b. The production cost is 65% variable
   c. The sales revenue is $18/case
   d. The tax rate is 35%
   e. The annual depreciation writeoff is $950K/yr
5. Two unrelated capital projects are being proposed to your boss. She has asked you to estimate their ROIs and to recommend whether or not either should be funded. Your company's hurdle rate is 10% and its tax rate is 32%. Assume a 10-year project life for both. What is your recommendation?
   - *Proposal 1*. Invest $11.9M to improve manufacturing productivity and yield. This will reduce net manufacturing costs* by $2.6M per year.
   - *Proposal 2*. Invest $5M for a new warehouse to reduce outside warehousing.
6. Calculate the NPV for the following cash flows. Use a 12% discount rate.

Year	AT Cash Flow ($K)
1	−190
3	10
4	17
7	−15
1–10	33

7. Calculate the AC for the cash flows in Problem 7.
8. If you invest $100K in Year 0 and it returns $18.5K (BT) for 10 years, what is the AT ROI? Assume a 28% tax rate.
9. For a product to have a minimum breakeven volume of 60K tons/yr (TPY) and selling price of $800/ton, what is the product cost in $/T at 60KTPY? Assume the product cost is 65% variable. The depreciation writeoff is 720K/yr. Use a tax rate of 33%.
10. Do a sensitivity analysis for the variation of project ROI versus:
    a. Capital: Assume a variation of +30% to −10%
    b. Product costs: Assume a variation of +20% to −10%
    c. Revenues: Assume a variation of +5% to −15%
    d. Introductory marketing expense: Assume a variation of +100% to −30%
    e. Startup expense: Assume a variation of +30% to −20%
    f. The most probable estimates for the project are:
       - ROI   18%

---

* See Appendix I for the definition of net savings.

- Project life   10 years
- Cash flows   See table below

Year	Capital	Working Capital	Startup Expense	Introductory Marketing Expense	Product Costs	Revenues
-2	-4	—	—	—	—	—
-1	-30	—	-3	—	—	—
0	-5	—	-2.5	-9	—	—
1	—	-16	—	-7	-60	50
2	—	—	—	—	-85	85
3	—	—	—	—	-95	111
4	—	—	—	—	-95	111
5	—	—	—	—	-95	111
6	—	—	—	—	-95	111
7	—	—	—	—	-95	111
8	—	—	—	—	-95	111
9	—	—	—	—	-95	111
10	—	16	—	—	-95	111

11. Calculate the NPV and ROI for the following project proposal. All cash flows shown are BT and are in $M. Use a discount rate of 12% and a tax rate of 32%. Assume a 10-year project life. Return working capital in Year 10.

Year	Capital	Working Capital	Startup Expense	Introductory Marketing Expense	Product Costs	Revenues
-2	-15	—	—	—	—	—
1	-82	—	-6	—	—	—
0	-16	—	-5	-19	—	—
1	—	-41	—	-12	-150	145
2	—	—	—	—	-225	238
3	—	—	—	—	-250	278
4	—	—	—	—	-250	278
5	—	—	—	—	-250	278
6	—	—	—	—	-250	278
7	—	—	—	—	-250	278
8	—	—	—	—	-250	278
9	—	—	—	—	-250	278
10	—	41	—	—	-250	278

12. Do a sensitivity analysis of NPV versus capital cost for:
    a. Most probable capital (spent in Year 0) of $8.3M
    b. Most probable annual AT cash flow (received in Year 1 through Year 15) = $1.3M

   c. Most probable NPV = $1.6M at a discount rate of 10%

   d. Other than those resulting from changes in the capital-ratioed expenses and depreciation, assume no other changes the AT cash flow

   e. Assume the expected variation in the capital is +35% to –5%

   f. Assume a tax rate of 35%

13. You are being offered a chance to win money by rolling a die. You must guess which face of the die will be up after you roll. If you guess correctly, you will win money — $10 for a 4 or 6 and $2 for any other number. To play, you must pay $5. Using a decision tree analysis, decide whether or not you should play. Explain your rationale.

5.6 ADDITIONAL TOPICS

5.6.1 Creating Guidelines for Rapid ROI Calculation

In the course of a project, especially in its early stages, one checks the ROI of a project many times to ensure it continues to make good business and economic sense. This section describes how to develop rapid ROI estimation guidelines. This is best explained by working an example.

Example AT5.1: Calculate a rapid ROI estimator based on a hurdle rate of 15%, a project life of 10 years, and a 35% tax rate.

To solve this problem, a rephrasing will help: what amount of annual savings or revenues (both BT and AT) will justify the investment of $1.00 of capital? By converting the $1.00 of capital into its equivalent annuity ($i = 15\%$, $n = 10$ years), we solve for the AT cash flow that will justify the $1.00:

$$A = P(A/P, 15\%, 10) = (1)(0.199) = \$0.199 \text{ per year}$$

Next, we translate this into the BT cash flow or BT annual net savings using Equation 2.6:

AT Cash flow = (Net BT earnings)(1 – Tax rate) + Depreciation

Substituting Capital/Life for depreciation and rearranging,

Net BT revenues = (AT Cash flow – Capital/Life)/(1 – Tax rate)

$$= (0.199 \ 1/10)/(1 \ 0.35) = \$0.152 \text{ per year}$$

Again using Equation 2.6 and using a factor of 11% to estimate the maintenance, insurance and taxes, operating supplies, and plant overhead costs, we can translate the AT cash flow into gross BT revenues:

AT Cash flow = [(Gross BT revenues) – (0.11)(Capital) (Depreciation)]
(1 – Tax rate) + Depreciation

$$0.199 = 0.65 \ (Gross\ BT\ revenues - 0.11 - 0.1) + 0.1$$

$$\textit{Gross BT revenues} = (0.199 + 0.037)/0.65 = \$0.362 \text{ per year}$$

Thus, to justify investing $1.00 of capital, one must be able to generate the following revenues:

- $0.199 per year of AT cash flow
- $0.152 per year of net BT revenues
- $0.362 per year of gross BT revenues

A more useful way to express these is that the ROI $\geq$ 15%:

- When the capital/AT cash flow ratio $\leq$ 5 (5 ~ 1/0.199)
- When the capital/net BT revenues $\leq$ 6.6
- When the capital/gross BT revenues $\leq$ 2.8

Using this methodology, one can calculate estimation guidelines for any set of circumstances – ROI, project life, and tax rate.

Example AT5.2: Is the ROI of a $22.8M capital project that returns $9.2M per year of gross BT revenues equal to or greater than 15%? Assume a 10-year life and a tax rate of 35%. Because these economic conditions are the same as in Example AT5.1, we will use those guidelines.

Calculate the Capital/Revenue ratio:

$$\textit{Ratio} = \$22.8M/\$9.2M = 2.48$$

Because that is less than 2.8, the project has better than a 15% ROI.

5.6.2 What to Use — Constant Dollars or Actual Dollars

So far, the calculations in this book have ignored the effects of inflation; they have used constant dollars. This section deals with whether or how one should should deal with inflation. First, we will define three terms:

- *Constant dollars* — These are dollars valued at Year 0. The effects of inflation are not considered in NPV, AC, or ROI calculations.
- *Actual dollars* — These are dollars valued at the time they are spent or received as income. Thus, they have been inflated versus Year 0.
- *Purchasing value dollars* — These are dollars expressed in terms of what they could buy in some reference year.

Table AT5.1 compares cash flows for these three types of dollars when inflation is 5%. Note that the cash flows in Year 0 (before inflation comes into play) are the same in all three scenarios.

As one can see, the BT cash flows are quite different from one another when inflation/deflation is brought into the picture. Whether to use constant or actual

Table AT5.1
Dollar Comparisons

	Constant $		Actual $*		Purchase Value $†	
Year	Capital Investment	Annual BT Cash Flow	Capital Investment	Annual BT Cash Flow	Capital Investment	Annual BT Cash Flow
0	1000K	—	1000K	—	1000K	—
1	—	200K	—	210K	—	191K
2	—	200K	—	221K	—	181K
3	—	200K	—	232K	—	173K
4	—	200K	—	243K	—	165K
5	—	200K	—	255K	—	157K

* Inflated at 5% per year.

† Purchase value expressed in Year 0 dollars.

dollars in financial calculations is a bit controversial. The key points in the discussion are:

- In the early stages of a design, when the basic financial data is uncertain, adding the refinement of inflation does not really increase accuracy of NPV or AC calculations.
- The ability to estimate future inflation rates is really guesswork, so calculations based upon future inflation guesses are of questionable accuracy and value.
- In the past, when inflation rates were low — around 5% or less — inflation effects were usually ignored. It was assumed that the competitions' costs and revenues would escalate at the same rate as they did in one's company, making it unimportant to deal with inflation. Additionally, it was assumed that prices increase at about the rate as costs, effectively erasing the impact of inflation.
- In industries with a lot of innovation (new and upgraded products), the failure rate of these products will overshadow inflationary effects.
- When comparing alternates, the absolute values of NPV or AC are less important than differences among the options. This permits less accuracy in the individual numbers, so I suggest ignoring inflation when performing alternate evaluations.
- When estimating the ROI of a proposal, the absolute value of the ROI is important, leading one to think a little differently about bringing inflation into these calculations. How a company sets its hurdle rates will help dictate whether or not to base ROI calculations on actual dollars. If rates

are set based upon the company's actual ROI experience, then inflation is dealt with in the experience base and need not be included in the calculations.

Thus, in most cases, I believe it is appropriate to ignore inflation in financial comparisons. However, if one does need to include inflationary effects in the ROI calculations, approximating the impact is acceptable in the early stages of design. An acceptable approximation for inflation rates up to 10% is:

$$ROI_{Actual} \sim (0.3) \ (Inflation \ rate) + ROI_{Constant \ dollars} \qquad (AT5.1)$$

Example AT5.3: For a project having the following cash flows — capital of $100K (spent in Year 0) plus revenues and product costs of $30K per year and $25K per year (Year 1 through Year 10) — find the ROI for four situations: constant dollars, 5% per year, 10% per year, and 15% per year inflation rates. Use a 35% tax rate and a project life of 10 years.

Compare ROIs calculated exactly to those estimated by Equation (AT5.1).

Constant dollars:
Cash flow$_{BT}$ = Revenues − Product cost = $30K/yr − $25K/yr = $5K per year

Cash flow$_{AT}$ = (Cash flow$_{BT}$)(1 − Tax rate) + Depreciation

= ($5K/yr)(1 − 0.35) + $100K/10 = $13.25K per year

Calculate P/A; and using the compound interest tables, find the ROI.

P/A = $100K/$13.25K = 7.547

(P/A, 5.5%, 10) = 7.538; (P/A, 5%, 10) = 7.722

Call the ROI 5.5%

Inflated dollars at 5% per year:
The calculations are summarized in Table AT5.2. Explaining a few calculations is in order. These are shown for Year 1.

BT Cash flow = (Revenues − Product cost) (1 + Inflation rate)^n

= (30K − 25K)(1 + 0.05)^1 = 5.25K

Depreciation = Capital/Project life = 100K/10 = 10K

AT Cash flow = (BT Cash flow)(1 − Tax rate) + Depreciation

= (5.25K)(1− 0.35) + 10 = 13.41K

P = F(P/F, 6%, 1) = 13.41K (0.9434) = 12.65K

TABLE AT5.2
Example AT5.3: 5% Inflation

Year	Capital	BT Cash Flow, 5% Inflation	Depreciation	AT Cash Flow=	$P, i = 6\%$	$P, i = 7\%$
0	−100	—	—	—	—	—
1	—	5.25	10	13.41	12.65	12.54
2	—	5.51	10	13.58	12.09	11.86
3	—	5.79	10	13.76	11.56	11.23
4	—	6.08	10	13.95	11.05	10.64
5	—	6.38	10	14.15	10.57	10.09
6	—	6.70	10	14.36	10.12	9.57
7	—	7.04	10	14.57	9.69	9.08
8	—	7.39	10	14.80	9.29	8.61
9	—	7.76	10	15.04	8.90	8.18
10	—	8.14	10	15.29	8.54	7.77
Total P =	-100				104.46	99.58
				NPV=	4.46	-0.42

To find the ROI, interpolate to find where the NPV = 0. The ROI = 6.9%.

Inflated dollars at 10% per year and 15% per year:
Using similar methodology, the ROI at 10% inflation is 8.7% and at 15% inflation is 10.7%.

Summary: The table below compares the exact calculations to those from Equation (AT5.1). It shows good agreement when the inflation rate is 10% or less.

	Actual ROI (%)	ROI from Equation AT5.1 (%)
Constant dollars	5.5	5.5
5% inflation	6.9	7.0
10% inflation	8.7	8.5
15% inflation	10.7	10.1

Section II

Economic Design

6 Economic Design: A Model

Companies and exist for one simple reason — to make money for their shareholders and investors. The good companies do this while operating legally, ethically, and responsibly. The ones that cannot make money or that will not operate properly will eventually cease to exist. These will go bankrupt or be purchased for their assets.

This chapter presents a method for creating economically viable designs — for producing economic designs. Recall that economic viability requires cost-competitive products and plant designs that economically balance capital and production costs. The method has three phases:

- Defining business and technical objectives for the project before starting work
- Creating a list of options to be studied
- Analyzing the options and selecting the most economic

Most of an engineer's impact on costs occurs in the early stages of a design — during process development, feasibility, and conceptual engineering. Thus, the analytical part of the process uses simple tools, ones designed for use when few design details exist.

6.1 DEFINING ECONOMIC DESIGN

One finds many definitions of economic design. These include lowest capital cost, lowest production cost, highest rate of return, and so on. All fall short of the mark. In this book, the definition is more comprehensive: it means finding the economic balance between capital costs and production costs. To create an economic design, one must answer two questions:

- Is it better to spend more capital and have lower production costs?
- Is it better to spend less capital and have higher production costs?

Two examples show how one might apply these questions.

Example 1: One must decide how much energy to reclaim in the process shown below. The process involves heating a feed stream to reaction temperature, reacting it adiabatically, and cooling the reacted stream to storage temperature.

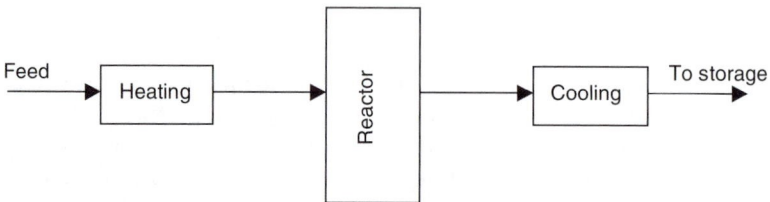

To avoid some of the energy use, one could add a heat exchanger before the heating step. The hot stream from the reactor would heat the feed and would be cooled. Adding the "interchanger" decreases utility usage, both heating and cooling, and reduces the sizes of the heater and cooler.

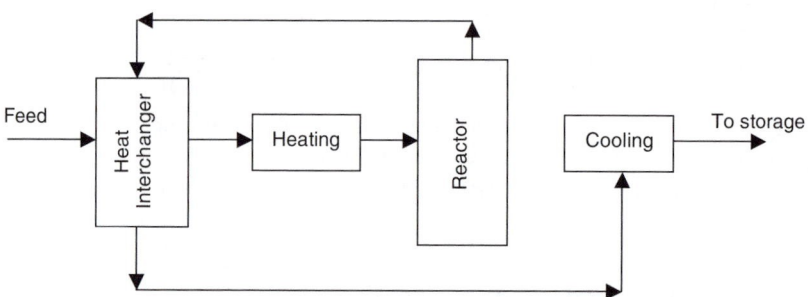

Adding heat interchange creates an almost infinite number of technically acceptable options. Each has a different size interchanger, heater, and cooler and reclaims a different amount of energy. One finds the more capital spent, the larger the energy saving or profit increase. Whenever one is balancing capital and production cost spending, annualized expenditures reach a minimum as shown in Figure 6.1. This is the economic design point. The curves are typical where a continuum of options is found.

Example 2: You are to select the type of reactor to use for a catalyzed gas-liquid reaction. Assume that three technically acceptable options exist and that the catalyst is a finely divided solid, slurried in the feed liquid. These are:

- A gas-sparged batch reactor with gas recirculation, gas purge, a 180-min reaction time, and a catalyst usage of 0.008%
- A well-agitated batch reactor with venting of the excess gas, a 120-min reaction time, and a catalyst usage of 0.01%
- A continuous, agitated, multistage reactor with venting of the excess gas, a 5-min reaction time, and a catalyst usage of 0.15%

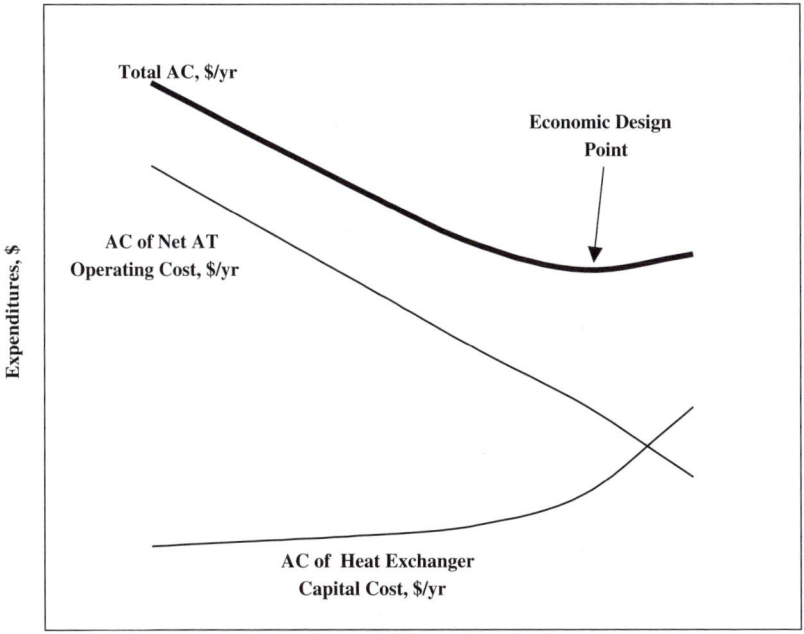

FIGURE 6.1 Balancing capital and production costs.

To select the most economic option, one first estimates the capital and production costs for each option. As each reactor system is quite different, the costs for each also will be different. The major design differences affecting capital costs are shown in the table below:

	Gas-Sparged	Well-Agitated	Continuous, Multistage
Reactor size	Large	Smaller	Smallest
Agitation	Gas-sparging ring	High rpm turbine with side-wall baffling	High rpm turbine with side-wall and interstage baffling
Other	Gas-recirculating system		

Production costs are also different for each option because gas losses, catalyst usage, operating labor, electric power, maintenance costs, insurance, taxes, overhead, and depreciation vary from option to option. After estimating the capital and production costs for each option, one calculates the NPV or AC for each. The most economic is the one having the largest NPV or AC.

Worth noting is a distinction between Example 1 and Example 2. Example 1 has a continuum of options, whereas Example 2 has a few distinct options. Both situations occur routinely.

6.2 THE ECONOMIC DESIGN MODEL

Figure 6.2 illustrates the economic design model. Use of the model is iterative; one cycles through it in each phase of engineering. Whereas this book only deals with the product development, feasibility, and conceptual phases of a design, one can use the model in all project phases.

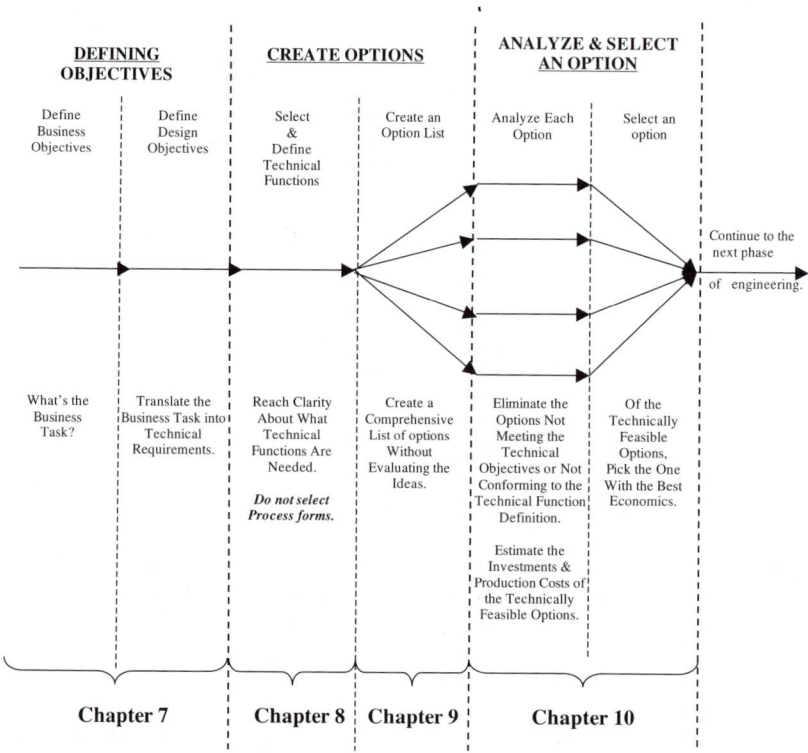

FIGURE 6.2 Economic design model.

6.2.1 DEFINING OBJECTIVES

It is vital to have management's business goals clear before any major amount of engineering begins. Once these are clear, they can be translated into the technical needs that will guide the design process. As a project moves through the process development, feasibility, and conceptual phases of design, objectives evolve and become more detailed and precise.

Often objectives are not clear because management is busy and has not had the time to figure out what they want the project to do for the company. If work begins without clear goals, management-driven changes often cause cost and schedule upsets. The changes occur when design features emerge that do not match what management really wants.

To help management become clear about the objectives, engineers can assist them by conducting short interviews designed to engage them in the project. A good way to engage the managers is to ask them a series of probing and discriminating questions. This helps the managers think through their views on the project, conflicting priorities, quality/schedule/cost tradeoffs, and so on. The level of the managers and number of interviews depends upon the project importance, size, and cost.

Following the interviews, the interviewer summarizes the results in a single text. This ensures that only one set of objectives is used by those on the project. If different managers have conflicting points of view, the interviewer can also use the single text to help resolve the differences.

6.2.2 CREATING OPTIONS

The heart of economic design is a thorough and complete list of options. Options identified in this step of the model will be studied in the next.

Too often, the list of options is incomplete, resulting in lost profit. The primary reasons option lists are incomplete:

- Engineers often converge quickly upon a design. Rapid convergence is most likely the result of the engineers' excellent problem solving skills. Their skills allow them to quickly develop a flowsheet and select unit operations and hardware.
- Engineers are often uncertain about how to go about creating a list of options and about what all the options might be. After putting a few items on the study list, they feel their list is good enough and they stop working on it.

Both reasons result in the exclusion of possible options. When the option list is incomplete, project economics will probably suffer. To illustrate, if one had to heat a process water stream, they might immediately select a shell and tube unit for the heating task. However, a steam injector could also do the job and is much less expensive. Had one initially selected the exchanger, s/he would not have considered the injector option. As a result, more capital would have been spent and more steam used.

6.2.2.1 Flowsheet Development

The method to offset rapid convergence is simple. It involves a disciplined, three-step process for flowsheet creation or development. It begins with a generic block flow diagram, converts this to a new type of block flow diagram (the technical function flowsheet), and ends with a process flow diagram. *Technical function flowsheets* focus on the functions that are needed in a process before deciding upon the unit operations that will be in the process. Keeping the focus on functions lets the engineer consider all the possible unit operations before selecting one. The difference between functions and unit operations is shown below.

Function	Unit Operation
Separate solids from a liquid stream	Filter
Heat a heat transfer fluid	Furnace
React three liquid streams	Jacketed reactor
Remove a component from a gas stream	Packed column absorber
Reduce the moisture content of granules	Rotary kiln

Figure 6.3 is an example of a technical function flowsheet. It is the first issue for a process to hydrogenate vegetable oil and would have been issued during the early part of process development. Note that only functions are shown in the different flow sheet blocks — heat oil, hydrogenate oil, remove catalyst, and cool oil. As flowsheet definition progresses, more details would be added such as the selection of unit operations, selection of equipment, addition of purge and recycle streams, addition of tanks, and so on.

1ST Issue

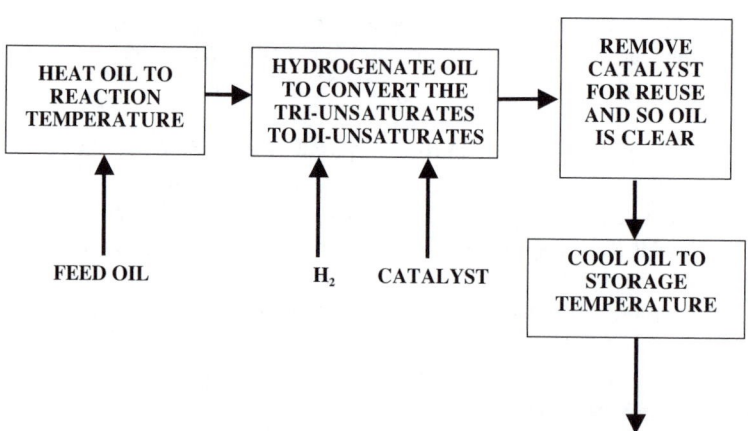

FIGURE 6.3 Technical function flowsheet, oil hydrogenation.

Not much detail is found on the flowsheet, so an associated "technical function definition" is needed. This definition covers key details for each block on the flowsheet. One usually develops these details in the laboratory, in the pilot plant, or by plant experimental testing. To illustrate, consider the first block of the flowsheet, oil heating. The definition statement might be as simple as stating, "Heat the oil is to the reaction temperature, 345°F to 355°F." If other conditions were vital to correctly making the specified product, such as a maximum skin temperature in the heating equipment, they would also be stated. One would also list the bases of the specifications, e.g., lab, pilot plant data, reports, and so on. This way, those having questions or needing clarification can easily refer to the source documents.

Technical function flowsheets and definitions are covered in detail in Chapter 8.

6.2.2.2 Creating an Option List

When done well, creating a process flowsheet involves exploring many options. The focus on technical functions sets the stage for doing this thoroughly. The first step is to produce a comprehensive list of options. Once the list is complete, analysis — first technical, then economic — begins. It is important to separate list creation and analysis. Combining the two results in the prejudgment of ideas, which gets in the way of idea generation. Several methods are detailed in Chapter 9 that one can use to help develop option lists.

- Brainstorming and the "6-3-5" method.[1] These are useful in many situations.
- Unit operation guides. The guides assist in the selection of unit operations by listing the most common unit operation choices for different technical functions. There are eight guides:
  - Blending-mixing
  - Drying (water removal only)
  - Heat transfer (this guide includes Evaporation)
  - Mass transfer (this guide includes Crystallization)
  - Material transport
  - Mechanical separation
  - Reactions
  - Size modification
- Questions for understanding. These are designed to increase one's knowledge about the process. Answering the questions will lead to an awareness of design options. To illustrate, several of the questions from Table 9.9 and Table 9.10 are:
  - Should a product be bought or made?
  - What different grades or sources of raw materials are available? What effects do these have on human or environmental safety and on process operation and costs?
  - Does the process require materials of construction more expensive than carbon steel? If so, how might the corrosive streams be removed or their concentrations changed to reduce corrosion problems and lower costs?
  - How can this unit operation be changed to control the final product attributes or the conditions of its output streams?
  - How do the upstream unit operations or processes affect this unit operation?
  - How does this unit operation affect the downstream unit operations or processes?

- How many sites are optimal? Where should they be located? Should an existing plant or process be modified or should a new facility be built? Should health, safety, and environmental (HSE) considerations be a key siting factor?
- Are materials used or made in the process that are hazardous for HSE reasons or that require environmental treatment? If so, can they be eliminated, used in reduced quantities, or replaced by less hazardous or nonhazardous materials? If these materials are reaction products, can reactor conditions, recycle amounts, or the catalyst be changed to reduce the amount generated?
- Should one or multiple units be used?
- Should there be surge between unit operations? How much?
- What is the heat recovery plan? Optimize the heat recovery system and the utilities.

Example 3: You are to develop a vegetable oil hydrogenation process. So far, you have learned the reaction will need to be catalyzed with a small amount of a finely divided solid catalyst. You also know this catalyst must be removed from the product stream for product quality reasons. Because the catalyst is so expensive ($350/lb), it must be reused. You have found it can be reused about six times. Your next step is to deal with the catalyst removal step.

Following the discipline suggested in this book, the technical function flowsheet would show the removal function as: *remove catalyst from the liquid.* You could also note why this is required: *for product quality reasons and to reclaim the catalyst for reuse.*

Having specified catalyst removal in function terms keeps the options open. It allows for later consideration of the many ways the catalyst could be removed from the oil. For example, using the unit operation guides, the options for removal are shown below.

Unit Operation Type	Equipment Category
Clarifiers/thickeners	Rectangular
	Circular
	Tilted plate units
Screens	Fixed
	Reciprocating/shaking
	Revolving
Flotation systems	
Expression presses	
Hydrocyclones	
Sedimenting centrifuges	Tubular bowl
	Scroll
	Disc

Unit Operation Type	Equipment Category
Filtering centrifuges	Peeler
	Pusher
	Worm-screen/scroll
Pressure filters	Plate and frame
	Vertical element
	Horizontal element
	Cartridge
Vacuum filters	Rotary drum
	Horizontal leaf
	Rotary disc

6.2.3 ANALYZING AND SELECTING OPTIONS

6.2.3.1 Technical Feasibility Comes First

The first step of analysis is to weed out options that are not technically feasible. These are the options that do not meet the needs defined in the Business and Technical Objectives and those that do not meet health, safety, and environmental requirements. Some of the weeding out will be quite simple and some may require detailed analysis, perhaps including bench-scale or pilot-plant studies.

6.2.3.2 Assess Costs of the Technically Feasible Options

Next, one uses economic criteria to select the best of the technically acceptable options. The first step in this analysis is to estimate the investments and production costs for each option. One then converts these cash flows into NPV or AC and selects the option having the best economics.

6.2.3.2.1 Estimating Investments

During process development, feasibility, and conceptual engineering, few design details are known. As a result, when defining the project investments one will have to use order-of-magnitude and study estimate methods (Chapter 3). These estimates are not very accurate in the absolute sense mainly because the design is incomplete. However, when comparing options one is only interested in the cost differences between options. For this, these methods provide acceptable accuracy — accuracy permitting good option selection.

One could use more precise estimating methods. However, that would require developing the design further before estimating costs and evaluating the options. This means one would have to design several options in parallel, a costly and time-consuming practice that would have little effect on decision making quality.

6.2.3.2.2 Estimating Production Cost

Again, one uses order-of-magnitude and study estimate methods because of the lack of design details. Recall that production cost includes raw material, packaging material, manufacturing, and product delivery costs. It does not include any costs for research and development, marketing, sales, or corporate administration.

6.2.3.3 Selecting an Option

This is the culmination of option analysis. Here one economically analyzes the technically feasible options to find which has the best economics. To do this, one converts the investment and production cost cash flows into an NPV or an AC. Of the technically correct options, select the one with the best NPV or AC.

There are times when the NPVs or the ACs of the options are so similar that one cannot use them to select one of the options. When this occurs, one can either develop the design in further detail and restudy the economics or use other criteria as the basis for selection.

6.3 SUMMARY

Engineers contribute to their company's profits by economically optimizing their designs, balancing capital and production costs. The three-part Economic Design Model (Figure 6.2) is a disciplined process for doing this. It involves:

- Defining business and technical objectives (Chapter 7).
- Creating options. When developing the process flowsheet, one avoids premature design convergence via technical function flowsheets. This keeps the focus on technical functions rather than on process forms (Chapter 8). Chapter 9 discusses several tools that will stimulate and expand one's creative thinking during option list development.
  - The Brainstorming method
  - The 6-3-5 Method
  - The Unit Operation Guides (Table 9.1 to Table 9.8)
  - The General Process/Process Interaction Questions (Table 9.9)
  - The Feasibility/Conceptual Checklist (Table 9.10)
  - It is important not to evaluate options during list creation so there will be an atmosphere conducive to idea generation
- Analyzing and selecting an option (Chapter 10). One first decides which options are technically feasible. When more than one option is technically feasible, one uses economics to select the best of them.

REFERENCE

1. Ulman, D.G., *The Mechanical Design Process*, New York: McGraw-Hill, 1977, 147–148.

7 Defining Objectives

7.1 WHY DEFINE OBJECTIVES?

Simply put, you set objectives to help you get wherever you want to go. On engineering projects, where more than one person is involved, clear goals are especially important. They help get everyone working toward the same endpoint. However, I have noted quite a few engineering projects begin before there is a good set of project objectives. What often happens is:

- The project leader believes everyone knows what is supposed to be done and so does not take the time to formally set goals.
- The people working on the project talk about their objectives but do not write them down.
- The key members of the project team set objectives before starting their work. They get no input from their bosses or from business management.
- Before starting their project and after getting input from their bosses and from business management, the project team formally set goals. They do not regularly measure or track results.

Each of these situations leads to problems. Because goals and results will be sorted out by trial and error as the design progresses, the trial-and-error approach will always result in design changes. These are often made at inopportune times, causing schedule delays and cost increases.

7.2 WHAT DO GOOD OBJECTIVES LOOK LIKE?

Many people are associated with a project — business management, technical management, the project leader, and the project team. Objectives are used to ensure everyone knows and agrees on what the project is to accomplish from a business and a technical standpoint. The key features of good objectives are:

- *They are developed using input from the principle stakeholders: business and technical management and the project team.* Because project objectives have two important facets — business and technical — both business and technical people must be involved in setting goals. Simply because of their positions in a company, the project leader, their team, and management have different points of view about a project. All are important when setting good objectives. If one's company is working on a project for a client, client representatives also should be involved.

- *They are written down and every stakeholder has a copy.* The rationale is simple: Different people hear and remember different things from the same conversation. During the objective-setting process, using a draft copy of the preliminary goals is a useful tool for comment collection. Once the stakeholders have agreed upon the goals, each should get their own copy. Just as people do not always hear the same thing during a conversation, months after they agree upon goals, people often forget to what they agreed. So the individual copies serve two purposes — everyone has a record of what was agreed upon and everyone can track project performance against the same standards.

- *They are measurable.* Someone once said to me, "In an organization, you get what you measure." By making sure you can measure an objective, you and others will be able to know whether you have done what you intended. Measures need not be — in fact, should not be — complicated. The simpler the better. This way, everyone is more likely to understand and agree upon performance. In his book, *Goal Analysis*, Robert Mager talks about selecting measures that one can measure.[1] He talks about how to get rid of unclear goal statements or "fuzzies." He suggests testing for fuzziness by finishing the phrase, "Hey, let me show you ..." with the substance of your goal statement. An example taken from Mager's book illustrates: "Hey, let me show you how I'm appreciating my deep sense of pride."[2] This is impossible to measure. Contrast that to something that is measurable: "Hey, let me show you that the cost of the plant will not exceed $25M."

- *They are realistic.* If all or much of the project team views its objective as impossible to achieve, project members will tend to lose their motivation to perform. Alternatively, if goals are too easy to reach, they will not stimulate the team to perform at a high level. Some organizations try to deal with this question by setting two levels of results — basic and outstanding. I have found that this two tier system focuses almost all of a team's energy on the basic level goal and that the outstanding goal is meaningless.

- *They are updated when conditions change.* A good time for updating is at the start of each project phase. As the work in one project phase ends, the work focus of the project team changes to what must be done in the next. For example, during process development the focus is on creating a process that can make the product as defined. When that is done, the focus shifts to finding out whether the process is feasible from an economic and schedule standpoint.

7.3 AN EXAMPLE

Assume you live in a small town in the South and you work out of your home. You have decided you want to move to another city, one in the Midwest. At this point, that is all you have decided. Before you pack your belongings, get in your car, and start driving (followed by the moving truck), you would really want to know to what state, city, neighborhood, and residence you will be moving. Deciding upon a state, city, and so on is a multiphase process. Each step in the process will have its own purpose or objective, objectives that build upon one another. Your first objective might be:

Decide where in the Midwest I wish to move.

To begin the process of city selection, you would probably define several criteria for the city in which you will want to live. These might include things like school system quality, collegiate and professional sports teams, transportation systems, safe neighborhoods, health facilities, the arts, housing prices, and so on. Using your criteria, you would most likely do research on Midwestern cities using the Internet and your local library.

Say your research locates four cities that seem to meet your criteria. At this point, you might make a brief visit to each city, selecting the best of the four. Say you pick Cincinnati, Ohio, and its metro area. Now you update your objective, building on what you have decided so far, making it more specific.

Select two or three great neighborhoods in the Cincinnati area where I can buy a four-bedroom house for no more than $250K.

Again, you would do a little preliminary research, selecting five or six probable neighborhoods based upon your own "neighborhood criteria" list. You would possibly visit each area, talk with some of residents, visit area schools, and so on. This would enable you to narrow your search to the two or three great neighborhoods. Now your objectives evolve into:

Select a house (new or used) in one of my great neighborhoods.

Decide whether to buy a used home or to build a new house.

At this point, you might do some research on the Internet and find a realtor who would help you find what you want. Once you have decided whether to buy or build and have selected your house, you would write your final objectives:

Build a house in Mason, Ohio (just north of Cincinnati), on Lot 1020 in the selected neighborhood for not more than $225K.

Move in no later than mid-August 2004 so my children can start school at the beginning of the school year.

Note that as you learned more, your objectives changed from the very general (deciding upon a city) to the highly specific (building a house in Mason, Ohio, for less than $225K). In the same way, as the design develops, objectives for engineering projects evolve, becoming more and more specific.

This example simply illustrates how objectives focus efforts. Say your first objective, set before you did any research on different towns in the Midwest, was to buy a house in Chicago. That objective would have produced a far different outcome than the evolution of objectives described above. Two things might have occurred if you had followed the "Chicago" objective:

- You would have discovered partway through your move planning that you were working toward the wrong objective and your needs were not going to be met. At that point, you would probably start over and work toward "Decide to what city I wish to move." As a result, you would have lost time and probably wasted some money.
- You would have moved to Chicago, a very different city than the Cincinnati area. Given how different the cities are, you most likely would not meet all of your "desirable city" criteria.

The same thing happens in industry.

7.4 PROJECT OBJECTIVES

If project objectives are not clear when work starts, the usual result is a series of design changes in the middle of a project. Most often, these cause cost or schedule disruptions. The primary cause of unclear starting objectives is that management is not clear about what they want the project to accomplish. Typically, this occurs because management just has not had or has not taken the time to figure out what a project must do for their business. This is not meant to imply that management does not do its job; rather, they just have not finished the task of defining project needs.

I suggest setting two types of project objectives — business and technical. Both are needed for a project to be successful. Having a clear statement of a project's business and technical objectives before starting work will ease most of the problems described in the last paragraph. Business (general) management can best define a project's business purpose, which takes the form of business objectives. However, business objectives are not sufficient to guide engineering work. The business objectives must be translated into technical objectives. Engineering management, working with the engineering leader and team, help and guide the translation. The "technical translation" becomes the statement of technical objectives.

7.4.1 BUSINESS OBJECTIVES

Business objectives will be quite different from project to project. For example, objectives for a new product introduction will be quite different from those for a project that is to reduce costs. The following are the topics business management should consider when developing their objectives.

- What is the project to accomplish from a business standpoint? This is the ultimate intent of the project. It is the reason the project is to be funded and completed; all else flows from it. Some reasons to fund a project are:
  - Introduce a new product
  - Upgrade or improve an existing product
  - Increase capacity
  - Reduce costs
  - Comply with some new governmental regulation
  - Solve an existing problem — i.e., low reliability, environmental, safety, community relations
- What are the schedule needs? These generally focus on the long-term business schedule. Timing of upstream work flows from the long-term schedule and is dealt with in the technical objectives. Some typical considerations are:
  - Is any production for consumer testing or customer sampling needed? When will it be needed?
  - When is the new or improved product to be introduced? Does a critical competitive situation exist requiring a short schedule?
  - When does the new capacity need to be online, and at what volume?
  - When do the costs need to be at the new lower level?
  - When does the environmental or safety problem need to be eliminated?
- What are the capacity needs?
  - How much product is needed for consumer or customer testing?
  - What production volume is needed? Will it phase in over time? Can it be built in stages?
  - Is the new process (and the product it makes) expected to cannibalize volume from some other process or product?
- What are the important economic factors?
  - Does a minimum acceptable rate of return or a hurdle rate exist? Rates of return usually take into account a company's use of capital policies and the risk level of the project. Generally, the higher the project risk, the higher the expected rate of return.
  - Does a capital spending limit exist? A company's capital budgeting plan or its cash flow position sometimes require capital spending limits for a project.
  - Does an upper limit for production costs exist? Production costs are usually set based on what a company believes it can competitively charge and upon the profit margin it wants.
  - Do cash flow restrictions exist? Cash flow might be restricted so that a company would not have to borrow money to fund a project.
  - Is the project risky enough to require capital investment be minimized until the project has proven itself to be successful?
- Do other items important to the business exist? These could include items such as:
  - Special security needs
  - Reemphasizing an existing or a new policy the project is to follow

Figure 7.1, Figure 7.2, and Figure 7.3 are examples of business objectives. They are for the same project as it moves through the process development, feasibility, and conceptual phases.

Business plan: There appears to be a market opportunity for an extension of four present product line of oils. Develop a process for the product (code named "Product X")

Projected volume: At this stage, potential volume is very uncertain. Estimates range from 200M to 700M lb/yr. We will need to do further consumer testing to more accurately estimate volume.

Timing: Complete the development work so the national introduction of product X can begin by late 2006.

Product for consumer testing will be needed as defined by the consumer testing schedule.

Economic factors: We expect we will have to sell Product X at the same price as our existing products. Therefore, finished product production costs cannot exceed $1.27/lb. Develop the process accordingly.

FIGURE 7.1 Business objectives (Process Development Phase).

Business plan: Determine the economic feasibility for product X. Include the cost of test market facilities in the study. The feasibility response should include a preliminary project schedule starting with the end of feasibility through the start of national production. Assume a one = year test market.

Projected volume: National volume is estimated at 400M to 600M lb/yr. Base the feasibility on a volume of 600M lb/yr. Test market volume is estimated at 6M lb/yr.

Timing: Complete the study within 3 months. If the project is feasible, begin test market shipments within 9 months. We wish to begin national shipments by late 2006.

Economic factors:
  • Return of investment at least 15%.
  • National production cost for the hydrogenated part of the product is not to exceed $0.292/lb. Finished product production cost is not to exceed $1.27/lb
  • Capital spending for national and test market facilities: not to exceed $6M.

Other: Follow the company's Health, Safety and Environmental Policy.

FIGURE 7.2 Business objectives (Feasibility Engineering Phase).

Business plan: Begin engineering and procurement for product X so it can be introduced nationally in September 2006.

Projected volume: Full national volume is projected at 500M lb/yr. The introduction will be phased and is expected to build as follows:

Month	% National Volume
9 & 10/06	25
11 & 12/06	60
1,2 & 3/07	80
4/07 →	100

Economic factors:
  • Return on investment: at least 15%.
  • Production cost for the hydrogenated oil is not to exceed $0.292/lb. Finished product production cost is not to exceed $1.27/lb.
  • Capital spending (excluding test market facilities): not to exceed $5M.

Other: Follow the company's Health, Safety and Environmental Policy.

FIGURE 7.3 Business objectives (Conceptual Engineering Phase).

Note how objectives evolve and develop as the engineering progresses by comparing the business plan statements for the different phases.

Project Phase	Business Plan
Process Development	Develop a process for the new oil product (code-named "Product X")
Feasibility Engineering	Determine the economic feasibility and develop a preliminary schedule for Product X
Conceptual Engineering	Begin engineering and procurement for Product X so it can be introduced nationally in September 2006

These shifts in business plans underscore the need to update objectives at the end of each project phase.

7.4.2 Technical Objectives

As is the case with business objectives, technical objectives will vary from project to project. One should consider some common topic areas when translating a set of business objectives into technical terms. Most flow directly from the business objectives but often additional requirements will be critical to the design. These would be included in the technical objectives.

- What is the business need? This section restates the business plan and often adds important technical details.
- What is the scheduled endpoint for the project as dictated by the Business Objectives?
  - If the project involves upstream work (anything prior to actual design and construction), what completion date is required for this phase of the project so that the final endpoint can be met?
  - When must in-specification production begin? How much inventory should be on hand when shipments begin? What production rates are needed during the introduction?
  - When must the new capacity be online at full production? What rate of startup is acceptable?
  - When must the cost reduction project be online to deliver the needed cost savings on time?
- What are the important economic factors? If additional detail on the economic factors is needed, they are included here.
- What are the important technical factors for the project? This section will often discuss:
  - Important sources of technical information such as pilot plant reports, process development reports, or previous engineering studies.
  - Siting requirements and considerations. Will the facility be a grassroots installation or an expansion of an existing facility? Where will it be built?

- Ancillaries. What's to be done for:
  - Raw material supply and storage
  - Product handling, storage, and shipping
  - Maintenance shops
  - Laboratories, offices, break areas, and so on
- Utility needs and considerations. What's to be done for:
  - Steam
  - Water-process, cooling, and so on
  - Refrigeration
  - Electrical power
- Health safety and environmental. For example:
  - Limitations on materials, byproducts, or product streams to limit HSE exposure
  - Environmental or safety hazards that must be dealt with, either eliminated or mitigated
  - Including funds in the project to upgrade a plant's environmental systems because of regulations about to take effect
- Other items. For example:
  - A special location for pilot plant facilities or for engineering personnel to ensure security is maintained.
  - Can construction be done only during specially arranged plant shutdown times to maintain production?
  - Operational schedule. How many hours per day, days per week, and weeks per year should the plant be operated?
  - Must a specific engineering or construction contractor be used?

Figure 7.4, Figure 7.5, and Figure 7.6 show the technical objectives corresponding to the business objectives in Figure 7.1, Figure 7.2, and Figure 7.3. As is the case with business objectives, they evolve and become more specific.

Business need: Develop raw material specs and a hydrogenation process (including catalyst selection) for the new product, "Product X".

Schedule: Pilot plant construction must be completed within 6 months. Thus should permit process development to be complete by 1/2004, enabling a start of national production in late 2006.

Sample product for consumer testing must be available by 7/2003.

Economic factors: The production cost of the hydrogenated oil cannot exceed $0.292/lb. The finished product production costs is not to exceed $1.27/lb.

Technical factors:
- See the Product Research report for product characteristics.
- The projected capacity (200M–700M lb/yr) would indicate a continuos process. However, you should work with the plant design engineers to determine whether a batch or continuous process will be best from an economic standpoint. This will probably involve deciding how many process location are best.
- Plant operation will be 24hr/day, 7 days/wk and 50wk/yr. The other 2 weeks will be used for maintenance shutdowns.
- Health, Safety and Environmental: all regulations and Company Policy will be followed. Since the hydrogenation catalyst will most likely contain heavy metals, your work must consider how to properly dispose of and/or reclaim the catalyst.

FIGURE 7.4 Technical objectives (Process Development).

Business need: Develop a feasibility grade design and estimates for the test market and national manufacturing facilities needed to Produce X. Assume volumes of 6M lb/yr for test market and 600M lb/yr for national production. Provide this data to the Financial Department so they can determine whether Product X is feasible.

As a part of the study, develop a milestone schedule for the funding, design, construction and startup of the test market and national facilities. Assume a one = year test market.

Assuming Product X meets the economic factors below and is economically feasible, you should also prepare appropriation requests to fund test market construction, national conceptual engineering and long lead-time equipment purchase.

Schedule: Complete the feasibility and appropriation requests within 3 months.

Economic factors:
- The return on investment must be at least 15%.
- Production cost for the hydrogenated oil is not to exceed $0.292/lb and for the finished product is not to exceed $1.27/lb.
- Capital spending for test market and national production cannot exceed $6M.

Technical Factors: Follow the company's Health, Safety and Environmental Policy.

- Production X contains a specially hydrogenated blend of 70/20/10-soybean/cottonseed/safflower oils. Operating conditions and the hydrogenation endpoint are specified in the pilot plan report.
- Operation will be 24hr/day, 7days/wk, and 50wk/yr. The other 2 weeks will be used for maintenance shutdowns.
- Health, Safety and Environmental: All regulations and Company Policy will be followed. There are no health hazards. H_2 handling is the only special safety risk. Environmentally, spent A3 catalyst will be returned to the manufacturer for reclaiming. Ensure the fat settling traps have sufficient capacity to keep the amount of oil in the wastewater at levels treatable by the sewage treatment plant.

FIGURE 7.5 Technical objectives (Feasibility Engineering).

Business need: Complete a conceptual design and estimate for the production of Product X. Purchase only equipment having long delivery times to protect the startup schedule. National shipments should begin in September 2006. The expected Product X volume is 500M lb/yr.

Schedule: The new capacity will have to start up 3 months in advance of the introduction, or in June 1999. Based upon the volume projections, the new hydrogenation system will have to produce the following:

Month	Production
9 & 10/06	5.2M lb/mo
11 & 12/99	12.5M lb/mo
1, 2 & 3/07	16.7M lb/mo
4/07 $\rightarrow$	250M lb/yr

Economic factors:
- The return on investment must be at least 15%.
- Production cost for the hydrogenated oil is not to exceed $0.292/lb.
- Capital spending: not to exceed $5M.

Technical factors:
- Hydrogenation is the only process or packaging system not having the capacity to produce Product X.
- Product X contains a specially hydrogenated blend of 70/20/10-soybean/cottonseed/safflower oils. Operating conditions and the hydrogenation endpoint are specified in the pilot plant report.

  Our existing facilities do not have the capacity to make the specially hydrogenated blend, Since 50% of *Product X* is the blend, 250M lb/yr of hydrogenation capacity is to be installed. Operation will be 24hr/day, 7days/wk and 50wk/yr. The other 2 weeks will be used for maintenance shutdowns.

- Hydrogenation ancillaries: Existing facilities at all plants are sufficient for feed oil storage and blending, for hydrogenated oil storage, for catalyst supply and for 50 psig H_2 supply. Use these systems making only the needed pump and piping changes.
- Product finishing: Only a few piping changes in the finishing tank farms are needed. Make no other changes in the Finishing Departments.
- Utilities: TBD
- Health, Safety, and Environmental: All regulation and Company Policy will be followed. There are health hazards. H_2 handling is the only special safety risk. Environmentally, spent A3 catalyst will be returned to the manufacturer for reclaiming. Ensure the fat settling traps have sufficient capacity to keep the oil levels in the wastewater at levels treatable by the sewage treatment plant.

FIGURE 7.6 Technical objectives (Conceptual Engineering).

To help understand how the focus shifts when the business objectives are translated into technical objectives, compare the business need statements from the conceptual phase objectives:

Business objective: "Begin engineering and procurement for Product X."

Technical objective: "Complete a conceptual design and estimate for the production of Product X. Purchase only equipment having long delivery times to protect the startup schedule."

Also note that for the conceptual phase, a section entitled "Technical Factors" was added to the technical objectives. Although none of the items addressed here were mentioned in the business objectives, they are definitely important clarifying details for the engineering team. Compare the technical factor statements for all three phases. This will show how the details expand as the process design progresses.

7.5 HOW TO GET INPUT FROM THE KEY STAKEHOLDERS (AND ENSURE IT IS CORRECT)

Getting input is a two-phase process, with as much recycle as is needed to ensure correctness. The first phase is to get the views of the managers in a way that helps them decide what is important for the project to achieve. During the interviews, the interviewer will uncover opinion differences among the managers. Major differences will need to be resolved.

The second phase is to write down what has been agreed upon. Because writing down the objectives makes misunderstandings much more apparent, it helps ensure that everyone is thinking about and agreeing to the same things.

7.5.1 GETTING MANAGER INPUT

One first gets the business managers' input (for the business objectives). This is followed by working with the technical managers (for the translation of business objectives into technical objectives). Obtaining input can be done in a number of ways — a face-to-face interview, a phone interview, a phone call followed by a written request, a written request, and so on. A good set of objectives requires manager engagement. The best way to do this is the face-to-face interview. Second best is a phone interview but nothing beats face-to-face contact. I believe written requests are inadequate because they do not force engagement.

7.5.1.1 Business Manager Interview Structure

Before starting, one must decide how to structure the interview. The two crucial questions are: Who will do the interviewing and what managers will be interviewed?

7.5.1.1.1 Who Will Do the Interviewing?

Whereas a number of people could do the interviewing, the project leader, his/her manager, or both are probably the best choices because they have the most at stake. If the project is small and process development work is being done, the project leader most likely would be the lead process development engineer. For a large project in feasibility, the project leader might be a group leader or section head from the organization responsible for plant design.

If the project leader is inexperienced, it would probably be best if both s/he and his/her supervisor do the interview. A joint interview brings the boss' skills into play and provides good training for the inexperienced person. A more seasoned project leader would probably handle the interview without his/her boss — unless major organizational conflicts or politics were involved. The project leader might include another key engineer from the project in the discussion to provide added perspective or for training.

7.5.1.1.2 What Managers Should Be Interviewed?

I define a business manager as a person having line management (as opposed to staff) responsibility for total business results.

The organizational level of the business manager depends on the size of the project, on its significance to the company, and on the size of the company. For example, in a small company one might interview the president for most of the projects. However, for a small project in a large company one would probably interview a lower-level business unit manager. In general, I suggest interviewing the lowest-level manager who has responsibility for business results. If a project is particularly complex or crucial to a company's success, one should consider adding the business manager's boss to the interview process. When that is done, consider interviewing them together. Joint interviewing generally improves the quality of the discussion by bringing several viewpoints together at the same time.

7.5.1.2 Technical Manager Interview Structure

The same two questions posed for business manager interviews apply to those for technical managers.

7.5.1.2.1 Who Will Do the Interviewing?

Again, I recommend the project leader always be a key player. Using the guidelines in the business manager interviewing section, one can sort out whether it would be helpful for someone to go with the project leader.

7.5.1.2.2 What Managers Should Be Interviewed?

As above, project size, complexity, and importance plus the manager's level are important considerations. I suggest always interviewing the project leader's boss. If the process were complex or crucial to the company's business, one might also interview the boss' boss.

If a company has organized its technical people by technical specialty, one must consider this when deciding which technical managers to interview. To guide the selection, follow the principle of picking only the technical managers whose people play pivotal roles in the project. By key managers, I mean those whose engineers will set the tone, technically and economically, of the project. Generally, no more than three to five tone-setters will be on a project. While limiting the list to no more than three to five people can eliminate some managers whose engineers have large roles in a project, it will result in the identification of the truly important decision makers. Some examples:

- An HSE manager would be involved if significant environmental or safety issues were included.
- For a project involving a large amount of controls work, interviewing the control engineer's boss would be appropriate.
- If a large amount of interaction exists between the process design and the structural design and if these interactions have major financial implications, the structural engineer's boss would be interviewed.

7.5.1.3 Conducting the Interview

Before starting, the interviewer must develop a plan so the interview will be efficient. Typically, an experienced interviewer can do a good job in 30 to 45 minutes. The plan should take into account a company's culture and norms.

Recall that the purpose of the interview is to help managers become clear about what a project is to accomplish. Their lack of clarity is usually a result of things such as:

- There has not been enough time to sort out the objectives
- Some uncertainty about what is really needed
- Concern about goals that seem to be in conflict with each other

Asking a series of questions designed to help the manager express their thoughts about the project is a good way to conduct the interview. Questions that probe and discriminate are most effective. The process should also look for conflicts among the manager's objectives. Except when checking for understanding, one should limit the number of questions needing only a yes/no answer. These are leading questions and do not probe for what is on the manager's mind.

Besides asking good questions, the interviewer must listen well. Good listening will help counteract people's tendency to hear what they want to hear and hear what confirms what they already believe. Good listening begins with a mindset of wanting to understand. In his book, *The 7 Habits of Highly Effective People*, Covey calls this empathic listening; others call it active listening.[3] Either way, good listening is much more than just listening to a person's words. It involves:

- *Paying attention to the entire conversation.* Watch the manager's body language; note their tone of voice and listen for what is not being said.*

---

* What is not said can provide insight into what is important and where the "watch-outs" are. Knowing what is not acceptable is often just as crucial as understanding what is important. Both define the playing field for a project.

- *Putting oneself in the manager's shoes.* By the nature of their work, people at different levels in an organization have different perspectives. To better understand what the manager is saying, listen from their perspective.
- *Paraphrasing what the manager says.* This shows the interviewer was listening and it provides the manager with a chance to find out how they came across. If the manager did not convey the intended message, they can clarify at that point.
- *Summarizing.* Often a conversation about an objective can be long and complex. When that is the case, it is important to summarize to ensure the objective is clear and is what the manager intended. To be effective, an objective must be simply stated; otherwise, confusion will rule.
- *Clarifying.* If the manager was not clear, one could say, "I'm not sure I really understood what you just said. Could you go over that again?" If the interviewer is not sure of the implications of something the manager said, they must ask for further clarification. If one feels the answers do not make sense, keep probing and asking questions until things are clear.
- *Taking notes.* This does two things: It captures the key thoughts and it demonstrates that the interviewer is paying attention to what's being said.

The following examples illustrate the kind of probing and discriminating questions one might ask. Note these examples require the manager to explain or clarify their thinking and to define their performance boundaries.

- What are the three to five most important things you want this project to achieve? Why are these important?
- How do you assess the business risk for this project? How do you believe that should affect the project?
  - How might this risk level impact the use of new technology that could result in a slower than desired startup?
  - What are your thoughts on using contract manufacturing until the market is certain? When we know what the volume really is, we could install our own system, end contact manufacturing, and reduce our costs.
- You said "_____" was important. How would you go about deciding if it has been done as you wanted? What level of performance do you want?
- What is the competitive situation relative to this project? How might this affect the project?
- What is it worth, if anything, to move the project startup forward?
- You said you did not want to spend any more than $_____ on this project. However, would you consider spending more than that amount if we could reduce costs and have a rate of return above the company's hurdle rate?
- Does advancing the schedule have financial value? Can you quantify this?
- Is there a priority order to your objectives that should be used to manage objective tradeoffs? What is it and how would you like us to manage tradeoffs if that becomes necessary?

7.5.2 WRITING DOWN THE OBJECTIVES

After the interview, the project leader puts the objectives in writing. This ensures s/he has captured what the managers said and gives them an opportunity to review and adjust their thoughts. In general, the process is to draft the objectives based upon the interviews, get comments from the managers on the draft, rewrite the objectives based upon the comments received, get comments again, and so on. The commenting and redrafting continues until an agreed-upon set of project objectives is produced.

If the project manager interviewed several managers, there will most likely be several points of commonality, some points of conflict, and some points unique to a given manager. When compiling the objectives from this kind of input, the interviewer musy choose how to go about drafting the objectives from the conflicting points of view. They can write draft objectives that:

- *Make the most sense to them.* This means the project manager will select some points of view, discard some, and combine others in a new way. This kind of draft will most likely contain something that will surprise some of the managers and can include an item or two of conflict. The project manger then uses drafting and commenting to resolve differences. The managers may need to meet with each other (either at the invitation of the interviewer or on their own initiative) to reach concurrence.
- *Identify the points of agreement, the points of conflict, and the unique points.* This most clearly spells out all the areas of disagreement, the areas needing resolution by the managers. Following this path may require more interviews or meetings with the managers in which they resolve their differences.
- *Develop a position they feel will be acceptable to all.* This may well contain some compromises and may be less than the "best" for the company.

Which path the project manager takes will depend on their sense of what may be required for the managers to come to an agreement. For example:

- If the managers' points of view are not too far apart and if they are not solidly entrenched in their opinions, getting agreement should be relatively easy. This would allow the project manager to choose to draft objectives that make the most sense to them (the first option listed above).
- However, if the points of view are quite different and the managers are entrenched in their positions but are willing to get together to resolve their differences, the project manager might choose the second option listed above. When the managers are willing to work toward agreement, having the project manager identify their similarities and differences will quickly focus them. To speed the resolution process, they may wish to use a facilitator. The project manager could fill that role if s/he has the skills.

- When the differences are significant and when it is expected the managers will not be able to work together to resolve them, the project manager might pick the third option listed above.* In this situation, based upon the input gathered in the interviews they work to find some middle ground — something they feel to which everyone might agree. This method is a variation of "single text negotiation." (In their book, *Getting to Yes*, Fisher, Ury, and Patton provide an easy-to-understand description of this technique.[4]) In single text negotiation, the parties in conflict employ a neutral third party to help them come to agreement. After talking with each shareholder, the third party composes a draft of a possible agreement, one that tries to find a meaningful middle ground to which everyone might agree. The mediator uses this draft to collect comments and suggested improvements. From this input, s/he prepares another draft and again gets suggestions for improvement. This continues until the parties have found common ground for agreement.

7.6 SUMMARY

Before a project starts, make sure it has clear business and technical objectives to guide those working on the project. Good objectives have five key features:

- They are developed using input from the principal stakeholders
- They are written down and every stakeholder has a copy
- They are measurable
- They are realistic
- They are updated when conditions change

Business objectives state what kind of business results the project must achieve. When developing these goals, think about questions like: What is the project to accomplish from a business standpoint? What are the timing or schedule needs? What are the capacity or production requirements? What are the important economic factors? Examples of business objectives and their evolution through the phases of a project are shown in Figure 7.1 to Figure 7.3.

Once the business goals are set, translate them into technical terms to create the technical objectives. These must address all the items found in the business objectives and should also deal with any important technical factors. Figure 7.4 to Figure 7.6 are examples of the technical objectives corresponding to the business objectives in Figure 7.1 to Figure 7.3.

When developing objectives, start by getting input from the key managers involved with the project. Interviewing is the best way to get their input.

---

* We would like to believe the managers in a company will always work toward mutual agreement, but this is not always the case. There are instances are encountered where significant, and at times, open conflict exists between two managers or their departments. In these cases, it is doubtful whether the managers will be able to resolve a conflict in project objectives until the underlying problems are resolved.

- Ask probing and discriminating questions that help the managers clarify their thinking and then confirm what has been said
- Actively listen to what the managers are trying to convey:
  - Pay attention to the entire conversation, both the verbal and nonverbal messages
  - Put oneself in the managers' shoes
  - Paraphrase, summarize, and clarify what's been said
  - Take notes

After the interviews, put the objectives in writing. The recording and finalizing of goals is multistep process. First, prepare a draft set of objectives based on what the managers said. Use this draft to collect comments; then modify the goals based upon the comments. The revised draft is used to collect a second round of comments. This draft/comment process continues until a set of agreed-upon project objectives is produced.

7.7 PROBLEMS AND EXCERCISES

1. Describe the characteristics of good objectives.
2. Write one of your present objectives or goals down and test it against the characteristics of a good objective.
3. If a student, write the objectives:
   a. For your senior design project, both business and technical.
   b. For one of your co-op or intern assignments. If you were never given objectives, hypothesize what they actually were. Consider both business and technical objectives.
   c. For finding a job.
   d. For your next 12 months in school.
4. If a practicing engineer, write the objectives:
   a. For a project you are working on currently. Consider both business and technical objectives.
   b. For your next 12 months of work.
5. For the objectives you wrote in Problem 4 or Problem 5, test them against the characteristics of a good objective.
6. Interview a professor, your advisor, a local businessperson, one of your parents, a boss from a previous co-op or intern job, or your present boss. First, select the topic of the interview and schedule it. Plan to spend between 20 and 30 minutes interviewing. Second, develop your interview plan, which would include a list of potential probing and discriminating questions. Summarize the interview, listing and describing the key points. Then give the summary to the person you interviewed to see if you correctly captured what they said.

REFERENCES

1. Mager, R.F., *Goal Analysis*, Belmont, CA: Fearon Publishers, 1972.
2. Ibid, 22.
3. Covey, S.R., *The 7 Habits of Effective People,* New York: Simon and Schuster, 1989.
4. Fisher, R., et al., *Getting to Yes*, New York: Penguin Books, 1991.

8 Creating Options: Flowsheet Development

This and the next chapter present methods that help one create a comprehensive list of design options. This chapter explains a process for developing flow sheets that reduce the tendency to converge rapidly, thus keeping open possible options. It involves focusing on technical functions before considering specific unit operations or equipment. Chapter 9 describes tools that will stimulate one's thinking about possible options. Good list creation involves considering all the options before selecting one. Thus, before selecting an option, one must answer questions like:

- What are all the different unit operations that might be used for a technical function?
- How do the unit operations in a process interact with each other?
- What are all the kinds of equipment that might be used for a unit operation?

Once the option list is complete, analysis begins. One first weeds out the options that are not technically feasible. After that s/he analyzes the technically feasible options to find which is economically best.

8.1 DESIGN (OR PROJECT) PHASES

Before moving to option list creation, we will briefly review what takes place in the design phases dealt within this book — Process Development, Feasibility, and Conceptual. This is summarized in Table 8.1. (For more detail, see Appendix V.) By comparing the outputs from each of the phases, one can see how a design develops from process development to conceptual engineering.

In Table 8.1, unit operations (UO) and equipment are described using the terms "category" and "type." The meaning of these terms is shown in Table 8.2. Note that as one moves from left to right in the table, the level of specificity increases. The next level of detail would be an equipment specification.

8.2 FLOWSHEET DEVELOPMENT

One of the main process development products is a block flow diagram showing all process steps in sequence and all the major flow streams. (See the "Additional Topics" section at the end of this chapter for a description on the types of flowsheets.) When specifying the process steps, use the following guideline: specify as

TABLE 8.1

Design Output Progression — Process Development through Conceptual Engineering

	Process Development	Feasibility Engineering	Conceptual Engineering
Flow diagrams	Block flow diagram showing technical functions or UOs for all major process steps and showing all major flow streams	Preliminary process flow diagram showing all major equipment (by equipment category), some minor equipment, and all major flow streams	Final issue of the process flow diagram
Unit operation and equipment selection	Selection of UOs if choices must be made (May have multiple choices if no single choice must be made)	If not specified during process development, selection of UOs Preliminary selection of equipment categories for major equipment	Selection of equipment types for all major equipment and equipment categories for most minor equipment
Process operation	Set operating ranges or boundaries for each process step	Set operating conditions for major UOs and equipment categories	First draft of the Process Description*
Material/energy balances	Pilot plant material balance	Preliminary process or plant material and energy balances	Second or third draft of the process or plant material and energy balances
Health, safety, and environmental	Identification of potential process and environmental hazards and of emissions and waste streams	Preliminary identification of process and environmental hazards, emissions, and waste streams	Preliminary risk prevention and mitigation plan

* A process description contains details about each piece of equipment in the process, such as: its purpose, flows, operating conditions, materials of construction, operating points and controls, safety/environmental considerations, and so on. The document serves two main purposes: it requires the engineer designing the process to thoroughly think through their design and it conveys design intent to others. "Others" would include controls engineers, safety engineers, designers or engineering contractors, construction contractors, operating personnel, and so on.

little as possible while ensuring the process will operate as intended. That means some process blocks will not be defined beyond the technical function, some will spell out several choices of UO, and some will require a certain UO or equipment category. This allows those working later in the project phases to bring current technology and economics to the selection of UOs and equipment.

To enable complete option investigation, this book proposes a three-step process for flowsheet creation or development:

TABLE 8.2
Unit Operation (UO) and Equipment Terminology

Unit Operation	Equipment Category	Equipment Type
Pressure filter	Plate and frame filter	Self-cleaning plate and frame filter
Heat exchanger	Shell and tube heat exchanger	U-tube heat exchanger
Pump	Centrifugal pump	ANSI standard centrifugal pump

- *The generic block flow diagram.* This is the where development begins. All processes are a more specific form of this flow diagram.
- *Technical function flowsheets and technical function definition.* During process development, the generic flow sheet is transformed into a technical function flowsheet. This and the corresponding technical function definition are the added step between the generic block flow diagram and a process flowsheet. They are designed to counter the tendency to rapidly converge upon specific UOs and equipment.
- *Process flow diagram.* During the feasibility phase, the technical function flowsheet develops into a process flow diagram.

8.2.1 THE GENERIC BLOCK FLOW DIAGRAM

The generic block flow diagram, adapted from Turton et al., is shown in Figure 8.1. One begins to transform this generic diagram by first defining the reactor block, then the separation and recycle systems, the environmental systems, the heat exchanger network, and last the utility systems.[2] Of course, design is not as linear as this sounds. Rather, it is circular and iterative; each system affects those upstream and downstream and is in turn affected by those. For example, when one is defining the reactor block, one must consider how the raw materials affect the reactor and how the reactor affects the separation block. These "upstream and downstream" considerations will most likely result in a series of options to be studied.

Even if one is designing only a part of a process, say the separator, it is wise to look at the entire process. For example, consider how the reactor affects the separator

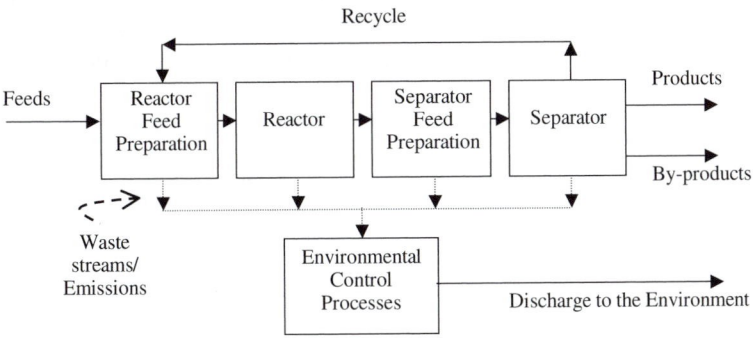

FIGURE 8.1 Generic block flow diagram.

design and vice versa. When thinking this through, one might find it is best to make changes in the reactor operation to simplify the separator design.

Process synthesis books cover questions like this. They discuss how one creates a process design; four good books are:

- *Analysis, Synthesis and Design of Chemical Processes* by Turton et al.
- *Conceptual Design of Chemical Processes* by Douglas
- *Process Design Principles* by Seider et al.
- *Chemical Process Design* by Smith

These books rely heavily on heuristics for flowsheet decision making. This book offers a different approach, one more economically sound. Whereas heuristics offer a path to rapid decision making, they have one major weakness — the analyses that led to their creation never exactly fit one's present situation. As a result, they never quite apply. In addition, the engineer applying the heuristic almost never knows its basis and therefore cannot adjust it to the current situation. Whenever a lot of money is at stake, I suggest doing one's own technical and economic evaluation using the methods in this book.

That said, heuristics have a place — for making decisions where the economic impact is small or for roughing in designs for use in a study. They are also quite helpful when reviewing a design to see if it makes sense.

8.2.2 TECHNICAL FUNCTION FLOWSHEETS

When one translates the generic block flow diagram into a process flow diagram, the tendency to converge rapidly and select specific UOs and equipment is pervasive. Earlier, we discussed the problems caused when one converges too rapidly. The books listed above also influence one to rapidly converge. With its technical function definition, the technical function flowsheet is designed to counter these forces.

Technical function flowsheets focus on first deciding what functions must be performed in a process. Only after that should one decide on the form of that function. Keeping the focus on functions allows the engineer to consider all the applicable types of UOs before selecting one. The difference between functions and forms is illustrated in the following list. Note that functions are verb-object combinations showing an action (verb) to be taken on the object, e.g., separate (verb) solids (object). Forms are nouns, often modified by an adjective. For example, filter (noun) and pressure (adjective).

Function	Form
Separate solids from a liquid stream	Pressure filter
Heat a heat transfer fluid	Furnace
React three liquid streams	Jacketed reactor
Remove a component from a gas stream	Packed column
Reduce the moisture content of granules	Rotary dryer

Example 1. Assume you are responsible for developing a drying process for a heat-sensitive material. You have quantified the heat-sensitivity as follows:

Maximum Product Temperature	Time at Temperature
200°F	15 min
250°F	8 min
275°F	4 min

Following the discipline suggested in this book, you would show the drying function on the technical function flowsheet as: Dry the product to 2.5% moisture without causing heat-damage. Your next step would be to develop the technical function definition. You would spell out how dry the product stream must be, the conditions to ensure no heat-damage, and any other conditions vital to the drying step. For example, that the product not come in contact with combustion gases (as would happen in a spray dryer) might be important. You would also list the rationale for each of the specifications, e.g., laboratory or pilot plant data or reports. That way, whoever uses the flowsheet and its related definition can refer to the source documents if questions or if clarification are needed.

Having specified drying in "function" terms keeps the options open. It allows later consideration of the many ways the water could be removed from the product. These are:

Unit Operation	Equipment Category
Batch-direct heating	Tray dryer
	Through-circulation dryer
	Fluid bed
	Tumbler
	Double-cone
Batch-indirect heating	Agitated atmospheric dryers
	Pan
	Horizontal cylinder
	Agitated vessel
	Freeze dryer
	Vacuum tray
	Agitated vacuum dryer
Continuous-direct heating	Spray dryer
	Flash dryer
	Belt conveyor
	Vibrating conveyor
	Pneumatic conveyor
	Rotary dryer
	Tray
	Through-circulation dryer
	Fluid bed
	Spouted bed
	Turbo-tray
	Tunnel dryer

Unit Operation	Equipment Category
Continuous-indirect heating	Drum (roll) dryer
	Rotating tray with wiper
	Rotary dryer
	Screw conveyor
	Vibrating conveyor
	Horizontal trough-paddle dryer
Infrared and dielectric drying	

This illustrates the power of the technical function flowsheet. If one had selected a specific drying method, no other options could have been considered, one of which might have been the most economic. Whereas this can appear to be an overwhelming number of options to study, many of the options will be quickly eliminated during technical analysis.

Figure 8.2 shows the first issue or draft of a technical function flowsheet for the oil hydrogenation process that is described in the Business and Technical Objectives in Chapter 7 (Figure 7.1 and Figure 7.4). Note that only technical functions are shown in the different flowsheet blocks — heat oil, hydrogenate oil, remove catalyst, and cool oil. (And note all are verb-object combinations.) This flowsheet would have been issued early in the process development. As the process is further developed, more details will be added.

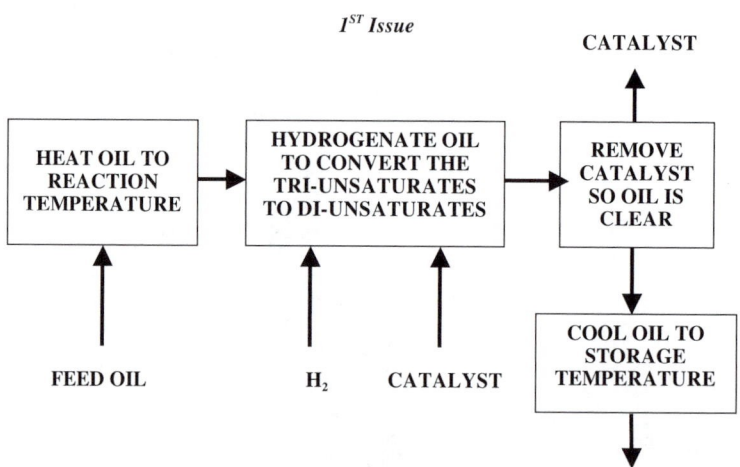

FIGURE 8.2 Technical function flowsheet, oil hydrogenation.

8.2.3 THE TECHNICAL FUNCTION DEFINITION

Because not much detail is on the flowsheet, the technical function definition fills the gap by defining each block on the flowsheet. The details come from the laboratory, pilot plant, and plant testing data and conclusions. A good format is shown in Figure 8.3. This is the technical function definition for the flowsheet in Figure 8.2. Note that for each function on the flowsheet, it:

- Defines the purpose of the function.
- Describes the operating conditions that must be met.
- Lists the basis for the operating conditions. Because of the large volume of information usually needed to define and explain the conditions, as a rule this section refers to detailed reports or studies.
- Lists the possible choices of unit operation types. This requires the developer to have weeded out the technically unacceptable options from those identified for study. (Remember that this weeding out should happen in the analysis phase of economic design, not in the "create options" phase.)

1ˢᵗ Issue

FUNCTION	PURPOSE	QUANTIFICATION	BASIS	UO TYPE
Feed oil	To provide oil for the reaction.	• 115-125°F	• This is the storage temperature of the oil. • See raw material specs.	
Oil heating	To bring the oil to reaction temperature.			TBD
Reactor	To hydrogenate the fatty acid chains to di-unsaturated chains.	• 99% conversion of tri-unsaturates to di-unsaturates • Limit hydrogenation of the di-unsaturate and oleic chains to < 1 and < .5%.	• Testing has shown this type of oil is preferred by 70% of consumers.	TBD
H₂ addition	To provide H₂ for the hydrogenation reaction.		• See raw material specs.	TBD
Catalyst addition	To catalyze the reaction.		• See raw material specs.	TBD
Oil-catalyst separation	To remove catalyst from the oil so that the oil is clear	• Oil clarity so that < 3ppm of catalyst is left in the oil.	• Oil is cloudy at concentrations > 3ppm and this drops consumer preference to <10%. See consumer testing results.	TBD
Oil cooling	To bring the oil to storage temperature.	• 115-125°F	• Experience has shown oil flavor is preserved at these temperatures.	TBD

FIGURE 8.3 Technical function definition, oil hydrogenation.

The technical function definition is also the precursor of the Process Description, which is a detailed description of the process.

As process development continues, one develops more data and makes more decisions. As that happens, more details are added to the technical function flowsheet and its technical function definition becomes more complete. Figure 8.4 through Figure 8.7 show how the technical function flowsheet and the technical function definition change as the design evolves. This is summarized in Table 8.3.

Also note in Figure 8.6 and Figure 8.7 that specific UOs were not selected for the oil heating and oil cooling blocks, and that choices were given for the reaction and catalyst removal blocks. During either Feasibility or Conceptual Engineering, one will again go through the economic design model to select these UOs.

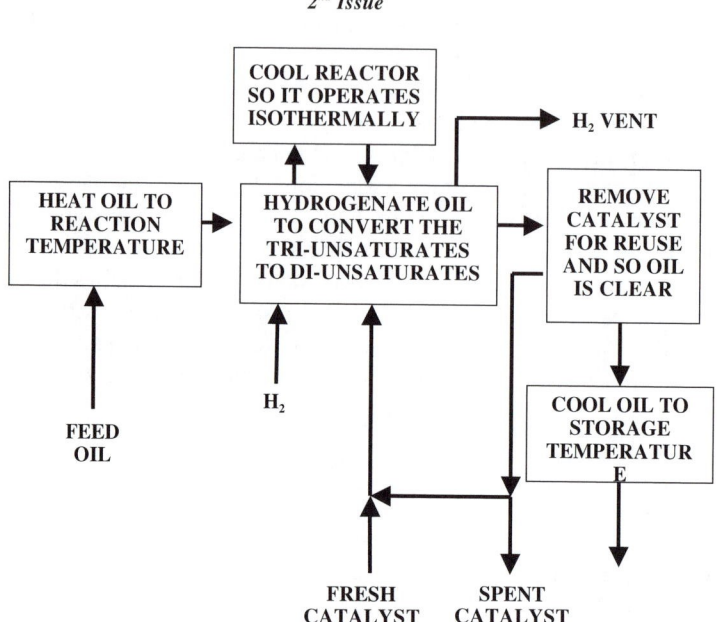

FIGURE 8.4 Technical function flowsheet, oil hydrogenation.

8.3 SUMMARY

Creating a list of options for study is a two-part process. Both are designed to help one's thinking diverge in the early stages of a design. The first, covered in this chapter, is a process for flowsheet creation based upon identifying the technical functions needed in a process. The second, covered in the next chapter, explains some tools designed to spark the generation of ideas, ideas that become the list of options to be later studied.

Flowsheet creation begins with identifying the technical functions, not the UOs or the equipment, needed in a process. Technical functions are verb-object combinations, e.g., separate (verb) solids (object of the verb). Design forms are nouns, usually modified by an adjective, e.g., pressure (adjective) filter (noun).

After one has identified the technical functions, one then creates a technical function flowsheet by arranging the technical functions in their proper order and adding the flow streams in and out of each function. As the design progresses, this flowsheet evolves into a block flow diagram that shows the UOs selected for each function. Figure 8.2, Figure 8.4, and Figure 8.6 show this evolution.

Because limited information exists on the technical function flowsheets, other important data is detailed in a technical function definition. For each function, it:

- Defines the purpose of the function
- Describes the operating conditions that must be met

FUNCTION	PURPOSE	QUANTIFICATION	BASIS	UO TYPE
Feed oil	To provide oil for the reaction.	• 115-125°F	• This is the storage temperature of the oil. • See raw material specs.	
Oil heating	To bring the oil to reaction temperature.	• 345 - 355°F	• See pilot plant report.	TBD

Reactor	To hydrogenate the fatty acid chains to di-unsaturated chains.	• Oil characteristics--
		  – 99% conversion of tri-unsaturates to di-unsaturates.
		  – Limit hydrogenation of the di-unsaturate and oleic chains to < 1 and < .5%.
		• Reaction conditions--
		  – Reaction endpoint: iodine value of 114
		  – Temperature: 345 - 355°F, Pressure: 30 - 40psig, H_2: up to 20% excess, Catalyst: < .5% by weight.
		  – The reaction rate is directly proportional to catalyst concentration.
			• Testing has shown this type of oil is preferred by 70% of consumers.
			• See pilot plant report for reaction conditions and reaction and process data.
				TBD

Reactor cooling	To operate the reactor isothermally.	• See reaction conditions.	• See pilot plant report.	TBD
H_2 addition	To provide H_2 for the hydrogenation reaction @ the reaction pressure.	• See reaction conditions. • Because of impurities in the H_2 feed stream, it is more economical to vent rather than recycle it.	• See raw material specs. • See pilot plant report. • See H_2 vent study.	TBD
Catalyst addition	To catalyze the reaction.	• Catalyst may be reused 6 times.	• See raw material specs. • See pilot plant report.	TBD
Oil-catalyst separation	To remove catalyst from the oil so that • The oil is clear. • The catalyst can be reused.	• Oil clarity so that < 3ppm of catalyst is left in the oil.	• Oil is cloudy at concentrations > 3ppm and this drops consumer preference to <10%. See the consumer testing report.	TBD
Oil cooling	To bring the oil to storage temperature.	• 115-125°F	• Oil flavor will degrade at higher temperatures.	TBD

FIGURE 8.5 Technical function definition, oil hydrogenation.

Final Issue

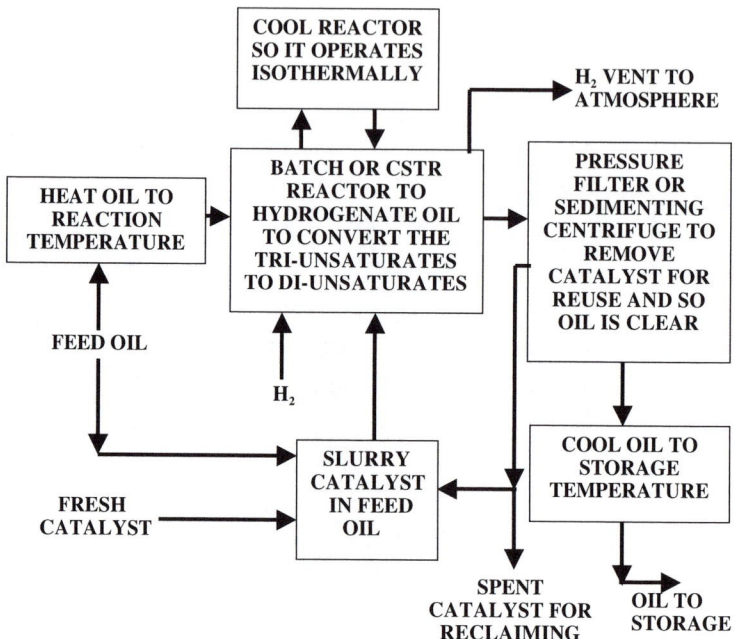

FIGURE 8.6 Technical function flowsheet, oil hydrogenation.

- Lists the basis for the operating conditions
- Lists the possible choices of UO types

The technical function definitions corresponding to the flowsheets in Figure 8.2, Figure 8.4, and Figure 8.6 are shown in Figure 8.3, Figure 8.5, and Figure 8.7, respectively.

Simply stated, one wants to begin a design by focusing on technical functions. This permits one to consider all the unit operations that could be used for a function. This is in contrast to quickly selecting a unit operation or a type of equipment. When that is done, no more options would be considered and no optimization could be done.

FUNCTION	PURPOSE	QUANTIFICATION	BASIS	UO TYPE
Feed oil	To provide oil for the reaction.	• 115-125°F	• This is the storage temperature of the oil. • See raw material specs.	
Oil heating	To bring the oil to reaction temperature.	• 345 - 355°F	• See pilot plant report.	TBD

Reactor	To hydrogenate the fatty acid chains to di-unsaturated chains.	• Oil characteristics--
		  – 99% conversion of tri-unsaturates to di-unsaturates.
		  – Limit hydrogenation of the di-unsaturate and oleic chains to < 1 and < .5%.
		• Reaction conditions--
		  – Reaction endpoint: iodine value of 114
		  – Temperature: 345 - 355°F, Pressure: 30 - 40psig, H_2: up to 20% excess, Catalyst: < .5% by weight.
		  – The reaction rate is directly proportional to catalyst concentration.
			• Testing has shown this type of oil is preferred by 70% of consumers.
			• See pilot plant report for reaction conditions and reaction and process data.
				• Use either an agitated and gas sparged batch reactor or a continuous CSTR reactor. See unit operation selection report.

Reactor cooling	To operate the reactor isothermally.	• See reaction conditions. • Keep the temperature in the cooling system above 275°F so the reaction will not be quenched.	• See pilot plant report.	TBD
H_2 addition	To provide H_2 for the hydrogenation reaction @ the reaction pressure.	• See reaction conditions. • Because of impurities in the H_2 feed stream, it is more economical to vent rather than recycle it. Venting to the atmosphere is ok.	• See raw material specs. • See pilot plant report. • See H_2 vent study.	TBD
Slurry catalyst	To make a catalyst slurry so catalyst can be added to the reactor	• Catalyst, fresh and for reuse, can be slurried in feed oil (up to 50% by weight).	• See pilot plant report.	TBD

Catalyst addition	To catalyze the reaction.	• Catalyst may be reused 6 times.
		• < .5% by weight usage.
		  – Continuous reactor: for 3 min reaction time, catalyst use = .2% by weight
		  – Batch reactor: for 60 min reaction time, catalyst use = .1% by weight
			• See raw material specs.
			• See pilot plant report.
				TBD

Oil-catalyst separation	To remove catalyst from the oil so that
	  • The catalyst can be reused.
	  • The oil is clear.
		• Oil clarity so that < 3ppm of catalyst is left in the oil.
		• At temperatures above 140°F, oil must not be exposed to the air (inert gas blanketed or enclosed equipment is ok)
			• Oil is cloudy at concentrations > 3ppm and this drops consumer preference to <10%. See the consumer testing report.
			• Oil will oxidize rapidly above 140°F.
				• Use either a pressure filter or a sedimenting centrifuge. See unit operation selection report.

Oil cooling	To bring the oil to storage temperature.	• 115-125°F	• Oil flavor will degrade at higher temperatures.	TBD

FIGURE 8.7 Technical function definition, oil hydrogenation.

TABLE 8.3
Design Evolution during Process Development

	Changes from First Issue to Second Issue	Changes from Second Issue to Final Issue

Technical Function Flowsheet	• Reactor cooling function added
	• Venting and recycle functions added
	  – Vent H2
	  – Reuse catalyst
	  – Added spent catalyst stream
	• For the Remove Catalyst function, added catalyst reuse
		• The Reactor function specifies either a batch reactor or CSTR
		• The Remove Catalyst function specifies either pressure filtration or centrifugation
		• H_2 vent now discharges to the atmosphere
		• Added a function: Slurry Catalyst
		• Spent catalyst now goes to be reclaimed

Technical Function Definition	• All the changes in the flowsheet were added to the definition • For the Oil Heating function, added the outlet temperature • For the Oil Hydrogenation function, the reaction conditions were added • For the H_2 Addition function, venting of excess H_2 was added. • For the Catalyst Addition function, catalyst reuse was added • A Reactor Cooling function was added • More references were added in the "Basis" column to explain the items in the "Quantification" column	• All the changes in the flowsheet were added to the definition • For the Oil Hydrogenation function, UO choices were added • For the H_2 Addition function, H_2 venting to the atmosphere was specified • For the Reactor Cooling function, a minimum coolant temperature was added • A Slurry Catalyst function was added • For the Catalyst Addition function, catalyst usage was added for the batch reactor and CSTR • For the Oil-Catalyst Separation function, UO choices were added and the maximum temperature for oil exposure to the atmosphere is specified • More references were added in the "Basis" column to explain the items in the "Quantification" column

8.4 PROBLEMS AND EXERCISES

1. Why would one want to show a technical function on the final issue of a block flow diagram rather than specifying a particular UO? Under what conditions might one specify a UO?
2. In your own words, explain the purpose of the technical function definition. What kinds of information does it contain?
3. Your company is looking into making soap from tallow. They have done some bench-scale testing that has indicated that the tallow can be hydrolyzed at temperatures between 460°F to 500°F. That reaction is:

Tallow + excess H_2O → Fatty acid (crude) + Glycerin (crude) + H_2O

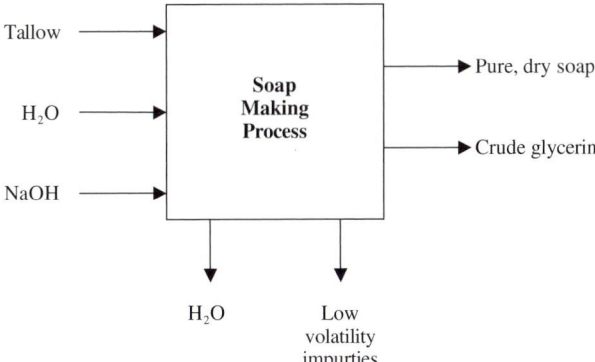

FIGURE 8.8 Problem 3: the soap-making process.

The crude fatty acid contains low volatility impurities plus dissolved water. Both must be removed before the fatty acid can be used to make soap. The soap making reaction is:

$$\text{Fatty acid} \rightarrow \text{NaOH} \quad \text{Soap} + H_2O$$

The water is in solution with the soap. It must be removed before the soap can be sent to finishing. The inputs and outputs for the process are shown in Figure 8.8.

Draw the first issue of a technical function flowsheet for the soap-making process.

4. In the orange juice-processing industry, about 50% by weight of the incoming product is orange peel. Because a typical processing plant processes over 1000M lb per year of oranges, the peel stream is too large to be sent to a landfill. With some further processing, the peel can be made into dry cattle feed pellets. The peel mainly contains water, fiber, and sugars and is about 80% water. Because the dry feed should contain only about 10% water, most of the water in the peel must be removed and disposed of.

   Bench-scale testing has shown that the water will be released from small pieces of peel when CaO is added. After treatment with CaO, the water in the peel stream is about 75% to 80% "free water" and the remainder is bound water. The "free water" stream contains the sugars, so this stream must go to a sewage treatment plant. Thus, water removal from the peel stream may have to include two steps — free-water removal followed by peel drying. The inputs and outputs for the process are shown in Figure 8.9.

   Draw the first issue of a technical function flowsheet for the making cattle feed from orange peels.

5. Figure 8.10 is a process flowsheet showing the steam reforming method of making H_2 from natural gas. In this process, methane — the main component of natural gas — is reacted with water at high temperatures. This yields CO_2 and CO. The reactions are:

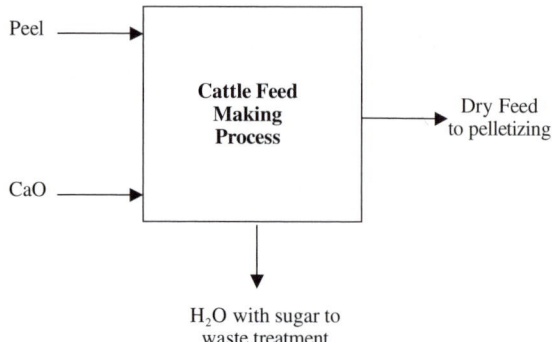

FIGURE 8.9 Problem 4: Citrus cattle feed making.

$$CH_4 + 2H_2O \rightarrow 4H_2 + CO_2$$

$$CH_4 + H_2O \rightarrow 2H_2 + CO$$

The CO is later reacted with H_2O, yielding CO_2 and H_2. The CO_2 is then removed from the H_2-CO_2 mixture by absorption with monoehtanolamine (MEA). Any trace amounts of CO are lastly converted back to CH_4 because CO is a poison for hydrogenation catalysts.

The following table explains more about each piece of equipment in the process:

V-1   The first step in process is a vessel filled with an adsorbent that removes sulfur from the natural gas. Sulfur is a poison for the reformer catalyst.

C-1   Mounted in the furnace stack, this coil uses the hot flue gases to superheat the steam used the reforming reaction.

R-1   In the reformer, CH_4 and water are reacted, forming H_2, CO_2, and small amounts of CO. The reaction is endothermic. It is a gas fired furnace containing several high-alloy tubes, which are filled with the catalyst. Superheated steam is added to the natural gas at the inlet of the tubes. The reformer operates at around 1500°F.

R-2   This reactor converts the CO made in the reformer to H_2 and CO_2 by reacting it with steam at 750°F to 800°F. The vessel contains two fixed beds of converting catalyst. This reaction is exothermic.

HE-1   This heat exchanger interchanges heat between gas streams, leaving the converter and the absorber.

HE-2   This heat exchanger cools the gas stream to about 100°F, the operating temperature of the absorber.

V-2   This vessel is a packed column in which the CO_2 in the gas stream is absorbed by a MEA solution. The gas stream is now essentially CO_2-free.

R-3   The methanator is another fixed catalyst bed reactor. Its purpose is to react trace amounts CO with H_2 to form CH_4 and water. This is done because CO is a poison to most hydrogenation catalysts. The reaction is exothermic, so the gas stream enters the methanator at about 350°F and exits at about 750°F.

HE-3   This exchanger cools the gas to about 100°F.

Draw a technical function flowsheet for this process.

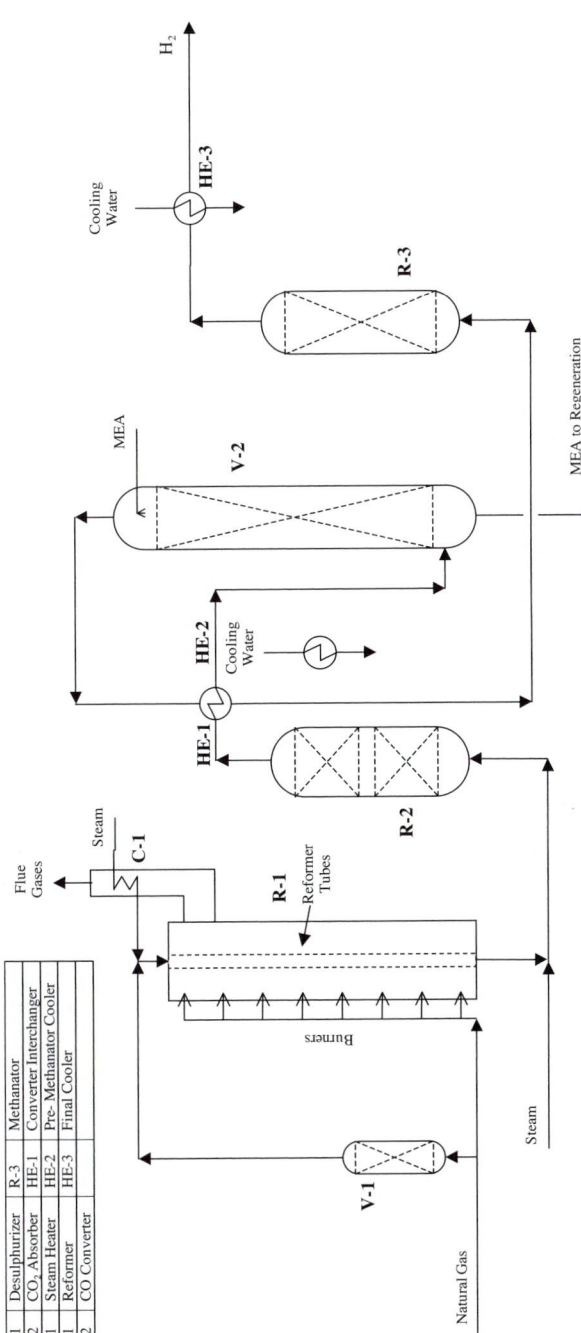

FIGURE 8.10 Problem 5: H_2 via steam reforming of natural gas.

6. You are about to begin developing a process to make benzene from toluene. Bench-scale work has shown the key reactions are:

$$\text{Toluene} + H_2 \rightarrow \text{Benzene} + CH_4$$

$$2 \text{ Benzene} \leftrightarrow \text{Biphenyl} + H_2$$

Data:
- The reactions must take place in the gas phase, at 1150°F to 1300°F and about 500 psig. Reacting at the low end of the temperature range minimizes the formation of biphenyl. It also slows the benzene reaction.
- Practically, an excess of H_2 (about 500%) should be used to stop coking. The benzene reaction does not go to completion, so toluene will be leaving the reactor. Both toluene and H2 are valuable enough that they should be removed from the reactor exit stream and recycled. Some methane in the H_2 stream is acceptable; preliminarily, up to 25% molar is acceptable.
- The boiling points of benzene, toluene, and biphenyl are 176°F, 231°F, and 493°F, respectively.
a. Draw the first issue of a technical function flowsheet for this process.
b. Draw two technical function flowsheets for different ways to structure the separation of H_2, CH_4, benzene, toluene, and biphenyl.

8.5 ADDITIONAL TOPICS

8.5.1 TYPES OF FLOWSHEETS

Three basic kinds of flowsheets are used — block flow diagrams, process flow sheets, and piping and instrumentation diagrams (P&ID). Block flow diagrams are used early in a project when the process is being developed. P&IDs are at the other end of the spectrum. They are used in the later stages of a project and contain a great deal of information, such as pipe sizes and materials and instrumentation details. Table 8.1 and Appendix V deal with the relationship between design status and these flowsheets.

8.5.1.1 Block Flow Diagrams

Broadly speaking, two types of block flow diagrams are used — one for a process where each block represents a piece of equipment or a unit operation, and one for a plant where each block represents a complete process. When complete, a process-level diagram includes:

- A block for each major unit operation — these are labeled with the name of the operation
- Lines for all major flow streams that indicate the flow direction

- Significant flows, temperatures, and pressures, shown on the flowsheet or in an attachment such as a "technical function definition"
- A material balance on the diagram or an attachment — these would usually be based upon a pilot plant material balance

See Figure AT8.1 for an example of a process-level block flow diagram.

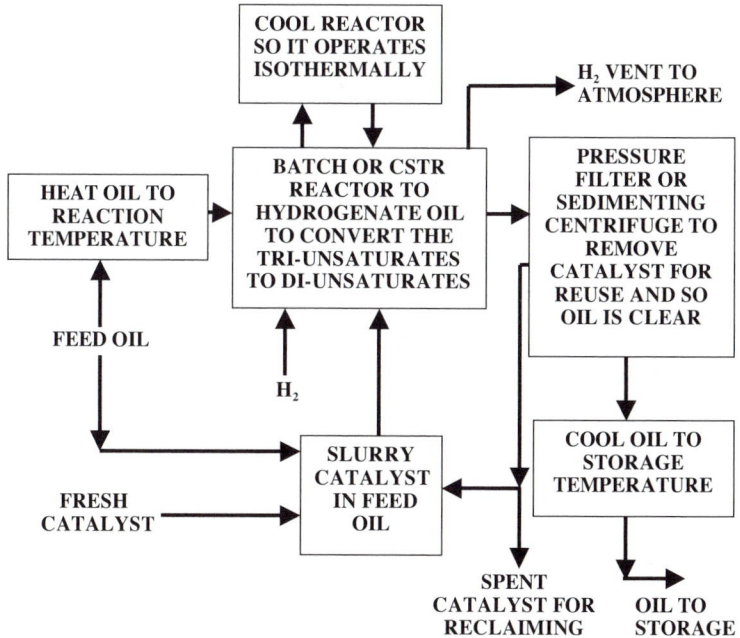

Material Balance & Energy Balance

Basis: 100 lb of hydrogenated product:	100 lb
Feed oil to the reactor:	100 lb
Catalyst slurry, per pass (catalyst can be used 6 times)	
Dry catalyst	0.001 lb
Feed oil to the catalyst system	0.005 lb
H_2 with 20 % excess	25.6 scf
Steam, saturated @ 150 psig	15.7 lb
Cooling water @ 85°F	60 gal
Electrical power, 220 V	0.65 kW

FIGURE AT8.1 Process-level block flow diagram, hydrogenation and filtration.

Figure AT8.2 for an example of a plant-level diagram. Note that Figure AT8.1 is the process-level diagram for the "Hydrogenation & Filtration" block in the plant-level diagram.

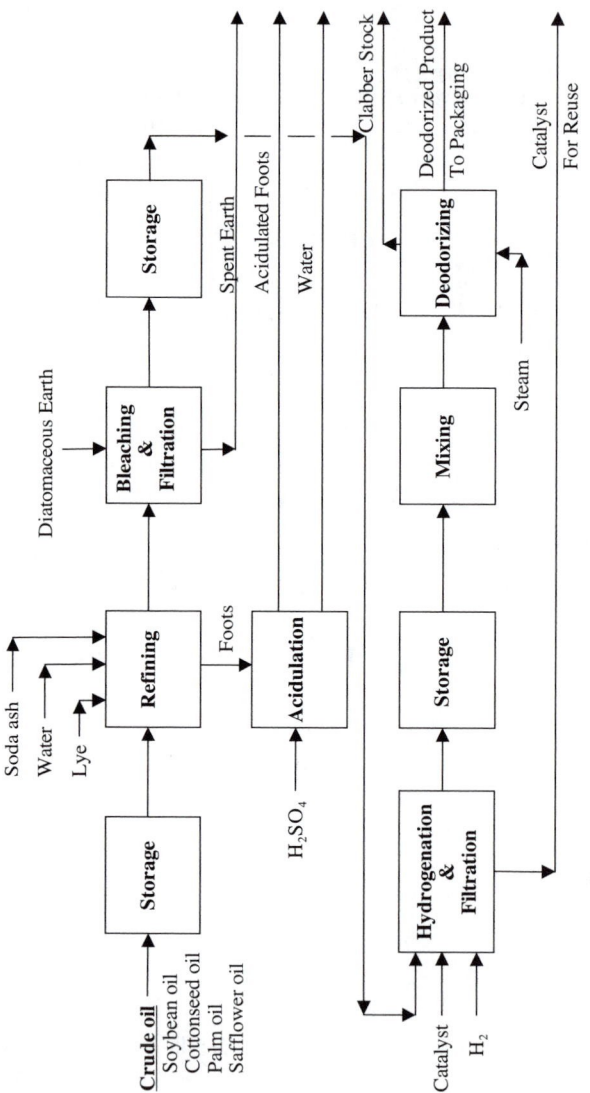

FIGURE AT8.2 Plant-level block flow diagram, edible oil processing.

8.5.1.2 Process Flowsheets

The process flowsheet contains quite a bit more information than the block flow diagram. When one draws these flow sheets, one has selected the type of major equipment, developed a preliminary material and energy balance, and so on, information that is reflected in the flowsheet. When complete, a process flowsheet includes:

- A symbol for all major and some minor equipment. The symbol is specific to a type of equipment. (Appendix B in Ulrich and Vasudevan's book shows some flowsheet symbols.[3]) Equipment is named or is given a number (an equipment list showing numbers and names is shown on or attached to the flowsheet.)
- All major and some minor flow streams, with the direction of flow, are shown.
- Significant flow rates, temperatures, and pressures are shown either on the flowsheet or in an attachment.
- Some control loops may be shown.

See Figure AT8.3 for an example.

8.5.1.3 Process and Instrumentation Diagrams (P&ID)

P&IDs are the most detailed of all the flowsheets; they show all the equipment, piping, and instrumentation that will be installed. When complete, a P&ID includes:

- Symbols and names for all equipment.
- All piping including valves, orifice plates, rupture discs, and so on. Size, schedule, and pipe material is shown for each line.
- All instrumentation, including instrument connections.
- All safety devices with sizes, such as a 4-in relief valve or a 9 ft^2 explosion-relief panel.
- All insulation, including types and thicknesses.

The flowsheet is based upon detailed material and energy balances and is often backed up by a process description.

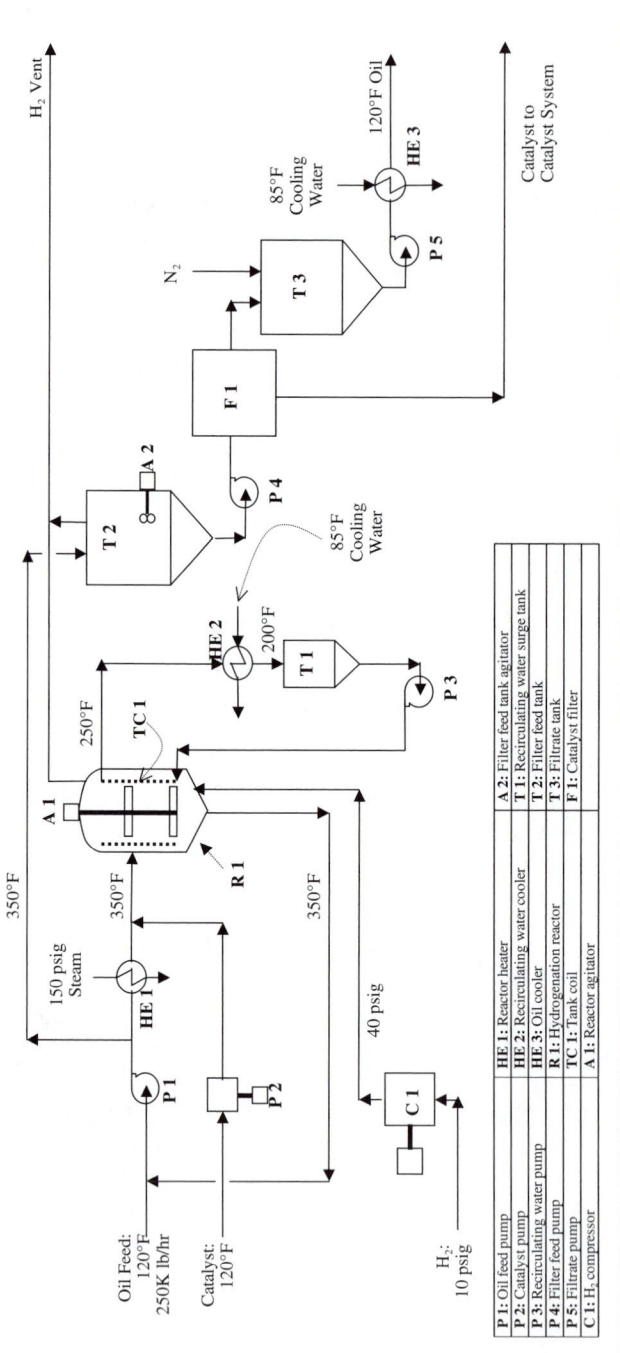

FIGURE AT8.3 Batch oil hydrogenation and filtration.

P 1: Oil feed pump	**HE 1:** Reactor heater
P 2: Catalyst pump	**HE 2:** Recirculating water cooler
P 3: Recirculating water pump	**HE 3:** Oil cooler
P 4: Filter feed pump	**R 1:** Hydrogenation reactor
P 5: Filtrate pump	**TC 1:** Tank coil
C 1: H_2 compressor	**A 1:** Reactor agitator

A 2: Filter feed tank agitator	
T 1: Recirculating water surge tank	
T 2: Filter feed tank	
T 3: Filtrate tank	
F 1: Catalyst filter	

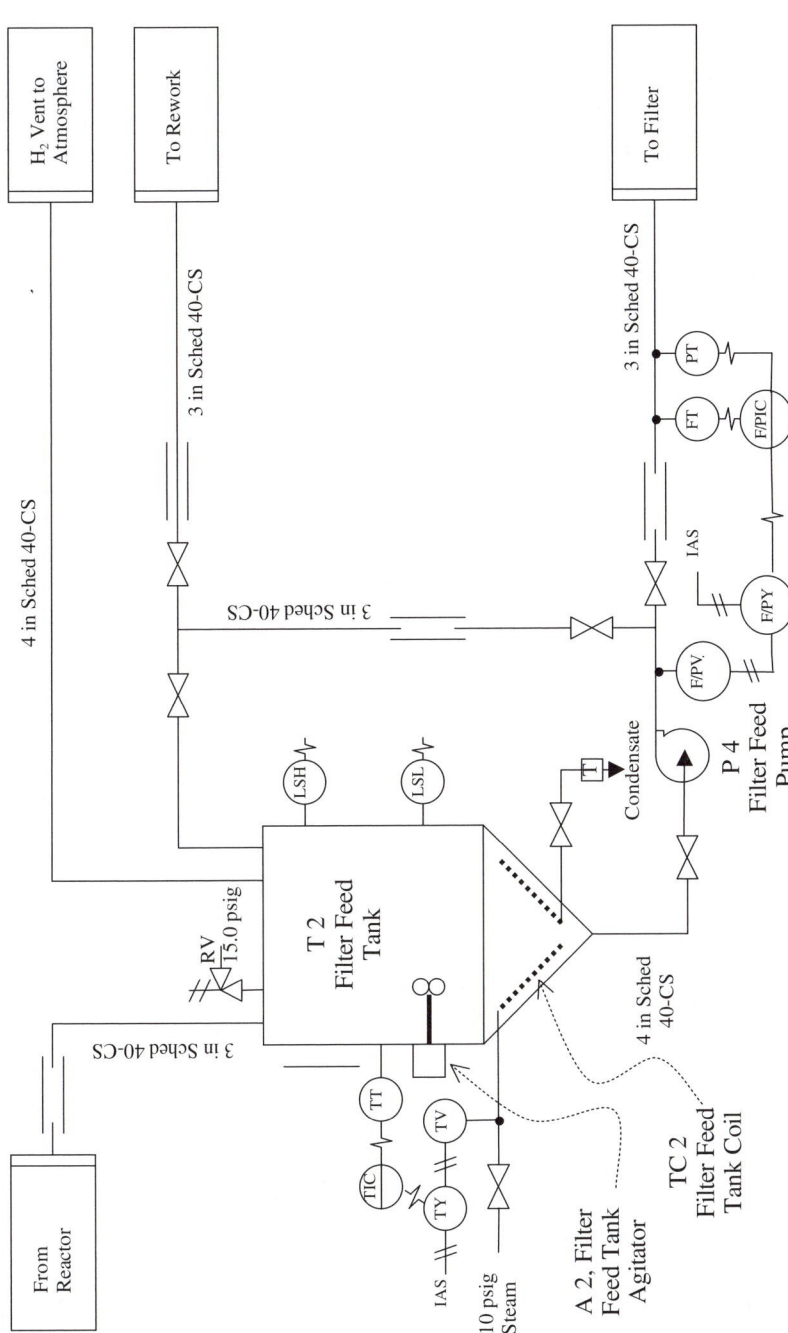

FIGURE AT8.4 P&ID, filter feed tank.

REFERENCES

1. Turton, R., Bailie, R.C., Whiting, W.B., and Shaeiwitz, J.A., *Analysis, Synthesis, and Design of Chemical Processes*, Upper Saddle River, NJ: Prentice Hall PTR, 1998, 184–186.
2. Smith, R., *Chemical Process Design*, New York: McGraw-Hill, 1995, 8–9.
3. Ulrich, G.D. and Vasudevan, P.T., *Chemical Engineering Process Design and Economics: A Practical Guide, 2nd Edition*, Durham, NH: Process Publishing, 2004, 563–571.

9 Creating Options: Creating the List

This chapter discusses how to create the list of options that one will evaluate in the analysis phase of the economic design model. The methods used for list creation depend upon what kind of work one is doing and in what project phase is current.

- *Process development work must be done.* The development might be needed for a new product, for a product upgrade, to reduce costs, to reduce energy use, to comply with a new safety or environmental regulation, to solve a raw or packaging material supply problem, and so on. Product specifications should be available before process work begins.
- *Process development work has already been done, sometimes years ago.* In this case, one is looking for opportunities to apply the process and must now do a feasibility study to see whether it is an economically viable project.

Each situation uses a different approach for list creation. We will deal with list development in three parts — general methods, those applying to process development, and those dealing with feasibility and conceptual engineering.

9.1 GENERAL METHODS

Two methods can be used in all situations — brainstorming and the "6-3-5" method.

9.1.1 BRAINSTORMING

Groups have used the brainstorming method for quite a while. Although mainly used by small groups (up to 10 or so people), individuals can also use it. Group use is best because ideas offered by one person often spark ideas from others. Whether used by a group or one person, a key ground rule is: Do not evaluate, compare, or criticize an idea or ask to have it explained when it is first proposed; that will be done later. The purpose of a brainstorming session is to create ideas, period. Here is how one conducts a brainstorming meeting:

- The person who called the group together clearly states the problem needing a solution. If necessary, everyone discusses the problem to ensure an understanding of the issue.
- The leader gives the group 5 to 15 min in which each member jots down ideas that might solve the problem. The leader should encourage wild,

seemingly impossible ideas because these sometimes lead to a useful solution.

- After the quiet preparatory time, the leader asks the group members to volunteer their ideas one at a time. Someone in the group records each as they are offered. Recording on a flip chart-easel is better than simply writing the ideas on a piece of paper because it visibly acknowledges each idea. It is best if each group member offers at least one idea. To maximize idea flow, just list ideas; do not discuss, criticize, or compare while they are being offered.
- Once the idea flow has stopped, each person explains their idea to make sure the group understands it. Recording the key explanatory points on the flip chart is also helpful.

When the meeting is over, the flip charts will contain a list of ideas plus some brief explanatory comments. The next step will be to do a preliminary assessment of the ideas. This will drop some ideas off the list. What remains will be a list of project options to be analyzed and evaluated later.

9.1.2 THE 6-3-5 METHOD

The following description is paraphrased from Ullman.[1] A 6-3-5 meeting is for a small group, say 3 to 8 people. Where a few people can dominate a brainstorming meeting, a 6-3-5 meeting requires that everyone participate.

- As with brainstorming, the person calling the meeting starts by clearly stating the problem needing a solution. Again, a discussion may be necessary.
- The leader then asks each member to divide a clean piece of paper into three columns. At the top of each column, the member writes an idea, adding whatever description or sketches are needed to explain the idea. Allow about five minutes to do this. No discussion occurs during this step.
- Next, each person passes their paper to the right and the next person studies the ideas on the page. After that, they add three more ideas. Again, allow five minutes for this step. Every five minutes, pass the papers to the right until everyone has seen and added to each paper. While passing, reading, and adding to the papers. there should be no discussion. This forces the members to interpret what is on the paper, potentially creating new insights and limiting critique and evaluation.
- After circulation of the papers, the group can discuss the results or the leader can take the papers, summarizing and evaluating them later.

9.2 PROCESS DEVELOPMENT METHODS

For Process Development, two methods are available:

- Unit Operation Guides. These help identify UO alternatives for the technical functions.
- A question list dealing with the overall process and the technical function interactions. Answering these questions will increase one's understanding of the process, leading to additional options.

As mentioned earlier, listing only possible options is important when working on list creation. Evaluation during list development will hinder the creation of a complete and comprehensive list. Because one will next do analysis, both technical and economic, no need exists for evaluation during list creation.

9.2.1 UNIT OPERATION GUIDES

The Unit Operation Guides are a series of tables that list the most common UO possibilities for different technical functions. There are eight guides:

1. Blending-mixing, Table 9.1
2. Drying (water removal only), Table 9.2
3. Heat transfer, Table 9.3 — this guide includes Evaporation
4. Mass transfer, Table 9.4 — this guide includes Crystallization
5. Material transport, Table 9.5
6. Mechanical separation, Table 9.6
7. Reactions, Table 9.7
8. Size modification, Table 9.8

These guides list the more common UOs. Atypical or unique options may not be included. As the guides are used, one can make them more useful by adding unique items specific to one's industry. Alternatively, I suggest removing nothing from the guides, even though certain UOs may not be used in one's industry. Removing items from the guides can create problems for future projects.

9.2.1.1 Use of the Guides

One uses the guides when identifying UO options. This work begins upon completion of the first issue of the technical function flowsheet (Figure 8.2). At this point, one has defined the technical functions needed in a process, their sequence, and the input and output flowstreams for each. Now one decides whether a UO must be specified. If specifying is not required, one need not develop a list of options for that function. If it is required, one simply finds the technical function in the guides and lists all the possible UOs. This then is the list of UO options, options that one will evaluate later.

To use the guides, find the one that correlates with the technical function currently being worked. Then identify the phase of each input stream — gaseous, liquid, or solid. Note the columns in the guides are classified by the phases of the input streams (Example 1 illustrates).

TABLE 9.1
Blending-Mixing Unit Operation Guide*

Gas–Liquid	Liquid–Solid	Gas–Gas	Liquid–Liquid	Solid–Solid
Gas dispersed in a liquid/foams	Propeller agitator	Fluid jet	Miscible liquids	Tumbler mixers
Turbine agitator with or without a draft tube	Turbine agitator	Orifice plate	Propeller agitator	- Double cone
Propeller agitator with a draft tube	Sparger	Static mixer	Turbine agitator	- Twin shell (V-mixer)
Sparger	Circulation by pump		Paddle-type agitator	- Horizontal drum
Static mixer	Static mixer		- Single paddle	Ribbon blender
Other	Fluid jet mixer		- Gate	Vertical screw mixer
- Impingement aerator	Sawtooth-blade mixer		- Intermeshing vanes	Muller mixer
- Aeration ejector	Closed-rotor mixer		- Anchor	Turbine trough mixer
- Wire whip	Rotor-stator mixer		Sparger	Rotor mixer
- Pipeline contactor	Orifice plate		Circulation by pump	Mills
- Fluid jet mixer			Static mixer	- Hammer
Liquid dispersion in a gas			Fluid jet mixer	- Impact
Atomizers			Orifice plate	- Cage
- Pressure-nozzle			Immiscible liquids	- Jet
- Two fluid nozzle			Propeller agitator	- Attrition
- Rotary atomizers			Turbine agitator	
Entrainment device			Fluid jet mixer	
Fog condenser			Orifice plate	
			High-shear mixer	
			Injector	
			Pastes and high viscosity liquids	
			Helical ribbon	
			Screw-type mixer	
			Kneader	
			Turbine trough mixer	

Extruder
Muller mixer
Static mixer
Roll mill
Banbury mill
Pug mill

[*] References:

Perry, R.H. and Green, D.W. (eds.), *Perry's Chemical Engineers' Handbook, 7th Edition*, New York: McGraw-Hill, 1997.
Sweeney, E.T., *An Introduction and Literature Guide to Mixing,* Cranfield, UK: BHRA Fluid Engineering, 1978.
Ulrich, G.D., *A Guide to Chemical Engineering Process Design and Economics*, New York: John Wiley & Sons, 1984.
Dietsche, W., Mix or match: Choose the best mixers everytime, *Chemical Engineering*, August 1998, 70–75.

TABLE 9.2
Drying Unit Operation Guide[*]

Gases	Solids (Including solid particles, rigid and flexible solids, pastes, slurries, sludges, and solutions)
Fixed bed adsorbers	<u>Batch drying</u>
Stream cooling to condense the water followed by entrainment eliminator	Direct heating (drying gas is the heat source)
	- Tray
	- Through-circulation dryer
	- Fluid bed
	- Tumbler
	- Double-cone
	Indirect heating
	- Agitated atmospheric dryers
	- Pan
	- Horizontal cylinder
	- Agitated vessel
	- Freeze dryer
	- Vacuum tray
	- Agitated vacuum dryer
	<u>Continuous drying</u>
	Direct heating
	- Spray dryer
	- Flash dryer
	- Belt conveyor
	- Vibrating conveyor
	- Pneumatic conveyor
	- Rotary dryer
	- Tray
	- Through-circulation dryer
	- Fluid bed
	- Spouted bed
	- Turbo-tray
	- Tunnel dryer
	Indirect heating
	- Drum (roll) dryer
	- Rotating tray with wiper
	- Rotary dryer
	- Screw conveyor
	- Vibrating conveyor
	- Horizontal trough-paddle dryer
	<u>Infrared and dielectric drying</u>

[*] References:

Perry, R.H. and Green, D.W. (eds.), *Perry's Chemical Engineers' Handbook, 7th Edition*, New York: McGraw-Hill, 1997.

Treybal, R.E., *Mass-Transfer Operations, 3rd Edition*, New York: McGraw-Hill, 1980.

Ulrich, G.D., *A Guide to Chemical Engineering Process Design and Economics*, New York: John Wiley & Sons, 1984.

Van't Land, C.M., Selection of industrial dryers, *Chemical Engineering*, March 5, 1984, pp. 53–61.

Walas, S.M., *Chemical Process Equipment*, Boston: Butterworth-Heinemann, 1990.

TABLE 9.3
Heat Transfer Unit Operation Guide[*]

Gas Heating/Cooling	Liquid Heating/Cooling	Solid Heating/Cooling
Direct contact (of the hot and cold streams)	Direct contact	**Particles/divided solids**
	Steam injector	Direct contact
Cooling/condensing		Rotating drum
- Spray chambers	Indirect contact	- Plain cylinder
- Packed columns	Shell and tube heat exchangers	- Flighted cylinder
- Tray columns	- U-tube	Fluid bed
- Pipeline contactors	- Fixed tube sheet	Vibrating conveyor
Heating	- Floating head	Belt conveyor
	Reboilers	Spouted bed
Indirect contact	- Kettle	Pneumatic conveyor
Shell and tube heat exchangers	- Thermosiphon	
- U-tube	- Forced flow	Indirect contact
- Fixed tube sheet	Double pipe heat exchangers	Rotating drum
- Floating head	Air cooled heat exchangers	- Tubed shell
Double pipe heat exchangers	Plate heat exchangers	- Plain cylinder
Air cooled heat exchangers	Spiral heat exchangers	- Flighted cylinder
Spiral heat exchangers	Scraped surface heat exchangers	- Deep finned
Process furnaces	Thermal fluid heaters	Spiral conveyors
	Process furnaces	- Jacketed solid flight
	Vessel heat exchange	- Large spiral hollow flight
	- Tank coils	- Small spiral large shaft
	- Bayonet heater	- Rotating paddle
	- Vessel jacket	
		Sheeted solids
	Evaporators	Cylinder heat transfer unit
	Natural circulation	
	- Short tube/calandria, basket	**Solidification of melted solids**
	- Long tube, vertical	Table type
	- Falling film	Agitated pan
	- Rising film	Belt type
		Rotating shelf
	Forced circulation	Rotating drum
	- Vertical tube	Vibratory type (caster)
	- Horizontal tube	
	- Agitated calandria	**Fusion/melting of solids**
	- Agitated thin film	Vertical agitated kettle
	- Horizontal spray film	- Screw conveyor
	- Plate type	- Ribbon blender
		Horizontal tank type
	Batch pan evaporator	Mill type
		- Paddle/screw
		- Banbury mixer
		- Multiple roll

[*] References:

Fair, J.R., Designing direct-contact coolers/condensers, *Chemical Engineering*, June 12, 1972, pp. 91–100.

Holt, A.D., Heating and cooling of solids, *Chemical Engineering*, October 23, 1967, pp. 145–166.

Mehra, D.K., Selecting evaporators, *Chemical Engineering*, February 3, 1986.

Perry, R.H. and Green, D.W. (eds.), *Perry's Chemical Engineers' Handbook, 7th Edition*, New York: McGraw-Hill, 1997.

Ulrich, G.D., *A Guide to Chemical Engineering Process Design and Economics*, New York: John Wiley & Sons, 1984.

TABLE 9.4A
Mass Transfer Unit Operation Guide

Gas–Liquid

Absorption
    Packed column
    Tray tower
    Wetted wall tower
    Spray tower

Stripping
    Packed column
    Tray tower
    Spray tower
    Flash drum/tank

Distillation
    Packed column
    Tray tower
    Sparged vessel
    Heated vessel

Humidification
    Packed column
    Tray tower
    Sparged vessel
    Spray contactor

Membrane separation
    Permeation

Evaporation will strip highly volatile
  materials from a liquid (see the Heat
  Transfer Guide)

Liquid–Solid

Leaching
  Percolation
    - Stationary bed
    - Moving bed
    - Bollman type
    - Horizontal basket
    - Kennedy extractor
    - Screw conveyor
  Dispersed solid leaching
    - Stirred tanks
    - Pachuca or Brown tank
    - Vertical plate unit
    - Gravity sedimentation

Adsorption
  Fixed bed
  Slurry systems-agitated tanks
  Pulsed columns
  Ion exchange

Crystallization
  From solution*
  Batch units
    - Agitated
    - Non-agitated
  Scraped-surface units
    - Trough
    - Double pipe
  Forced circulation[†]
  Draft tube[†]
    - Agitated
    - Non-agitated
  Fluidized bed
  Other
    - Cooling disc
    - Prilling tower
    - Rotating drum
    - Multistage horizontal vessel
    - Wetted-wall evaporative
    - Solar evaporation
  From the melt
  End-fed columns
    - Vertical with spiral conveyor
    - Vertical pulsed
    - Vertical with hydraulic movement
  Center-fed units
    - Horizontal with spiral conveyor
    - Vertical with spiral conveyor
    - Vertical vibrating ball and plate
    - Vertical, back-mixed

TABLE 9.4A
Mass Transfer Unit Operation Guide (continued)

Gas–Liquid	Liquid–Solid
	Layer crystallization
	- Falling film units
	- Rotating drum units
	- Horizontal belt units
	- Bubble column
	Other
	- Melt-spray units (prilling)
	- Sweat tank units
	- Freeze crystallization
	- Screw crystallizer
	<u>Sublimation/desublimation</u>
	* Methods to form crystals
	Evaporation (also see the Heat Transfer Guide.)
	- Internal coils/tubes
	- External heat exchanger
	- Vacuum
	Cooling
	- Internal coils/tubes
	- External heat exchanger
	- Direct contact with a refrigerant
	Chemical reaction
	- Neutralization
	- Precipitation
	- Electrochemical
	Salting out
	- With another solvent
	- With another solute
	† Classifying and nonclassifying

[a] See Table 9.4B for a list of references.

TABLE 9.4B[*]
Mass Transfer Unit Operation Guide

Gas–Gas	Gas–Solid	Liquid–Liquid
Membrane separations	Adsorption	Extraction (For settler options, see the
- Gaseous diffusion	Fluidized bed	Mechanical Separation Guide)
- Permeation	Fixed bed	Perforated plate tower
Sweep diffusion	- Horizontal	Packed column
Thermal diffusion	- Vertical	Pulsed column
Centrifugation	Moving bed	Baffle tower
	- Counter-current	Spray tower
	- Cross flow	Mechanically agitated vessels
	- Panel beds	- Agitated tank
	- Adsorbent wheels	- Multistage vessel
		- Disc contactor
	Regeneration	Centrifugal extractor
	Gas stripping	
	- Steam	Membrane separations
	- Air	Dialysis
	- Other gas	Electrodialysis
	Furnace regeneration	Osmosis/reverse osmosis
		Thermal diffusion
		Evaporation (see the Heat Transfer Guide)

[*] References:

Levenspiel, O., *Chemical Reaction Engineering, 3rd Edition*, New York: John Wiley & Sons, 1999.

McCabe, W.L., et al., *Unit Operations of Chemical Engineering*, New York: McGraw-Hill, 1993.

Mersmann, A., *Crystallization Technology Handbook, 2nd Edition*, New York: Marcel Dekker, 2001

Mullin, J.W., *Crystallization, 4th Edition*, Oxford, UK: Butterworth-Heinemann, 2001.

Nyvlt, J., Selecting a Suitable Crystallizer, *Proceedings of the 7th Symposium on Industrial Crystallization*, Warsaw, Poland, 1978, pp. 405–414.

Perry, R.H. and Green, D.W. (eds.), *Perry's Chemical Engineers' Handbook, 7th Edition*, New York: McGraw-Hill, 1997.

Treybal, R.E., *Mass-Transfer Operations, 3rd Edition*, New York: McGraw-Hill, 1980.

Rudd, D.F. et al, *Process Synthesis*, Edgewood Cliffs, NJ: Prentice-Hall, 1973.

Ulrich, G.D., *A Guide to Chemical Engineering Process Design and Economics*, New York: John Wiley & Sons, 1984.

Walas, S.M., *Chemical Process Equipment*, Boston: Butterworth-Heinemann, 1990.

TABLE 9.5
Materials Transport Unit Operation Guide[*]

Gas	Liquid	Solid
<u>Pressure units</u>	Centrifugal pumps	Pneumatic conveying
Fans	Single stage	- Pressurized
- Centrifugal	- Industry standard pumps	- Vacuum
- Axial	(ANSI, API, DIN, ISO)	- Vacuum/pressurized
Blowers	- Other process pumps	- Blow tank
- Turbo/centrifugal	- Canned pumps	Slurry transport
- Axial	- Sump pumps	Conveyors
- Rotary positive	Multistage	- Screw
displacement	Propeller pumps	- Belt
- Screw	Turbine pumps	- Bucket elevator
- Lobe	Regenerative pumps	- Vibratory
- Sliding vane	Positive displacement pumps	- Continuous flow
- Liquid-ring	Rotary	- Closed belt
Compressors	- Gear	- Flighted
- Centrifugal	- Lobe	- Bucket
- Axial	- Vane	- Apron
- Positive displacement	- Screw	
- Reciprocating piston	- Peristaltic	
- Rotary	Reciprocating	
- Screw	- Piston	
- Lobe	- Plunger	
- Sliding vane	- Diaphragm	
- Liquid-ring	Jet pumps	
<u>Vacuum units</u>	- Ejector type	
Steam ejectors	- Injector type	
Rotary vacuum pumps	Volumetric displacement	
- Liquid-ring	- Pressurized tank (blowcase)	
- Lobe	- Air lift	
- Sliding vane		
Reciprocating piston		
Diffusion pump		

[*] References:

Ulrich, G.D., *A Guide to Chemical Engineering Process Design and Economics*, New York: John Wiley & Sons, 1984.

Walas, S.M., *Chemical Process Equipment*, Boston: Butterworth-Heinemann, 1990.

TABLE 9.6
Mechanical Separation Unit Operation Guide*

Gas–Liquid	Liquid–Solid	Gas–Solid	Liquid–Liquid	Solid–Solid
Settling chambers	Clarifier/thickener	Settling chambers	Settlers/decanters	Screens
Cyclones	- Rectangular	Cyclones	- Horizontal	- Grizzlies
Impingement separators	- Circular	Mechanical centrifugal separators	- Vertical	- Reciprocating/shaking
- Wire-mesh	- Tilted-plate units	Fabric filters	- Tilted plate units (lamella)	- Vibrating
- Baffled devices	Screens	- Shaker cleaned	Hydrocyclones	- Revolving
- Jet impactor	- Fixed	- Reverse flow cleaned	Centrifuges	- Oscillating
- Wave plates	- Reciprocating/shaking	- Reverse pulse cleaned	- Disc	Wet classifiers
- Staggered channels	- Vibrating	Air filter types	- Tubular	- Cone type
- Vane-type	- Revolving	Scrubbers/wet collectors	- Disc cyclone	- Hydrocyclones
- Packed bed	Floatation systems	- Spray	Auxiliaries	- Sloping tank
Electrical precipitators	Expression presses	- Packed bed	- Coalescers	- Bowl
Venturi scrubbers	Hydrocyclones	- Plate units	- Membranes	- Hydroseparator
	Sedimenting centrifuges	- Mobile-bed scrubbers	- Electrical devices	- Scroll type sedimenting
	- Tubular	- Cyclone		- Centrifuge
	- Scroll	- Venturi		Jigging
	- Disc	- Sparged vessel		Tabling
	Filtering centrifuges	Granular beds		- Wet
	- Peeler	Impingement separators		- Dry
	- Pusher	Electrical precipitators		- Agglomeration
	- Worm-screen/scroll			Spiral concentrators
	Pressure filters			Dense media separators
	- Plate and frame			- Drum separation
	- Vertical element			- Cone Separation
	- Horizontal element			- Cyclone separation
	- Cartridge			- Dyna whirlpool
	Vacuum filters			Floatation
	- Rotary drum			- Dissolved air
	- Horizontal belt			- Dispersed air
	- Rotary disc			- Electrolytic
				Magnetic separators
				Electrostatic separators

* References:
Purchas, D.B., *Solid/Liquid Separation Technology*, Croyden, UK: Uplands Press, 1981.
Perry, R.H. and Green, D.W. (eds.), *Perry's Chemical Engineers' Handbook, 7th Edition*, New York: McGraw-Hill, 1997.
Svarovsky, L., *Solid-Liquid Separation, 4th Edition*, Boston: Butterworth-Heinemann, 2000.
Ulrich, G.D., *A Guide to Chemical Engineering Process Design and Economics*, New York: John Wiley & Sons, 1984.
Walas, S.M., *Chemical Process Equipment*, Boston: Butterworth-Heinemann, 1990.

TABLE 9.7A[a]
Reaction Unit Operation Guide

Gas–Liquid Reactants[b]	Liquid–Solid Reactants	Gas–Solid Reactants
Batch reactor[c]	Solid phase predominant	Rotating drums
- Static mixer	Rotary kiln	Moving bed
- Shell and tube heat exchanger, tube side	Fixed bed reactors	- Vertical vessels with trays
- Towers/columns	- Vertical	- Horizontal belt
- Packed column	- Horizontal	Kilns
- Tray tower	- Tubed	- Rotary
- Spray tower	Moving bed	- Vertical
- With catalyst beds	- Towers	Fluidized bed
- Fixed bed	- Belt conveyors	Spouted bed
- Trickle bed	- Augers	Pneumatic conveying
- Ebulating bed	Liquid phase continuous[a–c]	Furnaces
Thin film	Batch reactor	- Blast
- Rising film	Continuous well-mixed reactor	- Multiple hearth
- Falling film	Continuous stirred tank reactor (CSTR)	- Tubed
Continuous well-mixed reactor[c]	- Tanks in series	
Continuous stirred tank reactor (CSTR)[c]	- Vertical, multi-stage agitated vessel	
- Tanks in series	Static mixer	
- Vertical, multi-stage agitated vessel		
- Horizontal, compartmented, agitated vessel		
Miscellaneous		
- Venturi mixer		
- Pump impeller		
- Membrane reactor		

[a] See Table 9.7B for the list of references.

[b] Includes reactions using a solid catalyst.

[c] See the Blending-Mixing Guide for agitation options and the Heat Transfer Guide for heating and cooling options.

TABLE 9.7B[*]
Reaction Unit Operation Guide

Liquid-Liquid or One Liquid Reactant[a,b]	Solid-Solid or One Solid Reactant[c]	Gas-Gas or One Gas Reactant[a]	Catalyst Types
Single liquid and miscible liquids	Kilns	Pebble reactor	Homogeneous catalysts
Batch reactor	- Rotary	Tubed furnace	- Ions or metal
Continuous well-mixed reactor	- Vertical	Flame reactor	coordination
Continuous stirred tank reactor	Heaters/roasters	Fixed bed reactor	compounds (in
(CSTR)	- Rotating drum	- Vertical	aqueous solution)
- Tanks in series	- Plain cylinder	- Horizontal	- Enzymes
- Vertical, multi-stage agitated	- Flighted	- Tubed	- Immobilized or
vessel	cylinder	Cyclones	polymer-bound or
Static mixer	- Fluidized bed	Fluidized bed	hetrogenized (the
Shell and tube heat exchanger	- Spouted bed	Moving bed	catalyst is attached
Membrane reactor		Spouted bed	to a solid support)
		Pneumatic conveyor	- Phase transfer
Immiscible liquids		Jet impact	catalysis
Batch reactor		Membrane reactor	Solid catalysts
Continuous well-mixed reactor			- Strong acids
Continuous stirred tank reactor			- Base catalysis
(CSTR)			- Metal oxides,
- Tanks in series			sulfides, hydrides
- Vertical, multistage agitated			- Metals and alloys
vessel			- Transition metals
Static mixer			and organometallic
Shell and tube heat exchanger			catalysis
Raining bucket contactor			
Towers			
- Packed			
- Spray			
- Rotating disc contactor			
- Pulsed column			

[a] This includes reactions using a solid catalyst.

[b] See the Blending-Mixing Guide for agitation options and the Heat Transfer Guide for heating and cooling options.

[c] Most of these reactions involve using a gas as a heat source.

[*] References:

Cusack, R.W., A fresh look at reaction engineering, *Chemical Engineering*, October 1999, pp.134–146.

Levenspiel, O., *Chemical Reaction Engineering, 3rd Edition*, New York: John Wiley & Sons, 1999, pp. 134–146.

Perry, R.H. and Green, D.W. (eds.), *Perry's Chemical Engineers' Handbook, 7th Edition*, New York: McGraw-Hill, 1997.

Tominaga, H. and Tamaki, M. (eds.), *Chemical Reactions and Reactor Design*, Chichester, UK: John Wiley & Sons, 1997.

Ulrich, G.D., *A Guide to Chemical Engineering Process Design and Economics*, New York: John Wiley & Sons, 1984.

Walas, S.M., *Chemical Process Equipment*, Boston: Butterworth-Heinemann, 1990.

TABLE 9.8
Size Modification Operation Guide*

Solids Only

Size reduction
    Grinders/mills
    Stirred media
    Vibratory
    Mixer granulators
     - Pug mill
     - Paddle mill
     - Pin mill
    Disc attrition
    Tumbling
     - Rod
     - Ball
     - Tube
    Hammer
    Fluid jet
    Rolling compression
     - Bowl
     - Pan
     - Ring-roll
    Crushers
     - Jaw
     - Gyratory
     - Impact
     - Hammer
     - Rotor
     - Cage
     - Roll
    Cutters/slitters
Size enlargement
    Pressure compaction
    Presses
     - Tableting
     - Roll
     - Molding
    Pellet mill
    Extruders
     - Pug
     - Screw
    Agglomeration
     - Disc or pan
     - Drum
     - Prilling tower
     - Fluidized bed
     - Spouted bed

* References:
McCabe, W.L. et al., *Unit Operations of Chemical Engineering*, New York: McGraw-Hill, 1993.
Perry, R.H. and Green, D.W. (eds.), *Perry's Chemical Engineers' Handbook, 7th Edition*, New York: McGraw-Hill, 1997.
Ulrich, G.D., *A Guide to Chemical Engineering Process Design and Economics*, New York: John Wiley & Sons, 1984.
Walas, S.M., *Chemical Process Equipment*, Boston: Butterworth-Heinemann, 1990.

Example 1. You have just issued the first technical function flowsheet (refer to Figure 8.2) for the Product X process. In the laboratory, you found the hydrogenation reaction requires precise control so that the hydrogenation of di-unsaturate and oleic chains is minimized during the conversion of tri-unsaturates to di-unsaturates. As a result, you have decided to specify the reactor on the next issue of the block flow diagram. You have also found:

- A finely divided metal catalyst is needed to produce the proper reaction product. Only small amounts (less than 0.5% by weight) will be needed. (You used Table 9.7, the Reaction Guide, to help you create a list of catalyst options. Your analysis of these options led to the selection of the catalyst you plan to use.)
- For the reaction to proceed at a reasonable rate, a large volume of H_2 gas must be dispersed in the liquid oil. The volume ratio of H_2/oil is 13 at the reactor inlet.

Your next task is then to create a list of reactor options for later evaluation.

Because you are dealing with a reaction, you use Table 9.7, the Reaction Guide. You also know three input flow streams exist into the reaction block — liquid oil, gaseous H_2, and solid catalyst (which you plan to slurry in oil). Because the reactants are gas and liquid, you select the gas/liquid column in the guide. Note that footnote "b" mentions that the gas/liquid column "includes reactions using a solid catalyst." That column lists the possible reactor options:

- Batch reactor
- Static mixer
- Shell and tube heat exchanger, tube side
- Towers/columns
  - Packed column
  - Tray tower
  - Spray tower
  - With catalyst beds
    - Fixed bed
    - Trickle bed
    - Ebulating bed
- Thin film
  - Rising film
  - Falling film
- Continuous well-mixed reactor
- Continuous stirred tank reactor (CSTR)
  - Tanks in series
  - Vertical, multistage agitated vessel
  - Horizontal, compartmented agitated vessel
- Miscellaneous
  - Venturi mixer
- Pump impeller
- Membrane reactor

While developing the list, you became sure that quite a few of the items on the list above just would not work in your situation (e.g., a spray tower or a fixed bed reactor). You almost dropped them from the list but finally decided to leave them on the list because you know you will next do a technical evaluation of all the items.

Example 2. You are doing a portion of a feasibility study for glycerin production. The block flow diagram issued by the Process Development Department includes a block entitled "Concentrate Glycerin." The feed to this block is a 10% solution of glycerin in water and the outputs are an 85% glycerin solution and water. Your first task in the feasibility study is to select the method for concentrating the glycerin.

Because you are removing water from the glycerin, you go to Table 9.4, Mass Transfer, and use the liquid-liquid column. There you find four general methods — extraction, membrane separation, thermal diffusion, and evaporation. (Note that it refers you to Table 9.3, Heat Transfer, for evaporation.) You think extraction is not an appropriate method for your situation, but remembering the admonition not to evaluate during list creation, you leave this on the list knowing you will next do a technical evaluation. Your option list is:

- Extraction
  - Perforated plate tower
  - Packed column
  - Pulsed column
  - Baffle tower
  - Spray tower
  - Mechanically agitated vessels
    - Agitated tank
    - Multistage vessel
    - Disc contactor
  - Centrifugal extractor
- Membrane separations
  - Dialysis
  - Electrodialysis
  - Osmosis/reverse osmosis
- Thermal diffusion
- Evaporation
  - Natural circulation
    - Short tube/calandria, basket
    - Long tube, vertical
    - Falling film
    - Rising film
  - Forced circulation
    - Vertical tube
    - Horizontal tube
    - Agitated calandria
    - Agitated thin film
    - Horizontal spray film
    - Plate type

- Batch pan evaporator

Example 3. You are a project engineer in a plant. One of the key heat exchangers in your department needs replacement. The existing unit is a floating head exchanger that heats a low viscosity hydrocarbon to 250°F using 50 psig steam. You know there are no restrictions on what kind of unit to install, so you decide to use the Unit Operation Guides to help create a list of options. Near the heat exchanger other potential sources of heat are — 150 psig and 450 psig steam plus a condensate flash tank operating at 75 psig.

Because you are dealing with heat transfer, you use Table 9.3, Heat Transfer. You select the "Liquid Heating and Cooling" column, as the unit's purpose is heating, not condensing steam. Your list is:

- Direct contact
  - Steam injector
- Indirect contact
  - Shell and tube heat exchangers
    - U-tube
    - Fixed tube sheet
    - Floating head
  - Reboilers
    - Kettle
    - Thermoshiphon
    - Floating head
  - Double pipe heat exchangers
  - Plate heat exchangers
  - Spiral heat exchangers
  - Scraped surface heat exchangers
  - Thermal fluid heaters
  - Process furnaces
  - Vessel heat exchange
    - Tank coils
    - Bayonet heater
    - Vessel jacket

Because you are just heating the hydrocarbon, you do not include the options listed under Evaporation or air coolers. You leave all other options on the list.

9.2.2 GENERAL PROCESS AND PROCESS INTERACTION QUESTIONS

The general process and process interaction questions in Table 9.9 deal with the general nature of the process and with the interactions among its technical functions and flow streams. Answering the questions will increase one's understanding of the process and uncover additional design options.

9.2.2.1 The Questions — Further Thoughts

Two of the questions in Table 9.9 need added discussion.

TABLE 9.9
General Process and Process Interaction Questions

General Process

Should a product be bought or made? If it must be made, should a contract processor make it or should one's company produce it?

Should the process be batch or continuous?

What different grades or sources of raw materials are available? What effects do these have on human or environmental safety and on process operation and costs?

Does the proposed process use feed streams or create streams that are health or safety hazards or that require control or treatment for environmental reasons? If so, how might the process be changed to eliminate or minimize these issues and their associated costs?

Does the feed stream contain materials that should be removed before entering the process?

What reaction conditions maximize reactor yield and first-pass conversion?

Does the proposed process require operation outside the ranges of 0 psig and 150 psig or 100°F and 650°F? If so, how might the process be changed to permit operation within these ranges to lower costs?

Does the process require materials of construction more expensive than carbon steel? If so, how might the corrosive streams be eliminated or changed to reduce corrosion problems and lower costs?

What are the possible separation sequences for the removal of components from a process stream — gaseous, liquid, or solid?

Process Interactions

Should recycle or purge streams be used? Should the recycle streams be purified before reentering the process? Should the purge streams be treated or reclaimed?

What are the technical and economic tradeoffs among the main operations on the flow sheet — reactor (or dominant process), recycle and purge streams, separation systems, and heat recovery system? In general, the reactor design determines how the separator will be designed and what the need will be for recycle or purge streams. In turn, these affect the heat recovery plan and systems. All affect the utility needs.[2]

What attributes of the input streams will this UO change?

How can one manipulate this UO to control the final product attributes or the conditions of its output streams?

How do the upstream UOs or processes affect this UO?

How does this UO affect the downstream UOs or processes?

How could one change the attributes of the input streams to have a drastic impact on the performance or cost of this UO and those downstream?

Can the product specifications be relaxed, and would that make the process much less expensive to build or operate?

Does this UO have a significant impact on capital or production costs, either by itself or by its effect on downstream operations? If so, are there other options available, e.g., changes to the input streams, changes to the output streams' specifications, or changes to the final product streams' specifications?

9.2.2.1.1   Should a Product Be Bought or Made?

Can a supplier deliver the product to your plants more cheaply than you can make it? If yes, it makes more sense to buy the product. On the other hand, several instances can be found where a company might pay more for a material.

- If a supplier could supply the product sooner, it might make sense to buy the product and shorten the time to market.
- If the product has a high degree of market risk, it might make sense to buy the product — at least until the market is firmly established.

- If cash is limited, buying the product would spread the cash needs out over the life of the project as compared to a larger, one- or two-year cash flow for the capital investment.
- A company might also decide to make a portion of the product and to buy a portion. This could permit the company to fully load its own facilities and use the supplier for production swings.

The same rationale applies to intermediate materials used in one's process.

9.2.2.1.2 Should the Process Be Batch or Continuous?

Economics should be the key decision factor here. If the process must make a large number of products, changeover costs can be significant. Changeover costs include labor, system cleaning (supplies, labor, and costs to handle and treat the cleanout material), product losses or degradings, and startup and shutdown costs.

If production is seasonal so the system will not operate for long periods, one might want a flexible system that could run other products in the "off-season." This would more fully use the installed capacity, spreading capital and fixed costs over a larger production base.

9.3 FEASIBILITY AND CONCEPTUAL ENGINEERING

List creation for feasibility and conceptual engineering is combined here. Usually, one only does enough feasibility engineering to determine accurately whether the proposal is economically sound. Sometimes this requires a small amount of engineering and a lot of engineering at other times. As a result, the boundary between the two phases is fuzzy, making a combined list more practical.

In these phases, one often has UOs or equipment to select. When that is the case, use the Unit Operation Method described earlier. For the other types of options, use the checklist below.

9.3.1 THE FEASIBILITY/CONCEPTUAL CHECKLIST

Use the checklist in Table 9.10 to stimulate your thinking and discover options.

9.3.1.1 The Checklist — Further Thoughts

Several checklist items will profit from added discussion.

9.3.1.1.1 Siting Considerations

How many sites are economically optimal, and where should they be located?
Deciding how many sites are economically optimal requires balancing the capital cost of the process or plant, startup expenses, its manufacturing cost, in-freight costs, and distribution cost. In general, as the number of sites increases, capital costs, startup expenses, and manufacturing costs increase and distribution costs decrease. In-freight costs may increase or decrease depending on the location of raw material and packaging suppliers. Many texts and articles discuss the factors one considers when locating a site.[3-6] Granger lists the key ones as labor availability and conditions,

TABLE 9.10
Feasibility/Conceptual Checklist

Should a product be bought or made? (Refer to "The Questions — Further Thoughts" in the previous section.)

Siting considerations:

How many sites are economically optimal, and where should they be located?

Should HSE considerations be a critical siting factor?

Should an existing plant or process be modified or a new one be built?

Health, Safety, and Environmental (HSE) considerations: One will find more detail in the process synthesis books by Douglas, Smith, and Turton et al., plus *Plant Design and Economics for Chemical Engineers* by Peters and Timmerhaus and *Perry's Chemical Engineering Handbook*.

Are there materials used or made in the process that are hazardous for HSE reasons or that require environmental treatment? If so, can they be eliminated, be used in reduced quantities, or be replaced by less or nonhazardous materials? If these materials are reaction products, can reactor conditions, recycle amounts, or the catalyst be changed to reduce the amount generated?

Can hazardous materials, essential to the process:

Be contained to reduce exposure?

Have only minimal amounts kept in storage to reduce the extent of the hazard?

Have a stabilizing ingredient added, or can it be diluted? Both will reduce the consequences of a leak or a release.

Should the process be located:

In an unpopulated area or in an area not subject to natural catastrophes (earthquakes, tornados, and so on) to minimize community exposure to an HSE event?

In a remote area of the plant to reduce the in-plant consequences of an HSE event?

Does the process require added, oversized, or special equipment to permit safe and effective startup and shutdown? These items could affect the optimization studies. Consider:[8]

Reactor/reaction control

Equipment emptying or flushing or material disposal at shutdown

Heating or cooling sources needed to get the process operating

Equipment cool down or heat up after shutdown

System pressurization or depressurization

Optimize the process:

Optimize the tradeoffs among the reactor or dominant process, the recycle and purge streams, the separation systems, and the heat recovery system. (Refer to the "Process Interactions" questions in the previous section.)

Should the process be batch or continuous? (Refer to "The Questions — Further Thoughts" in the previous section.)

Should there be one or multiple units?

Should there be surge between UOs? How much?

Define the heat recovery plan. Optimize the heat recovery system and the utilities.

Optimize the equipment: Optimize operating conditions versus equipment cost/size and operating costs. Process design texts provide much more information on this subject. Several good references are *Plant Design and Economics for Chemical Engineers* by Peters and Timmerhaus, *Chemical Process Equipment* by Walas, *A Guide to Chemical Engineering Process Design and Economics* by Ulrich, and *Perry's Chemical Engineering Handbook*. See also Table 9.11 in this chapter.

Gas-liquid towers: diameter versus height for different internals, for different reflux ratios, for different L/Gs, and so on

Heat exchangers: area versus pressure drop and temperature approach

Pressure filters: filtering time, initial flow, and filtering temperature versus filter area

Reactors: reactor size (hold time) versus temperature, pressure, reactant concentrations, catalyst concentration, and recycle rate

Pipe: diameter versus pressure drop

Insulation: thickness versus heat loss

tax and financial inducements, utility needs and availability, site conditions and topography, local services availability, environmental considerations, freight/transportation availability and costs plus quality-of-life considerations for employees. Each company will weigh the importance of these factors differently.

To determine the optimal number of sites, one first decides the number of site options to be studied (one site, two sites, three sites, and so on) then roughly selects the locations for each set of options, estimates the costs for each, and then makes an economic comparison using the cost data. (See also "The Economics of Plant Siting" section in Chapter 10 and the "Optimal Number of Plants" case study in Chapter 11.)

Should HSE considerations be a critical siting factor? If the process or plant is handling or making a very hazardous material, HSE considerations should be critical. Consider the 1984 Bhopal, India, incident where methyl isocyanate was accidentally released from an insecticide plant. The release caused the deaths of 3800 people and disabled another 11,000.[7] Had the plant been located in an unpopulated area, few — if any — deaths or injuries would have occurred.

Should an existing plant or process be modified or a new one built? One can answer this question by economically comparing the capital cost, startup expenses, manufacturing costs, in-freight costs, and distribution costs for each option. Situations where one might spend more to install a process in an existing facility are:

- If building in an existing plant would shorten the time to market for your product, it might make sense to spend more.
- If cash is limited and the capital cost to build on an existing site is less than for a new site, it might sense to use the existing plant even if production costs are higher.
- To permit it to fully load its facilities, a company might also decide to make a portion of the product in an existing location and to build capacity for the rest at a new site. (See also "What to Choose: A Grass-Roots Plant or the Expansion of an Existing Plant" case study in Chapter 11.)

9.3.1.1.2   Should There Be One or Multiple Units?

This is an economic decision. Economies of scale drive the decision toward having one large unit. However, there are times when multiple units may be appropriate:

- If the process must run several different products, having more than one unit might allow smaller units to be dedicated to the smaller volume products. This would reduce or eliminate the changeover costs associated with one large unit. Changeover costs include labor, downtime, utilities, losses, and so on.
- If the process receives feeds from several different units or sends its product to multiple units, one might build several units, matching their size to the upstream or downstream systems. Intellectually, this symmetry is nice but be careful that the economics do not suffer.

9.3.1.1.3 Should There Be Surge between UOs? How Much?

This question is controversial. Strong proponents of not having surge exist as do strong proponents of having surge. Do not decide based upon principle; rather, examine the process in question for what seems right. Consider using surge:

- When a batch process has a continuous feed or when it feeds a continuous unit. Use surge to buffer the continuous units from the intermittent batch operation.
- When one part of the process cannot handle upsets and temporary shutdowns well. In this case, one may need to protect the process from upstream or downstream upsets. For example, the crystallizer in a freeze concentration system will freeze up if it cannot discharge crystals because of an upset in the downstream unit. Once a freeze-up occurs, the crystallizer takes about eight hours to get back online.
- When there are two or more units in series and:
  - The number of feed streams increase as one moves downstream. To illustrate, assume there are three processing steps (A, B, and C) followed by packaging. Step A produces 2 to 4 products. Step B uses them to make 6 to 8 other products. Step C uses these 6 to 8 products to make 20 to 40 different blends. The 20 to 40 blends are packaged in 2 to 4 different sizes. A plant like this needs surge between the operations to operate effectively.
  - The reliabilities of different parts of the process have an adverse effect on the system reliability. System reliability is equal to the product of the reliabilities of each part of the process. Assume four operations are in series and no surge exists between them. If the reliability of each operation is 98%, then the total system reliability will be (0.98)(0.98)(0.98)(0.98) = 0.92, or 92%. If surges existed between all the operations so down time in one seldom affected another, then the system reliability would approach 98%. If the value of reliability is high enough, one can justify adding surge.

9.3.1.1.4 Define and Optimize the Heat Recovery System

Processes and plants heat and cool many different process streams. If energy is not interchanged between some of these streams, it will be lost and utility costs will be higher than necessary. When defining a heat recovery system, one decides which hot and cold streams to use for energy interchange and about how much energy to transfer among them. S/he fixes the flow rates and estimates the inlet and outlet temperatures of the different streams in each heat exchanger. With a process as simple as the one shown in Example 1 in Chapter 6, this is fairly easy. In processes that are more complex or in plants, it is more involved. In these cases, one could use pinch technology to define the heat recovery plan. The process synthesis book by Turton et al. includes a good description of this method.[9] When using pinch technology, one must select a minimum approach temperature (ΔT_{min}); this defines the location of the pinch point. Most texts suggest using a ΔT_{min} of 10°C to 20°C

(18°F to 36°F). My experience is that when one uses the methods described in Chapter 10, ΔT_{min} should be more in the range of 60°F to 90°F.*

One optimizes the heat recovery plan by finding the optimum ΔT_{min}. This involves making heat transfer calculations at several different values of ΔT_{min}. The different values of ΔT_{min} redefine the pinch points and change the heat loads and inlet or outlet temperatures for the heat exchangers. With this information, one calculates the after-tax cash flows and the NPV or AC for each option. If utility capacity must be built to supply needed utilities to the process or plant, the engineer must include the costs — capital and operating — in the analysis.

9.3.1.1.5 Optimize the Equipment

During feasibility and conceptual engineering, one can choose to optimize operating conditions versus equipment cost or size for the major pieces of equipment. If this is not done, either capital or production costs may be too high. Table 9.11 contains examples for a few types of equipment. They are included to illustrate how to think about equipment optimization. For other types of equipment, one should refer to UO and equipment design texts and articles or consult with an equipment specialist.

9.4 SUMMARY

Developing a thorough list of options is key to creating an economic design. Without one, attention to economics will be haphazard and incomplete. It is important to separate creating the list from evaluating the options. If not separated, ideas will prejudged and that will get in the way of generating a comprehensive list.

When developing an option list use the following tools:

- Brainstorming and the 6-3-5 method can be useful in most any situations.
- When involved in process development work, use:
  - The Unit Operation Guides. They are: Blending-mixing, Drying (water removal only), Heat transfer (including Evaporation.), Mass transfer (including Crystallization.), Material transport, Mechanical separation, Reactions, and Size modification. These list the more common UOs.
  - The general process and process interaction questions. These questions deal with the process in general and how its different parts affect each other. Answering them will increase one's understanding of the process and lead to a better awareness of potential design options.
- When doing feasibility or conceptual engineering work, use the Unit Operation Guides when selecting UOs or equipment and the feasibility/conceptual checklist to increase your understanding of the process and project.

---

* This book uses complete after-tax cash flows to calculate either net present value or annual cost for economic comparisons. This is what makes sense in industry. Comparison methods in many books use before-tax cash flows, which distort one's conclusions.

TABLE 9.11
Equipment Optimization Examples

Pipe size	When one increases the pipe diameter:	Cost of the pipe increases
		Pressure drop through the pipe decreases, and the size and cost of the associated pump decrease
		Power usage by the pump and power cost also decrease
Heat exchanger size	When one increases shell-side or tube-side flow (and thus turbulence):	Size and cost of the heat exchanger decrease
		Pressure drop through the exchanger increases, and the size and cost of the associated pumps increase
		Power usage by the pumps and power cost also increase
	When one changes inlet and outlet temperatures so they approach more closely:	Heat exchanger area, and the cost, increase. For shell and tube units, the number of shells also increase which further increases the cost
		Utility usage and cost decrease
Evaporator configuration	When one increases the number of effects:	Cost of the evaporator increases
		Steam usage and cost decrease
	When one adds vapor recompression:	Cost of the evaporator system increases
		Steam usage and cost decrease
Batch pressure filter size[10]	When one increases the initial mass flow rate:	Size and cost of the filter decrease
		Filter cycle decreases and the labor costs for downtime operations (filling, precoating, emptying, and cleaning) increase
		Size and costs of the feed pump increase
	When one increases the filtering temperature:	Filtrate viscosity decreases and the filter size and cost decrease
		Filter cycle decreases and the labor costs for downtime operations increase
	When one increases the pressure drop across the filter (assuming an incompressible cake):	Size and cost of the filter decrease
		Filter cycle decreases and the labor costs for downtime operations increase
		Size and costs of the filter feed pump increase
		Power consumed by the pump and electrical cost increase

9.5 PROBLEMS AND EXERCISES

1. Figure 9.1 is a technical function flowsheet for a batch oil hydrogenation process. The batch size is dependent upon charging, heating, cooling, and discharge time. The reaction time is 1 hr per batch. Draw two possible process flow diagrams, each having different piping, pumping, and heat transfer arrangements. Show the reactor, all oil lines, pumps, and heat transfer devices. If it is important to convey your concept, indicate temperatures of the oil streams on the flow diagram and/or include a brief description of the option.

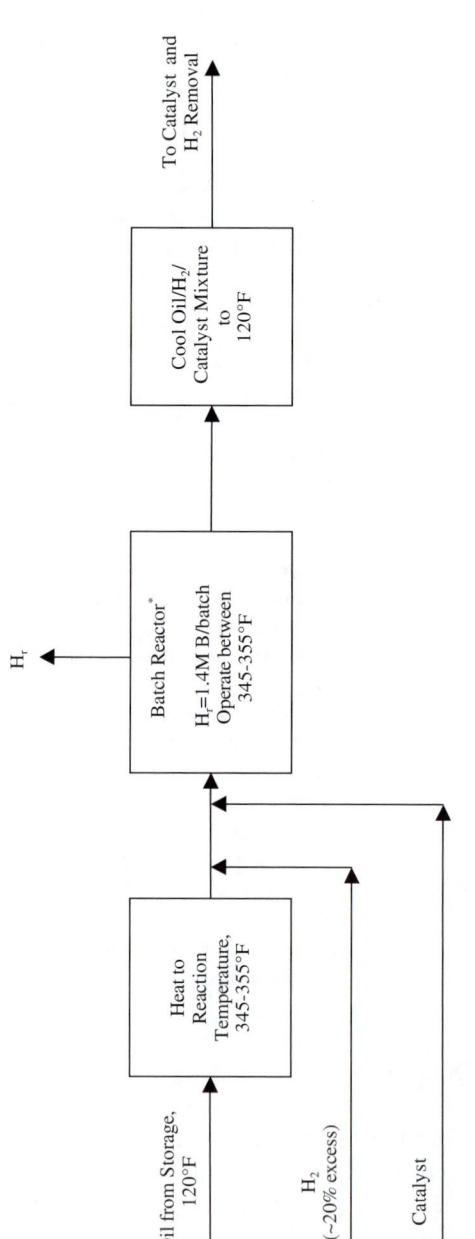

FIGURE 9.1 Problem 1: Technical function flowsheet for batch hydrogenation.

2. In a whiskey distillery, one of the waste streams is a slurry of spent grain in water. You wish to remove the water and send it to the sewer before processing the spent grain into cattle feed. What are all the UOs that would go on an options list?

3. Problem 4 in Chapter 8 presented information about processing orange peel into cattle feed. It mentioned the possibility of drying the peel to about 10% moisture after the free water had been removed. Assuming that peel drying was one of the technical functions, what drying UOs would you place on the options list? Both direct and indirect drying are technically acceptable.

4. Your company is considering expanding its business in Mexico by beginning to make a laundry bleach for consumer use. The general management is discussing locating the plant in or near Mexico City because most of the consumers will be there. You are considering three processes:
   a. Electrolyzing a salt-water solution into hypochlorite and hydrogen
   b. Reacting Cl_2 gas with NaOH to form hypochlorite, salt, and water
   c. Reacting calcium hypochlorite with sodium carbonate to form sodium hypochlorite, calcium carbonate, and salt
   Considering the questions in Table 9.10, Feasibility/Conceptual Checklist, comment on the HSE features of the three options.

5. During the Feasibility phase, you selected a vertical-tube, forced circulation evaporator to concentrate sodium hydroxide from 10% to 50%. As you optimize the evaporator, what options would you study?

6. You are beginning the feasibility study for a batch process that includes mixing three high-viscosity liquids. The technical function flowsheet shows the function as: mix liquids A, B, and C into a uniform mixture. All have viscosities similar to that of cold molasses. Create a mixing UO options list.

7. What UOs would you put on the options list for the following technical functions:
   a. Form crystals of material A from a 10% feed solution
   b. Remove trace amounts of benzene from water
   c. Create a vacuum in a packed stripping column
   d. Separate detergent dust from the airveying system exhaust (the airveying system is used to move detergent powder from the process to storage)
   e. React magnesium oxide (a solid) with chlorine (a gas) to form magnesium chloride

REFERENCES

1. Ullman, D.G., *The Mechanical Design Process*, New York: McGraw-Hill, 1997, 147–148.
2. Smith, R., *Chemical Process Design*, New York: McGraw-Hill, 1995, 3–8.
3. Granger, J.E., Plantsite selection, *Chemical Engineering*, June 15, 1981, 88–115.

4. Douglas, J.M., *Conceptual Design of Chemical Processes,* Boston: McGraw-Hill, 1988, 417–421.
5. Peters, M.S. and Timmerhaus, K.D., *Plant Design and Economics for Chemical Engineers 4th Edition*, New York: McGraw-Hill, 1991, 91–95.
6. Jelen, F.C. and Black, J.H., *Cost Optimization Engineering 2nd Edition*, New York: McGraw-Hill, 1983, 455–456.
7. Browning, J.B., Union Carbide: Disaster at Bhopal, in *Crisis Response: Inside Stories on Managing Under Siege*, Gottschalk, J.A. (ed.), Detroit, MI: Visible Ink Press, 1993.
8. Roodman, R.G., Operations: A critical factor often neglected in plant design, *Chemical Engineering*, May 17, 1982, 131–133.
9. Turton, R. et al., *Analysis, Synthesis, and Design of Chemical Processes*, Upper Saddle River, NJ: Prentice Hall PTR, 1998, 528–542.
10. Brown, T.R., Designing batch pressure filters, *Chemical Engineering*, July 26, 1982, 58–63.

10 Option/Alternate Analysis

There are almost always several technically acceptable answers to any design question. One can use economic factors to select which is best.

The analysis of options begins by weeding out the items on the option list that are not technically feasible. Only after that is done should one begin an economic analysis. In my experience, technical analysis removes most, but not all, of the options from the option list. This chapter covers how one goes about the economic analysis of options.

10.1 TECHNICAL FEASIBILITY COMES FIRST

A technically feasibility process, unit operation, or piece of equipment is one that can produce a product as intended. This includes making a product of the right quality and producing it at the expected capacity, reliability, and cost. In addition, the product must be made so that all environmental and safety requirements are met. Product quality is defined by the product and process specifications; project objectives generally spell out the capacity, reliability, and cost requirements; and environmental and safety provisions are defined by regulations — local, state, and national — and by company policy. During design, engineers convert these factors into process requirements, which are things such as: flow rates, operating pressure, operating temperature, properties of the feed streams, continuous or batch operation, output streams' properties (composition, purity, odor, flavor), and so on.

One analyzes for technical feasibility by comparing the process requirements to the capabilities of the unit operation or equipment. Determining technical feasibility is outside the scope of this book. The typical sources one would use to answer technical feasibility questions are:

- Bench-scale and pilot plant testing results
- Company operating and pilot plant experience
- Unit operations and process design texts and articles

10.2 ECONOMIC ANALYSIS METHODOLOGY

Once one has analyzed the option list for technical feasibility, usually several options are still on the list. In this part of the Economic Design Model, one subjects the remaining options to an economic evaluation to find which is best. Analysis moves through several steps:

1. Selecting the independent and dependent variables in the analysis.
2. Picking the initial values of the independent variables. These are the values for which one will do calculations. If these first picks do not define the best option, pick another set of values and recalculate.
3. Selecting and sizing the equipment for the different values of the independent variable and estimating the cost variables for each.
4. Calculating the NPV or AC for each option.
5. Selecting the economic design point.

10.2.1 VARIABLE SELECTION

In this first step of economic analysis, one must decide which is the independent variable. Recall that this is the variable whose value is specified first and that determines the values of the other variables. After selecting the independent variable, one figures out what are the dependent variables. There are two types — physical equipment-related and cost-related. The equipment-related are selected using the engineer's knowledge of the system or equipment being analyzed. Then one determines which costs will vary as a result of changes in the equipment-related variables. The cost checklist in Table 10.1 will help identify the costs that will change.

One uses it to ensure they have considered all the costs. Remember, study grade estimates are usually used when evaluating options and they use factors for many of the costs. For example, maintenance, insurance and taxes, operating supplies, and plant overhead are all expressed as a percentage of capital. Thus, whenever capital costs change from option to option, these costs will also vary.

Example 1: If one were trying to find the economic diameter of a pipe in a pipe/pump/motor system, pipe diameter would be the independent variable. When one changes pipe diameter, the pressure drop in the pipe will change, resulting in changes in the pump and motor size and the amount of power used to drive the motor. Using the checklist in Table 10.1, one can sort out which costs will vary as a result of changes in the pipe diameter. This process is shown in Table 10.2. The economic variables are in bold italics and are the only costs that must be considered.

10.2.2 EQUIPMENT SIZING — COST ESTIMATION

Once one figures out which are the variables, s/he selects the values of the independent variables that will be studied. Then for each, roughly size the equipment and estimate the capital costs. The order-of-magnitude and study grade methods described in Chapter 3 are the perfect tools when comparing most options.

Next, one would estimate the production cost variables using the study grade methods from Chapter 4.

Example 1 (continued): As described in the first part of Example 1, one would calculate the pressure drop, pump and motor size, and power usage for each pipe diameter being studied. Having that information, one next calculates the capital costs for the pipe and pump (with motor) and estimates the utilities, maintenance, insurance and taxes, operating supplies, plant overhead, and depreciation costs for each diameter.

TABLE 10.1
Cost Variable Checklist

Investments
    Capital
    Working capital
    Startup expenses
    Supplier advances and royalties
Production costs
    Raw materials
    Packaging materials
    Manufacturing
        Operating labor
        Employee benefits
        Supervision
        Laboratory
        Utilities
        Maintenance
        Insurance and taxes
        Operating supplies
        Plant overhead
        Depreciation
    Product delivery

10.2.3 COMPARE THE OPTIONS ECONOMICALLY

Knowing the variable costs for each option, one is ready to do the economic comparison. For this, I recommend using either NPV or AC because they allow one to compare many options at the same time. ROI can be used but it is cumbersome. It requires comparing pairs of options and justifying one versus the other. When multiple alternatives are involved, option pairing is difficult to use.

10.2.3.1 Use the Same Economic Life

When evaluating options, one must use the same economic life for each. When some options need replacement before others have worn out, this is not so straightforward.

Example 2: Consider the selection of materials for a pump in corrosive service where there are several possible materials of construction. The materials that extend the life of the pump are more expensive. The question is, what is the best economic choice? Say there are three options:

Option	Useful Life
A	2 yrs
B	4 yrs
C	5 yrs

TABLE 10.2
Example 1: Economic Variables

Cost Variables	Does the Variable Change?	Comments
Investments		
Capital	*Yes*	*Varies dependent upon pipe size and pump/motor size*
Working capital	No	Ignore
Startup expense	No*	Ignore
Supplier advances/royalties	No	Ignore
Production costs		
Raw materials	No	Ignore
Packaging materials	No	Ignore
Manufacturing	No	
Operating labor	No	Ignore
Employee benefits	No	Ignore
Supervision	No	Ignore
Laboratory	No	Ignore
Utilities	*Yes*	*Varies with power usage*
Maintenance	*Yes*	*6% of capital*
Insurance and taxes	*Yes*	*3% of capital*
Operating supplies	*Yes*	*1% of capital*
Plant overhead	*Yes*	*1% of capital*
Depreciation	*Yes*	*Capital/equipment life*
Product delivery	No	Ignore

* Chapter 3 suggested estimating startup expenses as a percent of capital. In spite of this, it seems logical that the startup expenses for one pipe/pump system would be the same as for another. Thus, no change is shown here.

When faced with this kind of situation, one looks for what might be called the "least common denominator." In this case, an economic life of 20 years is the shortest economic life that would work. That means Option A will need replacements in years 2, 4, 6, 8, 10, 12, 14, 16, and 18; Option B in years 4, 8, 12 and 16; and Option C in years 5, 10, and 15. These intermediate replacements become a key part of the analysis. The other part is the annual operating cost for each option.

If one does not use the same economic life for all options, Option A would automatically be selected because it is the cheapest of the options; however, it has the highest maintenance and replacement costs. If anything other than a 20-year comparison period were used, a significant portion of the cost differences would be ignored in the analysis, thus changing the conclusion.

10.2.4 THE ECONOMIC DESIGN POINT

The purpose of comparing options is to select the one having the best economics. The most economic option is the one having the greatest NPV or AC. Because the only costs included in the analysis are those that change as the independent variable changes, most or all of the cash flows will be negative. Thus, the highest NPV/AC may be negative. That is acceptable; it still defines the best option. Times will be found when the NPVs/ACs of the competing options will be so close that one cannot pick an option on the basis of economics. When that occurs, use some other criteria for selection.

Unless one has evaluated every possible value of the independent variable, it is best to plot the NPC/AC vs. the independant variable to ensure finding the true optimum. For example, if one were trying to find the optimum reflux in a distillation column, essentially an infinite number of options exist. The only way to find the maximum NPV/AC is to plot NPV/AC versus the percent reflux. On the other hand, for the three options in Example 2, a graph serves no purpose.

Example 3: You have a heat exchanger with a carbon steel shell and tubes in your department that is about to fail due to corrosion. You are considering two options — replacement in kind (carbon steel shell and tubes) or replacement with a unit having 314 stainless steel shell and tubes. The present unit had a life of 5 years and will cost $90K to replace. The 314 stainless steel unit is expected to have a life of 10 years and will cost $160K. Your company's tax rate is 37% and its hurdle rate is 15%. Assume maintenance, insurance and taxes, operating supplies, and plant overhead are 11% of capital for both options.

After using the cost checklist in Table 10.1, you determine the only costs that must be considered are capital costs, depreciation, maintenance, insurance and taxes, operating supplies, and plant overhead. Because the two units have different lives, we must pick a common life for comparison. In this case, a 10-year life will work for both. No capital investment will be made in year 10 for either option because that is the last year in the comparison period.

Carbon Steel Option:

Capital = $90K every 5 yrs

BT costs = Depreciation + Maintenance, etc.

$= -\$90K/5yrs + (-0.11 * \$90K) = \$27.9K/yr$

AT cash flow = Production cost difference (1 – Tax rate) + Depreciation (Equation 2.6)

$= (-\$27.9K/yr)(1 - 0.37) + \$18K/yr = \$0.4K$

Drawing the cash flow diagram:

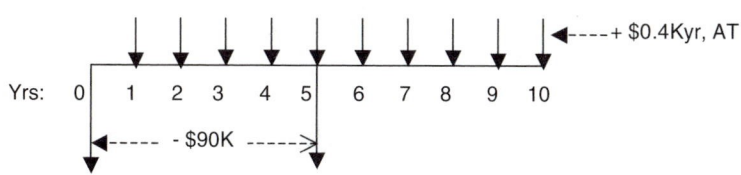

Find the NPV:

$$P_{capital} = P_{capital,yr\ 0} + P_{capital,yr\ 5}\ (P/F,\ 15\%,\ 5) = -90K + [-90K(0.497)]$$

$$= -\$134.7K$$

$$P_{AT\ cash\ flow} = AT\ cash\ flow\ (P/A,\ 15\%,\ 10) = 0.4K(5.019) = \$2K$$

$$NPV = -134.7K + 2K = -\$132.7K$$

314 Stainless Steel Option:

Capital = $160K every 10 years

BT costs = *Depreciation* + *Maintenance, etc.*

$$= -\$160K/10\ yrs + (-0.11 * \$160K) = -\$33.6K/yr$$

AT cash flow = $(-\$33.6K/yr)(1 - 0.37) + \$16K/yr = -\$5.2K$

Drawing the cash flow diagram:

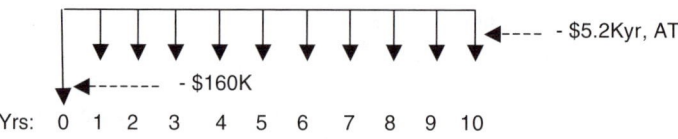

Find the NPV:

$$P_{capital} = \$160K$$

$$P_{AT\ cash\ flow} = AT\ cash\ flow\ (P/A,\ 15\%,\ 10) = -5.2K(5.019) = -\$26.1K$$

$$NPV = -160K + (-26.1) = \$186.1K$$

Because the carbon steel option has the larger NPV (−$132.7K versus −$186.1K), it is the better, more economic choice.

Example 4: You are part of a team designing a new plant and have responsibility for process B. You are considering three alternates. All meet the project's technical requirements and are sound from an environmental and safety standpoint. You have estimated capital and production costs for each option; they are summarized below:

Cost Category	Option 1	Option 2	Option 3
Capital ($M)	2.5	2.9	3.2
Raw materials, catalyst ($M/yr)	0.70	0.50	0.30
Manufacturing costs ($M/yr)			
Operating labor	0.25	0.20	0.17
Employee benefits, supervision, laboratory	0.18	0.14	0.12
Utilities	0.37	0.19	0.13
Maintenance	0.15	0.17	0.19
Insurance and taxes, operating supplies, plant			
overhead	0.13	0.15	0.16
Depreciation	0.17	0.19	0.21
Total ($M/yr)	1.25	1.04	0.98
Production cost differences ($M/yr)	1.95	1.54	1.28

Your company's tax rate is 40%, its hurdle rate for this type of project is 12%, and it specifies a 15-year project life for this type of project. Determine the most economic option. Use the annual cost method for comparing.

Find the AT cash flow for each option using Equation 2.6:

For Option 1:

$$AT\ cash\ flow = Production\ cost\ differences\ (1 - 0.4) + Depreciation$$

$$= -1.95(0.6) + 2.5/15 = -\$1.0M$$

Similarly for Option 2: $AT\ cash\ flow = -\$0.73M$

Similarly for Option 3: $AT\ cash\ flow = -\$0.56M$

Find the AC for the capital for each option. This involves converting the capital into its equivalent annuity.

For Option 1:

$$AC = P(A/P,\ 12\%,\ 15) = -2.5(0.147) = -\$.37M$$

Similarly for Option 2: $AC = -\$0.43M$

Similarly for Option 3: $AC = -\$0.47M$

Find the total AC for each option. Because the AT cash flow is already in annuity form, it is equal to $AC_{product\ cost}$.

$$AC_{total} = AC_{capital} + AC_{product\ cost}$$

For Option 1: $AC_{total} = -\$0.37 + (-\$1.00) = -\$1.37M$

For Option 2: $AC_{total} = -\$0.43 + -(\$0.73) = -\$1.16M$

For Option 3: $AC_{total} = -\$0.47 + -(\$0.56) = -\$1.03M$

The economic design point: All options are technically feasible, so we will pick the one having the best economics. Because Option 3 has the largest AC (the least negative), select this one.

Example 5 (Economic Pipe Diameter): Example 1 introduced finding the economic diameter of a pipe. This example will work the problem to completion. Consider the following pumping situation where 160 gpm of Newtonian fluid, having a viscosity of 20.6 centistokes (cs) and a specific gravity of 0.9, is pumped from Tank A to Tank B.

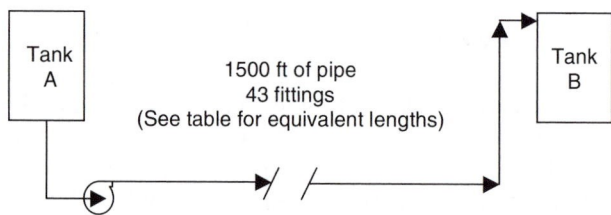

Using March 2005 costs, find the most economic pipe diameter for the following situation. Pumping occurs 10 hr/day, 5 days/wk, and 50 wk/yr. March 2005 electrical power costs are $0.047/kWh. The system is manually operated. The pipe is Schedule 40 carbon steel and the pump is a cast iron ANSI centrifugal pump. Your company's hurdle rate is 15% and its tax rate is 35%. Use a 10-year project life.

Pipe Diameter (in)	Equivalent Length of Pipe (ft)	P (ft of fluid/ 1000 ft of Pipe)[1]
$2\frac{1}{2}$	1688	281
3	1709	98.2
4	1820	27.2
6	1967	3.9

Example 1 identified pipe diameter as the independent variable. The dependent equipment variables were pump and motor size and the dependent cost variables were capital, utilities (electrical power), maintenance, insurance and taxes, operating supplies, plant overhead, and depreciation.

To solve this problem, start by sizing the pump. (We will ignore the $2\frac{1}{2}$-in. pipe because its ΔP exceeds 150 psig, which significantly increases the pipe and pump costs.) The following are the calculations for the 3-in. pipe.

Size the pump:

$$\Delta P_{ft} = (Equivalent\ length) * (\Delta P/1000\ ft) = 98.2/1000\ ft * 1709 = 168\ ft\ of\ head$$

Pump size: 160 gpm, required head of 168 ft

Electrical power costs (assume a pump/motor efficiency of 60%):

$$HP = \frac{\text{flow} * \text{head}}{\text{Motor Efficiency}}$$

In standard units, this becomes:

$$HP = \frac{\text{gpm} * \Delta P_{ft} * SG}{3960 * E} = \frac{160 * 168 * 0.9}{3960 * 0.6} = 10.2$$

$$\$/\text{yr} = 10.2HP * 10\frac{hr}{day} * 5\frac{day}{wk} * 50\frac{wk}{yr} * 0.746\frac{kWh}{HP-hr} * \frac{\$0.047}{kWh} = \$894/\text{yr}$$

Capital costs:

Pipe: Use the piping page in Appendix IV. The chart is based upon a CEPI of 460, which is a March 2005 index. This chart gives the installed cost of piping as:

$$\$K/100\text{ ft} = (0.1F + 0.924)\,D^{0.83}$$

where F = the number of fittings/100 ft of pipe = 43 fittings/1500 ft = 2.9/100 ft

D = nominal pipe diameter = 3 in

$$\$_{2005} = [(0.1 * 2.9 + 0.924) * 3^{0.83}] * (1500\text{ft}/100\text{ft}) = \$45.3K$$

Pump. Use the pump page in Appendix IV.

The "gpm * ft" for the pump = 160 gpm * 168 ft = 26.9K

$$\$_{2005} = 5K$$

Use Equation 3.3 to calculate the capital cost.
The Hand Factor (Table 3.2) is 4.

F_m = 1 because the pump is cast iron

F_i = 1 because the pump is manually operated

F_b = 1.06 — assume the system will be part of an expansion in an existing plant

F_p = 1 — assume the system will be in the U.S.

$$\$_{2005} = 5K * 4 * 1 * 1 * 1.06 * 1 = \$21.2K$$

$$Total\ capital = 45.3 + 21.2 = \$66.5K$$

Cash flows:

$$BT\ cash\ flow = Utilities + (Maintenance,\ etc.) + Depreciation$$

$$= -\$0.9K + (0.11 * -\$66.5K) + (-\$66.5K/10) = -\$14.9K$$

$$AT\ cash\ flow = (BT\ cash\ flow)(1 - Tax\ rate) + Depreciation$$

$$= (-\$14.9K)\ (1 - 0.35) + \$6.09K = -\$3.04K$$

NPV:

$$NPV = P_{capital} + P_{AT\ cash\ flow} = -\$66.5K + [-\$3.04K(P/A,\ 15\%,\ 10)]$$

$$= -\$66.5K + (-\$15.3) = -\$81.8K$$

Similarly, one calculates the NPVs for 4-in. and 6-in. pipe:

Pipe Diameter (in)	NPV ($K)
3	−81.8
4	−86.9
6	−106.2

The economic design point: Because there are discrete options, one need not plot NPV versus diameter. The maximum NPV (least negative) occurs for the 3-in. pipe. This is the economic diameter.

10.3 THE ECONOMICS OF SELECTING EQUIPMENT

The economics of equipment selection is a broad field and must be tailored to each type of equipment. This section deals with selection economics for the common types of pumps and heat exchangers. Each uses a different approach to the selection problem, the more exact being for heat exchangers. In his book, *Chemical Engineering Process Design and Economics, A Practical Guide, 2nd Edition*, Ulrich covers the economic selection for a broad range of equipment.[2] His approach is similar to the one used in this chapter for pumps. In keeping with the focus on the early phases of design, this section deals only with the economics of preliminary selection. Final selection requires further work.

10.3.1 PUMPS: PRELIMINARY SELECTION

Two principal types of pumps are used — centrifugal and positive displacement. All other types are designed to solve some unique pumping problem. *Centrifugal pumps* move liquids by the rotation of an impeller, which imparts kinetic energy to the fluid. The kinetic energy is converted into pressure energy. The typical centrifugal pump has an open, rather simple impeller and casing. *Positive displacement* pumps use close tolerances between gears, lobes, vanes, screws, or pistons and cylinders to positively push or force the fluid through the pump. The primary subtypes of these two types of pumps are:

Centrifugal	ANSI, horizontal mounting[*]
	Multistage
	Vertical, in-line mounting
	API, horizontal and vertical mounting
Positive displacement	Rotary
	Reciprocating

These six subcategories cover about 95% of all pump applications.

10.3.1.1 Technical Considerations

The guidelines below will help one decide between a centrifugal and a positive displacement pump.

Centrifugal pumps can handle all types of fluids including those that are non-lubricating and those with suspended particles. The discharge is steady and nonpulsating. Flow from the pump can be controlled by throttling. Use them for:

- Lower viscosity fluids — less than 650 cs (3000 SSU)
- Lower total developed head (TDH) requirements — single stage up to around 150 psi and multistage up to about 2000 psi
- Higher available NPSH: more than 5 ft

Positive displacement pumps can handle clean, lubricating fluids. The discharge is pulsating. Flow from the pump can be controlled by controlling the pump speed or by recycling fluid from the discharge to the suction. Use them for:

- Low to high[†] viscosity fluids
- Low to high TDH requirements
- Negative to high NPSH situations

---

[*] The basic difference between ANSI and API pumps are: ANSI pumps have a maximum operating pressure of 375 psig @ 700°F, whereas API pumps can operate up to 870 PSIG @ 800°F; and ANSI pumps having the same flow rate and head have the same outline dimension regardless of manufacturer, whereas API pumps do not.
[†] Greater than 650 cs.

10.3.1.2 Economic Considerations

The rough guidelines in Table 10.3 will help one use economics to select among technically feasible pump options.

TABLE 10.3
Pump Economics

	Type	Purchase Cost Ratio	Maintenance Cost
Centrifugal	ANSI, horizontal	1.0	Lowest
	Multistage	1.5	
	Vertical, in-line mounting	2.2	
	API, horizontal and vertical	3.5	
Positive displacement	Rotary	1.5	
	Reciprocating	3.5	Highest

10.3.1.3 Selection Guideline

If an ANSI centrifugal pump can be used, use it. This will result in the lowest capital and maintenance costs. If one cannot be used, balance capital and maintenance costs.

Example 6: You are doing a feasibility study for a process upgrade and are making a preliminary selection of a pump that will pump 130 gpm of a clean fluid, having a viscosity of 5 cp. The available NPSH will be more than 20 ft and the pump will have to develop a discharge head of 1050 ft (435 psig). You have determined a multistage centrifugal or either type of positive displacement pump are technically feasible. Which should you select?

Table 10.3 shows that the purchase cost ratio for both the multistage centrifugal and the rotary positive displacement pump is 1.5, whereas it is 3.5 for the reciprocating pump. The table also shows the maintenance for the multistage centrifugal is the lowest of the three. Given that, the multistage pump is the best economic choice.

Example 7: As a part of an option study, you must make a preliminary selection of a pump that will pump 300 gpm of a clean fluid that has a viscosity of 70 cp. The available NPSH will be well over 5 ft, and the pump will have to develop about 300 ft of head. From a technical standpoint, all the types of pump are feasible. Which should you select?

Because all the pump types are feasible, pick the ANSI centrifugal pump. It is the least expensive from both a purchase and maintenance cost standpoint.

10.3.2 Heat Exchangers: Preliminary Selection

The two most common types of heat exchangers for heating and cooling service are shell and tube and plate and frame exchangers. Three shell and tube units are common: U-tube, fixed-tube sheet, and floating head exchangers. The many other types of exchangers are specialty units and are generally more expensive to use. (For a list of the specialty types, see Table 9.3, Heat Transfer Unit Operation Guide.) When trying to select an exchanger, one immediately finds out that shell and tube units are less expensive on a $/ft^2 basis but their overall heat transfer coefficients (U) are lower. Conversely, plate exchangers have a higher $/ft^2 cost but their Us are also higher. If both types are feasible, selection of the most economic requires some analysis.

10.3.2.1 Technical Feasibility

Table 10.4 compares the uses, usual applications, and limitations of both types of exchanger. Table 10.5 further compares the three types of shell and tube exchangers. Using the data in these tables, one can determine technical feasibility. If one type is not feasible, drop it from consideration. If several are feasible, use economics to decide which type to use. If none are feasible, consider the specialty exchangers.

10.3.2.2 Economic Considerations

If both the shell and tube and the plate and frame exchangers are technically feasible, one must sort out which is the better economic choice. This is a multistep process that involves:

- Selecting the type of shell and tube exchanger to be used. If all are feasible, select the U-tube because it is the least expensive. The heat exchanger cost chart in Appendix IV shows fixed-tube sheet units are 5% more expensive than a U-tube and floating head units are 30% more.
- Rough sizing of both the shell and tube and plate units.
- Estimating the purchase prices using the heat exchanger pages in Appendix IV.
- Estimating the capital cost of each using Equation 3.3.
- Selecting the exchanger having the lowest capital cost. NPV or AC calculations are not needed because the only portions of production costs that will be different are those that vary with capital costs.

10.3.3 Examples Illustrate

Example 8: You are in the process of sizing and selecting a heat exchanger that will use 15-psig steam to heat a hydrocarbon from 150°F to 200°F. You already know both U-tube and plate exchangers are technically feasible. Either carbon or stainless steel metallurgies are acceptable. The U-tube is carbon steel but the plate unit has stainless steel plates because carbon steel plates are not available. The gasket material

TABLE 10.4
Technical Selection Guidelines — Heat Exchanger Type*

Shell and Tube	Plate and Frame

Typical uses

Liquid cooling	Liquid cooling
Liquid-liquid heat exchange	Liquid-liquid heat exchange
Steam heating	Low pressure steam heating
Condensing	
Low- to high-pressure gas heating and cooling (finned tubes are often used)	
Reboilers	

Usual applications

Services with no to slight temperature crossing	Services with no to high-temperature crossing
Low- to high-temperature services	Lower temperature services due to temperature limitations of gaskets
Low- to high-pressure services	Nitrile ≤ 275°F
Low- to medium-viscosity fluids (up to 75–100 cP)	EPDM: ≤ 300°F
Low fouling services	Viton: ≤ 350°F
	Lower pressure services[†]
	Low- to high-viscosity fluids
	Medium fouling services
	Where mechanical cleaning is required
	Where little space is available
	Where more costly metallurgies are required
	Where heat exchanger area increases are probable

Limitations

Less efficient and more costly with closely approaching or crossing outlet temperatures	Not feasible at pressures > 370 psig[†]
Poor shell-side flow distribution causes inefficiency and fouling	Gasket materials limit use to temperatures ≤ 350°F
Mechanical cleaning of the shell side is almost impossible	
Exotic metallurgies are costly	
Require a large amount of space	

* Adapted from: Brown, T.R., Use these guidelines for quick preliminary selection of heat exchanger type, *Chemical Engineering*, February 3, 1986, 107–108.

† Standard units are rated up to 150 psig. Special high-pressure designs for pressures up to 370 psig are available for an upcharge of 20% to 35%.

TABLE 10.5
Technical Selection Guidelines — Shell and Tube Heat Exchangers

	U-Tube	Fixed-Tube Sheet	Floating Head
Advantages	• Acceptable for differential thermal expansion between the shell and tubes • Shell side: Can handle fouling fluids because the tube bundle can be	• Tube side: Can handle fouling fluids because the tubes can be mechanically cleaned	• Acceptable for differential thermal expansion between the shell and tubes • Tube side: Can handle fouling fluids because the tubes can be mechanically cleaned
Disadvantages	• Tube side: Must have clean, nonfouling fluids because the tubes cannot be mechanically cleaned	• Limited to minimal differential thermal expansion between the shell and tubes unless an expansion joint is used. (These are subject to leakage and require maintenance)	• The packing gland limits operating temperatures and pressures.
Cost	• The least expensive shell and tube	• The intermediate cost shell and tube	• The most expensive shell and tube

for the plate exchanger is nitrile. You have sized and estimated the purchase cost for both exchangers.

	U (BTU/h-ft²-°F)	Area (ft²)	Cost ($/ft²)	Purchase Price ($K)
U-tube (CS)	60	2000	9.50	19.0
Plate (SS)	150	800	23.10	18.5

Note the significant differences in Us and $/ft² between the shell and tube and the plate units. In this example, they balance each other so that the purchase price of both is about the same. The capital costs of the two options are calculated as follows.

The Hand factor for heat exchangers is 3.5. We will assume typical instrumentation or $F_i = 1.35$, $F_b = 1.06$, (assuming an expansion in an existing plant) and U.S. construction. F_m for both options is 1.0.* Using Equation 3.3:

*Capital cost = Σ[Equipment purchase cost * (Hand factor * F_m)] * F_i * F_b * F_p*

$\$_{S\&T} = [\$19.0K\ (3.5 * 1)] * 1.35 * 1.06 * 1 = \$95.2K$

---

* Because the Hand factor for plate units having stainless plates is 3.5, F_m would be 1.0.

$$\$_{Plate} = [\$18.5K\ (3.5 * 1)] * 1.35 * 1.06 * 1 = \$92.7K$$

The capital cost for the plate unit is the lowest, so it is the best economic choice.

Example 9: You are sizing and selecting a heat exchanger to interchange heat between two fatty acid streams. You will cool the hot stream from 290°F to 220°F and heat the cold stream from 120°F to 240°F. Both shell and tube and plate exchangers will work in this service. They must be made from stainless steel and will have a pressure rating of 140 psig. Because the cold steam is "dirty," you expect the tube sheet in the shell and tube exchanger will have to be mechanically cleaned. Because of this, you select a fixed-tube sheet unit. You have roughly sized and priced both.*

	U (BTU/h-ft²-°F)	Area (ft²)	Cost ($/ft²)	Purchase Price ($K)
Fixed tube sheet	30	2280*	16.14	36.8
Plate	90	670	25.37	17.0

As in Example 8, note the large counterbalancing Us and $/ft². In this case, they do not offset each other because the crossing outlet temperatures require that the shell and tube exchanger have three shells. The capital costs of the two options are as follows.

The Hand factor is 3.5. Assume typical instrumentation or $F_i = 1.35$, $F_b = 1.06$ (assuming an expansion in an existing plant), U.S. construction, and $F_m = 1$ for the plate unit. (See the footnote in Example 3 for why $F_m = 1$.) For the fixed tube sheet the alloy/CS ratio is 2.0 (from the heat exchanger chart, Appendix IV), so $F_m = 0.69$. Again, use Equation 3.3:

$$\$_{S\&T} = [\$36.8K\ (3.5 * 0.69)] * 1.35 * 1.06 * 1 * = \$127K$$

$$\$_{Plate} = [\$17.0K\ (3.5 * 1)] * 1.35 * 1.06 * 1 = \$85.1K$$

Because the capital cost of the plate exchanger is lower, select it.

Example 10: You are sizing and selecting a heat exchanger to heat a process stream to 450°F using 600 psig steam. You rule out the plate unit because of high steam pressure and high temperature. Because only the shell and tube unit is feasible, you select it.

10.4 THE ECONOMICS OF PLANT SITING

The methodology described here is different from that found in the literature in that it first uses economics to decide on the number of plants to be built and their approximate locations. An approximate location is one identified by a circle having a 50 mi to 100 mi radius. The second step would be to adjust this plan if any sites do not meet some of the important siting criteria. One might also adjust the initial

---

* Because the outlet temperatures of the two streams cross, three shells are needed @ 760 ft²/shell.

plan to place the new construction in an existing plant rather than build a new site. The third and last step would be to do a detailed site search within the approximate location circles.

This section will deal primarily with the first step. In addition, it will briefly review the factors to consider when locating a new process or plant.

10.4.1 HOW TO SET UP THE ECONOMIC STUDY

Deciding how many sites are economically optimal requires balancing the costs that vary as the number of sites change — capital, startup expenses, manufacturing costs, in-freight costs, and distribution costs. In general, as the number of sites increase, capital, startup expenses, and manufacturing costs increase whereas distribution costs decrease. In-freight costs may increase or decrease depending on the location of raw material and packaging suppliers.

The work to find the optimal number of plants begins with deciding how many plants one wishes to consider. Once that is done, the engineer would subdivide the customer distribution area according to the number of sites in each option. For example, if one were considering options of two, four, and six sites, they would divide the distribution area into two, four, and six parts, one part for each site in the option. Subdividing is best done with Sales or Marketing. In some companies, one might also involve those responsible for getting the plant's product to its customers.

Whereas it is usually best to locate plants on the basis of being near customers, instances may well exist where it is more appropriate to locate on the basis of where one's suppliers are. The rest of this section will discuss only customer-based siting but the methodology described applies to either situation.

10.4.2 HOW TO GO ABOUT ROUGH SITING

Rough siting involves two steps: subdividing the distribution area and roughly locating a site in each subdivision.

10.4.2.1 Subdividing the Distribution Area

The distribution area is the area in which a company's customers are located. This could be as big as the entire world or as small as a state or county. Consider a hypothetical situation where the distribution area is the continental United States. Assume the product to be made is a packaged food that is expected to have a volume of 150M lb/yr. Table 10.6 lists the company's 18 largest customers, which represent 87.5% of the total U.S. volume.

If the engineer were studying one, two, and three sites, they would subdivide the distribution area into two and three regions. To illustrate, Figure 10.1a and Figure 10.1b show two ways the U.S. could be split up for a three-plant option. Sales and Marketing could have specified the subdivision or the engineer doing the siting study could have generated both options. During the study, they would economically compare both three-plant options and pick the one having the best economics. Because shipping across the Rocky Mountains is so expensive, the split off of the West Coast would often be done for multiplant plans.

TABLE 10.6
Volume Distribution

Largest Metropolitan Market Areas	% of U.S. Distribution
New York	15.2
Los Angeles	11.7
Chicago	6.7
Baltimore/Washington D.C.	5.5
San Francisco	5.1
Philadelphia	4.4
St. Louis/Kansas City	4.4
Boston	4.3
Detroit	4.1
Cleveland/Columbus/Cincinnati	4.1
Dallas/Fort Worth	3.7
Houston	3.4
Atlanta	3.0
Seattle/Portland	3.0
Miami	2.9
Phoenix	2.4
Minneapolis/St. Paul	2.2
Denver	1.4
Total	87.5%

Once one has subdivided the distribution area, one must size the plants for the different options. Two things determine the capacities of the plants — the number of regions and how one has subdivided the distribution area. Using the data in Table 10.7 and Table 10.8, one would size the plants for both of the three-plant options as shown below. (Note the "% Volume" numbers are on a 100% basis.)

	Region I		Region II		Region III	
	% Volume	Capacity (M lb/yr)	% Volume	Capacity (M lb/yr)	% Volume	Capacity (M lb/yr)
From Fig.10.1a	49.7	74.5	24.9	37.4	25.4	38.1
From Fig.10.1b	58.2	87.3	16.4	24.6	25.4	37.5

This illustrates the impact of subdividing the distribution area on plant design. Next, one roughly locates the plants within each region.

10.4.2.2 Locating Plants Based on Customer and Supplier Locations

This section presents two ways to locate plants, one graphical and one analytical. Respectively, these are the "tie-line" and the "coordinate" methods. Both require knowing where a company's primary customers and suppliers are located and how

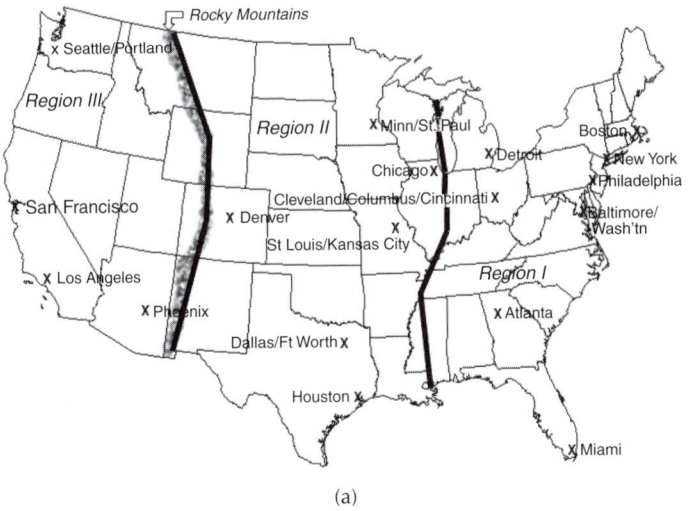

(a)

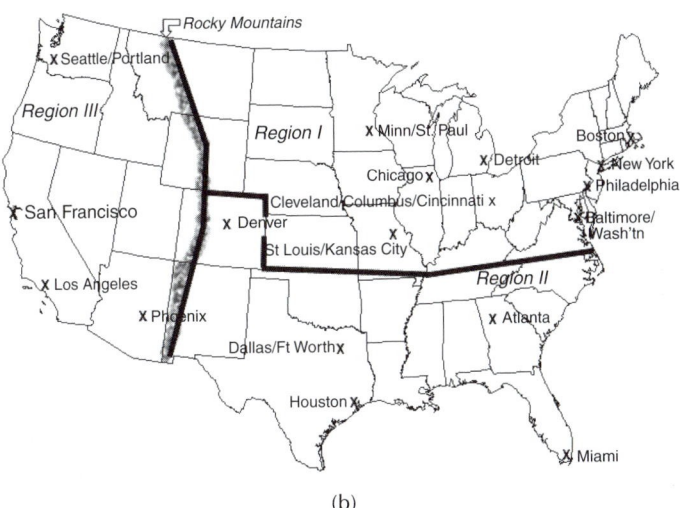

(b)

FIGURE 10.1 (a) Subdividing the U.S. distribution area. (b) Subdividing the U.S. distribution area.

much "volume" can be attributed to each. Customer volume is most often expressed in the dollar volume of sales or the amount of product shipped. Raw and packaging materials received by a plant are often so different that shipping costs per unit are also quite different. For example, a plant might get most of its raw materials in 20,000 gallon tank cars, whereas its primary packaging material is shipped in semi-trailers. When this is the case, volume would be expressed in a way that

TABLE 10.7
Volume/Plant Capacity for Figure 10.1a

	% of U.S. Distribution	% of Company Volume
Region I		
New York	15.2	17.4
Baltimore/Washington D.C.	5.5	6.3
Philadelphia	4.4	5.0
Boston	4.3	4.9
Detroit	4.1	4.7
Cleveland/Columbus/Cincinnati	4.1	4.7
Atlanta	3.0	3.4
Miami	2.9	3.3
Subtotal	43.5	49.7
Region II		
Chicago	6.7	7.7
St. Louis/Kansas City	4.4	5.0
Dallas/Fort Worth	3.7	4.2
Houston	3.4	3.9
Minneapolis/St. Paul	2.2	2.5
Denver	1.4	1.6
Subtotal	21.8	24.9
Region III		
Los Angeles	11.7	13.4
San Francisco	5.1	5.8
Seattle/Portland	3.0	3.4
Phoenix	2.4	2.8
Subtotal	22.2	25.4
Total	87.5%	100%

takes this variability into account. One way to do this is to express "volume" on the basis of:

$$(\text{lb shipped}) * (\$/\text{lb/mi shipped}) = \$/\text{mi shipped}$$

While not really "volume," this measure provides a way to balance shipments from different suppliers. It can also be used when finding the combined center of mass of customers and suppliers.

Most plants have many customers and suppliers. To consider all of them in the location decision is not necessary because the smaller ones have essentially no impact on the outcome. One only needs to take into account those representing 80 to 90% of the volume

TABLE 10.8
Volume/Plant Capacity for Figure 10.6b

	% of U.S. Distribution	% of Company Volume
Region I		
New York	15.2	17.4
Chicago	6.7	7.7
Baltimore/Washington D.C.	5.5	6.3
Philadelphia	4.4	5.0
St. Louis/Kansas City	4.4	5.0
Boston	4.3	4.9
Detroit	4.1	4.7
Cleveland/Columbus/Cincinnati	4.1	4.7
Minneapolis/St. Paul	2.2	2.5
Subtotal	50.9	58.2
Region II		
Dallas/Fort Worth	3.7	4.2
Houston	3.4	3.9
Atlanta	3.0	3.4
Miami	2.9	3.3
Denver	1.4	1.6
Subtotal	14.4	16.4
Region III		
Los Angeles	11.7	13.4
San Francisco	5.1	5.8
Seattle/Portland	3.0	3.4
Phoenix	2.4	2.8
Subtotal	22.2	25.4
Total	87.5%	100%

10.4.2.2.1 The Tie-Line Method

This method involves locating the "center of mass" of the customer and supplier locations. To show how it works, we will locate a single plant on the basis of just its customers. This plant supplies three major customers:

Customer Location	% of Company Shipping Volume
Houston, TX	15
Chicago, IL	45
Atlanta, GA	25

To begin, mark the three customer locations — Houston, Chicago, and Atlanta — on a U.S. map. See Figure 10.2; we will use it in the analysis.

FIGURE 10.2 The tie-line method.

Next, locate the center of consumer mass for the Chicago and Houston custom-ers. Connect them with a line. These two customers represent 60% of the company volume and Chicago receives 75% (45/60) of the combined volume. The center of consumer mass is a point 75% of the way from Houston to Chicago. Mark that point on the line and call this point "CH."

Now draw a line between point CH and Atlanta and find the center of consumer mass for these two locations. Point CH corresponds to 60% of plant volume (the combined volume of Chicago and Houston) and Atlanta to 25% for a total of 85%. Point CH would receive 70.6% (60/85) of the volume represented by these two points. The center of consumer mass for point CH and Atlanta is a point 70.6% of the distance from Atlanta to point CH. Note the centers of mass we have found are closer to the location getting the most volume. Because all three of the plant's major customers are now accounted for, the center of mass just located is also the location of the plant. Because this is an approximate location, draw a 50 mi to 100 mi radius around the location. Detailed site selection will focus on potential locations within that circle.

10.4.2.2.2 The Coordinate Method

Based a 1981 article by Granger, this method also locates the plants at its customer-supplier center of mass.[3] We will again locate a single plant using the data in the "tie-line" example. For this analysis, one draws a coordinate grid on the map. Use Figure 10.3 to follow the rest of explanation of this method.

After locating the customers on the map, one finds the x-y coordinates for each. Next, one converts the volume column, which only totals 85% to a 100% basis or to the percent of volume being considered in the analysis. Then multiply the coor-dinates by the "100% volumes." The sum of the x-coordinate products is the x-

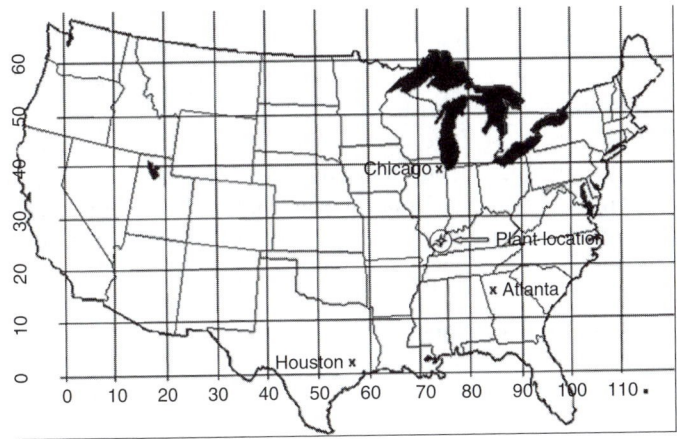

FIGURE 10.3 The coodinate method.

coordinate of the plant location, or 73.9. Similarly, the y-coordinate of the plant is 25.4. Note this is the same location as found by the tie-line method.

Customer Location	% of Company Shipping Volume	% of Volume Being Considered	X	Y	Volume * X	Volume * Y
Houston, TX	15	17.6	57	2	10.0	0.4
Chicago, IL	45	52.9	74	39	39.1	20.6
Atlanta, GA	25	29.5	84	15	24.8	4.4
Totals	85	100	—	—	74	25

10.4.3 SITING CRITERIA[4–8]

One uses the following factors when adjusting the rough siting plan and when doing a detailed site search within the 50 mi to 100 mi radius circle located during rough plant siting.

One can group the siting criteria into five major groups — proximity to suppliers and customers; labor considerations; utility availability and cost; health, safety, and environmental considerations; and locale and community considerations. Each company will assess the importance of the factors differently.

10.4.3.1 Proximity to Suppliers and Customers

Freight costs and the time between order and delivery are the key considerations for shipments to a plant's customers and for material shipments into the plant. Even when raw and packaging materials are priced delivered to the plant's unloading dock, one must take into account in-freight costs because they are an integral part of a supplier's pricing. If one can reduce in-freight by locating a plant near its suppliers, Purchasing should be able to negotiate lower material costs. If the receiving plant pays the in-freight charges, the cost effect of a location close to a supplier is easier to see but is no more significant. This same logic holds true for delivery charges. Here the advantage shows up in a plant's ability to price its product more competively.

Order-delivery response time is important because a rapid response allows a plant's customers to reduce the inventory in their plants. The same is true for one's own plant when its supplier has short response times. In addition, a plant will be better able to meet its customers' needs when it can respond more quickly to emergency orders.

10.4.3.2 Labor Considerations

One of the crucial labor factors is the availability of the skills needed to build the plant and then operate it. One would also consider:

- The work ethic and the attitude of the available work force.
- The labor climate. Is it union or non-union? Is it antagonistic or cooperative?
- What are the prevailing wage rates in the area? What will have to be paid to recruit the skills needed to operate the plant?
- Is the area one that will be amenable to the hiring and transfer of managers?

10.4.3.4 Utility Availability and Cost

In this category, one determines the availability of all the utilities needed by the plant and what each will cost.

- Electricity — consider quantity needed, interruption history, voltages
- Water — consider process, potable, cooling and any other needs
- Fuels — consider what type and quality will be needed (natural gas, fuel oil and coal)
- Other sources — is there a potential for these?

10.4.3.5 Health, Safety, and Environmental Considerations

Here one must consider:

- The availability of waste disposal availability and cost. This includes things such as sewage treatment, land fill (both sanitary and hazardous), and air quality.
- Permitting. How easy or difficult might it be? How long might it take? Consider the permitting history/record in that area, whether the local treatment plant presently has enough capacity for the proposed plant, and whether the region is in compliance with air quality standards.
- The hazard level of the plant and its proximity to a population center.

10.4.3.6 Locale

Here one considers items such as:

- The potential for natural disasters — earthquakes, tornados, hurricanes, or floods
- The climate — facility costs and operation can be affected hot or cold temperature extremes, heavy snows, excessive rainfall, high humidity, and so on
- The topography of the area — the terrain (flat, hilly, mountainous), soil characteristics (bearing strength, sandy, rocky), and so on

10.4.3.7 Community Considerations

This category includes many miscellaneous items related to the area where the plant might be built. Besides affecting plant operation, these considerations will have an effect on the plant's ability to recruit and transfer managers.

- Community services availability — fire, security, hazardous release response, waste disposal, and medical
- Financial — tax structure, financial inducements, training allowances
- Quality of life — general cost of living, housing quality and costs; shopping, transportation, medical, and educational facilities; cultural facilities/activities; civic organizations; outdoor activities (ocean, lakes, skiing, hunting, fishing); racial and religious balance; and so on

10.5 SUMMARY

10.5.1 ECONOMIC COMPARISONS

When making economic comparisons, follow these steps:

- Select the independent variable and determine the dependent variables in the analysis. Use Table 10.1 to help sort out which are the dependent cost variables.
- Pick the initial values of the independent variable — the ones for which one will perform calculations. If these first picks do not define the best option, pick another set of values and recalculate.
- Select and size the equipment for the different values of the independent variable and estimate the dependent variable costs for each. This will include both capital and production costs.
- Calculate the NPV or AC for each option:
  - Use your company's hurdle rate as the discount rate, or if a specific discount rate has been established for a project, use that rate.
  - Use the same economic life for options being compared.
  - Use obsolescence or company financial guidelines to set the economic life for the comparison.

- Establish year 1 as the year in which revenues begin.
- Use AT cash flows (Equation 2.6).
- Select the economic design point. The option having the maximum NPV/AC defines the design point. Because only the costs that vary from option to option are used in the NPV/AC calculations, this will often be a negative number. Unless one is considering only a limited number of discrete options, it is best to plot NPV/AC versus the independent variable to ensure finding the maximum. In the case of discrete options, select the one having the maximum NPV/AC.

10.5.2 ECONOMIC EQUIPMENT SELECTION

The economics of equipment selection must be tailored to each type of equipment. The economic part of the process begins as all option analysis do: determining what is technically feasible. After that, economics can be brought into play.

- For the two principle types of pumps — centrifugal and positive displacement — consider:
  - If an ANSI centrifugal is one of the technically feasible options, select it because it is the least expensive and has the lowest maintenance costs of all the pumps.
  - If considering both centrifugal and positive displacement pumps (but not an ANSI pump), use Table 10.3 to estimate the balance point between capital costs and maintenance costs. Pick the one having the best balance point.
  - If the data in Table 10.3 is not discriminating enough to find the balance point, do more in-depth estimating to find the most economic unit.
- For the two common types of heat exchanger — shell and tube and plate and frame exchangers — when both are technically feasible:
  - Select which type of shell and tube unit will be used. If all three types are feasible, pick the U-tube because it is the least expensive.
  - Roughly size both exchangers.
  - Estimate the capital cost for both and pick the one that has the lowest capital cost.

10.5.3 ECONOMIC PLANT SITING

When siting plants, first use economics to decide on the number of plants to be built and their approximate locations. Selecting the optimal number of plants involves finding the economic balance point among all the costs affected by the number of plant and locations — capital, startup expenses, distribution/in-freight costs, and manufacturing costs.

- The first step in the analysis is to decide the number of options one will study, i.e., the number of plants.

- Next, for each option subdivide the distribution area into as many parts as the number of plants in the option. This is usually done in consultation with Sales and Marketing.
- Again, for all options locate the plants in each subdivision on the basis of where the customers (or suppliers) are located. Use the tie-line or coordinate methods to find the best locations. Draw a 50 mi to 100 mi radius circle around that location.
- Size the plants in each subdivision based on its customer volume (on a 100% basis).
- Estimate costs and calculate the NPV/AC for each option. Select the most economic number of plants.

10.6 PROBLEMS AND EXERCISES

1. What are the five steps when performing an economic analysis?
2. In *a* through *e*, you are comparing options having different lives. For each, what economic life would you use to compare the options?

	Option A	Option B	Option C	Option D
a.	4 yrs	6 yrs	—	—
b.	3 yrs	8 yrs	12 yrs	—
c.	3 yrs	8 yrs	12 yrs	6 yrs
d.	5 yrs	20 yrs	10 yrs	—
e.	4 yrs	2 yrs	10 yrs	5 yrs

3. Make preliminary pump selections for the following services. Explain your rationale.
   a. A 70% propylene glycol solution: flow = 450 gpm; required head = 120 ft; specific gravity = 0.87; temperature = 70°F; viscosity = 23 cP; available NPSH ~ 20 ft; system design pressure = 150 psig
   b. Glycerol: flow = 100 gpm; required head = 89 ft; specific gravity = 1.27; temperature = 60°F; viscosity = 1070 cP; available NPSH ~ 20 ft; system design pressure = 150 psig
   c. A 5% slurry of diatomaceous earth on oil: flow = 1000 gpm; required head = 290 ft; specific gravity = 0.95; temperature = 85°F; viscosity = 1.5 cP; available NPSH ~ 10 ft; system design pressure = 150 psig
4. For the flowsheet in Figure 10.4, identify the heat exchanger types that are technically feasible. Both the feed and the stream from the reactor are clean, nonvolatile, nonhazardous organic liquids. Their viscosities are 10 cP to 20 cP. Explain your rationale.
5. You are heating a hydrocarbon oil stream from 130°F to 165°F with 200°F water. Both shell and tube and plate heat exchangers are technically feasible. You have roughly sized both exchangers — the U-tube has three

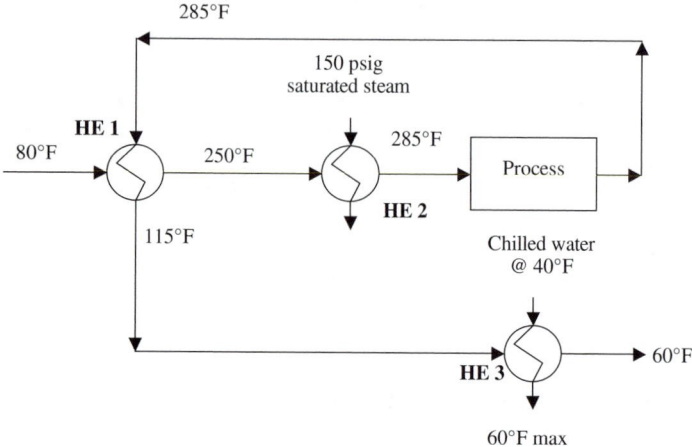

FIGURE 10.4 Problem 4: Heat exchanger selection.

shells, each with an area of 1250 ft², and the plate unit has an area of 2500 ft². Either carbon steel or stainless steel is acceptable. Use a Hand factor for stainless steel/nitrile plate units of 3.5. For both units, assume typical instrumentation and construction in an existing site. Preliminarily select the heat exchanger for this service.

6. You are removing the heat of reaction from a reactor. The reaction fluid is circulated through a heat exchanger where it is cooled from 305°F to 295°F. The coolant is a glycol solution that enters the heat exchanger at 180°F and exits at 200°F. Both shell and tube and plate heat exchangers are technically feasible. The U-tube has a single shell of 365 ft² and the plate unit has an area of 220 ft². Either carbon steel or stainless steel is acceptable. Use a Hand factor for stainless steel/nitrile plate units of 3.5. For both units, assume typical instrumentation and construction in an existing site. Preliminarily select the heat exchanger for this service.

7. You are working for a company that wants to build a small plant. The company has three major customers and one major supplier, as shown below. The in-freight and product shipping costs per mile are about equal. Preliminarily locate the plant.

Customer Locations	% of Volume Shipped
Atlanta	40
Nashville	30
Birmingham	20

Supplier Location	% of Volume Shipped
Chattanooga	85

8. Select a preliminary plant location for a new plant that will ship to customers in the following areas:

Metropolitan Area	% of Shipping Volume
New York/Newark	20.2
Boston	9.2
Chicago/Milwaukee	8.5
Cincinnati/Indianapolis	7.6
Dallas/Houston	6.9
Detroit/Cleveland	6.5
South Florida	6.1
Atlanta/Birmingham	5.9
Memphis	5.9
St. Louis/Kansas City	5.2
Denver	2.2

For this problem, ignore in-freight costs.

9. You are trying to decide how to replace a 15K gal carbon steel tank that is failing due to corrosion. Its life, installation to removal, will be 5 years. You can extend this life to 10 years by installing an epoxy-lined carbon steel tank. Which type of tank should you install and why? Use a 13% discount rate and a 32% tax rate.

10. You must decide what thickness of insulation to put on a 3-in. Schedule 40 carbon steel pipe. The pipe contains a fluid at 700°F. You want the exterior surface of the insulation to be at 100°F. The pipe is used 6000 hrs/yr. What is the economic insulation thickness?

Data/Assumptions:
- Ignore the resistance of the pipe wall, assuming the inside surface of the insulation will be at 700°F (the resistance of the fluid film and the pipe wall are so small that this will have little effect on the answer)
- The film coefficient at the external surface of the insulation is 0.8 B/hr-ft^2-°F
- The thermal conductivity of the insulation is 0.42 B-in/hr-ft^2-°F
- The cost of energy is $6.50/M Btu
- The insulation comes in $1/2$-in. thick increments with an installed cost of:

Thickness (in)	Installed Cost ($/ft)
$1/2$	3.00
1	4.10
$1\,1/2$	5.10
2	6.80
$2\,1/2$	8.90
3	11.00
4	19.40

- Use a 20-year economic life, a 10% discount rate, and a 32% tax rate.

11. You are deciding what kind of filtration system to install to remove catalyst from the discharge stream of a continuous oil hydrogenation reactor. The system will be built in the Chicago plant. The stream from the reactor is at 300°F. You are evaluating several options:
    - A CS candle filter that can operate at 300°F because it is an enclosed filter. This type of unit has an automated, self-cleaning cycle and requires almost no operator attention. The clean oil stream will be cooled to 120°F after filtration. You have sized this filter using a method by Brown that minimizes the area of the filter.[9] The filter size and purchase cost are 38 ft² and $85K.
    - A CS filter press. The oil stream must be cooled to 120°F before filtration because the press must be manually cleaned. Press cleaning requires two operators, who are each paid $15/hr. You have sized the filter two different ways:
      - Using Brown's "minimum area" method, the size and purchase cost are 181 ft² and $20.6K. This filter must be cleaned every 3.4 hrs and cleaning takes 55 min for two people. The operating labor cost for cleaning is $67K.
      - Using a method that finds the economic balance between cleaning costs and capital cost, the size and purchase cost are 400ft² and $34K. This filter must be cleaned every 14.3 hrs and cleaning takes 2 hrs for two people. The operating labor cost for cleaning is $35K.
    Which of the three options is best and why? Use a 10-year project life, a 15% discount rate, and a 35% rate.
12. In the process illustrated in Figure 10.5, the feed of 15K lb/hr is first heated using a CS coil inside the Boiling Water Heater/Cooler. Condensate from around the plant is flashed into this tank, where the temperature is 212°F. The rest of the heating of the feed is done in a CS shell and tube exchanger using 150 psig steam. In the process there is a 50°F temperature rise. The 300°F product is first cooled in the Boiling Water Heater/Cooler, where it passes through an SS coil and is partially cooled by boiling the condensate in the tank. The remainder of the product cooling is done in an SS plate and frame exchanger using 85°F cooling water. Find the economically optimal outlet temperature for the outlet temperature of the feed stream from the Boiling Water Heater/Cooler.

    Data/Assumptions:
    - The process flow rate is 15K lb/hr.
    - U for the heating coil in the Boiling Water Heater/Cooler is 40 B/hr-ft²-°F.
    - U for the steam heater is 110 B/hr-ft²-°F.
    - Assume C_p for the process fluid is constant at 0.65 B/lb-°F.

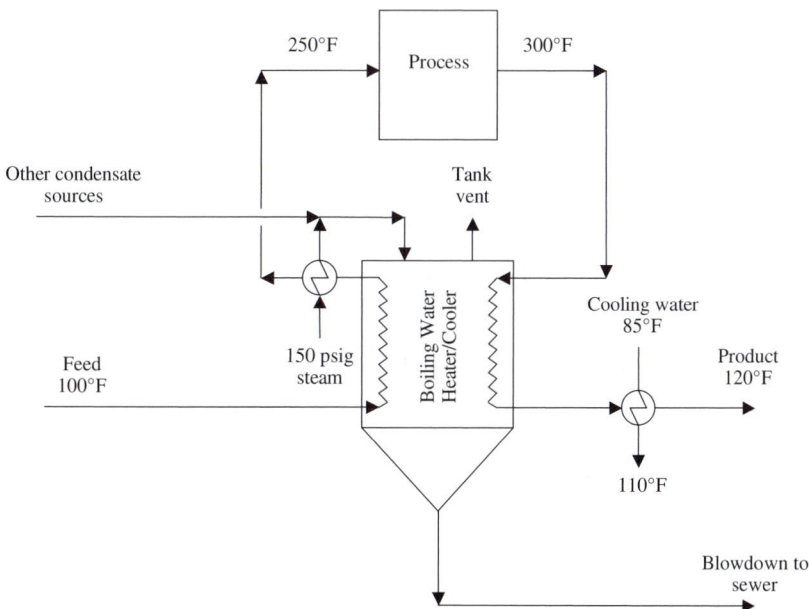

FIGURE 10.5 Problem 12: Boiling water heater/cooler.

- Tank coil costs: a 500 ft² coil costs $6K (@ a CEPI of 460). The size factor is 0.96.
- There is no instrumentation on the Boiling Water Heater/Cooler. The heater is typically instrumented.
- The process operates 6000 hr/yr.
- Steam costs (March 2005 dollars) are $6/1000 lb.
- Use March 2005 economics, a 15-yr project life, a 10% hurdle rate, and a 35% tax rate.

13. You are specifying a tank for a corrosive service. You have worked with the materials engineers and have come up with three options.
    - Option 1: CS
        - Capital cost = $130K
        - Estimated life = 4 yrs
    - Option 2: 304 SS
        - Capital cost = $156K
        - Estimated life = 12 yrs
    - Option 3: Fiberglass
        - Capital cost = $158K
        - Estimated life = 12 yrs

    Which option do you recommend and why do you recommend it? Your company's hurdle rate is 10% and its tax rate is 32%.

10.7   ADDITIONAL TOPIC

10.7.1   How to Decide Whether to Install Capacity for an Assumed Future Need

In 1977, T.R. Brown published an article on this subject in *Chemical Engineering*.[10] This section is a major update of that material.

Whenever one is designing a new system, one should consider whether to oversize new equipment for future expansion needs. The decision is often made using judgment rather than analytic methods. This creates a risk of improperly spending capital. This section presents an analytic decision-making method that is adjustable to the economic policy of one's company.

10.7.1.1 The Two Options

When one is working to decide whether or not to install oversized or higher-capacity equipment, one is choosing between two options:

- Installing the higher-capacity system, even though only part of the capacity will initially be used
- Installing the lower-capacity system and investing the capital cost difference between the larger and the smaller system

The pros and cons of the options are:

	Pros	Cons
Buy the higher capacity system	The larger system costs less to install now. Later, it will cost more because the cost will have inflated. No money is lost having to scrap equipment when the larger system is needed.	During the time the additional capacity is not needed, the added capital spent for the larger system cannot be invested and earn profits.
Buy the lower capacity system	The capital difference between the large system and the small system can be invested and earn profits.	When more capacity is needed, it will cost more to install. This capacity might be added in two different ways: Remove the original equipment and replace it with the larger equipment. Keep the original equipment in service and add more capacity in parallel. The lower-capacity system may have to be removed for its salvage value (before it is worn out).

10.7.1.2 The Breakeven Point and Decision Guidelines

One can use a future worth analysis to develop guidelines for deciding which option is more economic. How the future worth of the two options change over time is illustrated in Figure AT10.1. The breakeven point is the point where the future worths are equal.

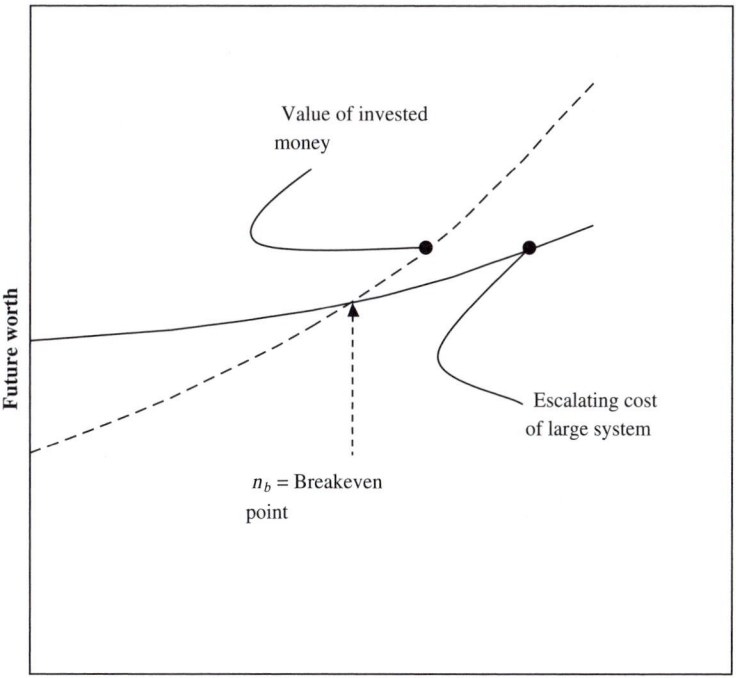

Years after the decision

FIGURE AT10.1   Breakeven point.

10.7.1.2.1 The Decision-Making Guidelines

To decide which option is the better economically, consider the following situation:

- The small unit has been purchased
- The cost difference between the large and small unit has been invested
- Extra capacity will be needed before the breakeven point

Examination of Figure AT10.1 shows that prior to the breakeven point, the projected future worth of the large system is greater than the projected value of the invested money. Therefore, it is the more economic choice.

From this situation, we can create decision-making guidelines:

- If the extra capacity will be needed before the breakeven point, buy the larger system
- If the extra capacity will be needed after the breakeven point, buy the smaller system

10.7.1.3 Calculation of the Breakeven Point, n_b

Using the equation for future worth, we will derive an equation for n_b. First, we will define terms:

Nomenclature

C_L	Capital cost of the larger system, in today's dollars
C_S	Capital cost of the smaller system, in today's dollars
C_P	Capital cost of the parallel system, in today's dollars
e_c	Annual rate of capital cost escalation
F_L	Future capital cost of the larger unit
F_{L-S}	Future worth (after-tax) of the capital cost difference between the larger and smaller systems when invested at r_{AT}
n	Years after the decision
n_b	Breakeven point (years)
r_{AT}	Rate of return, after-tax
S_s	Salvage value of the smaller system

For the capital cost of higher capacity system, the equation is:

$$F_L = C_L (1+e_c)^n$$

and for the increasing value of the invested capital cost difference between the higher and lower capacity options:

$$F_{L-S} = (C_L - C_S)(1+r_{AT})^n$$

At the breakeven point, $n = n_b$ and $F_L = F_{L-S}$. If the salvage value of the smaller system is insignificant, one sets these two equations equal to each other, takes logarithms of each side, and solves for n_b.

$$n_b = \frac{\ln\left(\dfrac{C_L - C_S}{C_L}\right)}{\ln\left(\dfrac{1+e_c}{1+r_{AT}}\right)} \qquad \text{(AT10.1)}$$

If the added capacity would be installed by adding more equipment in parallel, Equation AT10.1 becomes:

$$n_b = \frac{\ln\left(\dfrac{C_L - C_S}{C_P}\right)}{\ln\left(\dfrac{1+e_c}{1+r_{AT}}\right)} \qquad \text{(AT10.2)}$$

If the salvage value of the smaller system is significant compared to $F_{L\text{-}S}$, solving for n_b requires a trial-and-error solution of:

$$F_L = F_{L\text{-}S} + S_S$$

$$C_L(1 + e_c)^{n_b} = (C_L - C_S)(1 + r_{AT})^{n_b} + S_S \qquad \text{(AT10.3)}$$

10.7.1.4 Examples Illustrate

Example AT1: An engineer is sizing a batch reactor system. A high probability exists that 50% more capacity will be needed in about five years. In today's dollars, the capital cost of the smaller system is $395K, and of the larger unit, $550K. If the larger unit is not purchased, whenever the extra capacity is needed the small system will be removed and the larger reactor installed. Because the reactor is so specialized, the salvage value of the smaller system will be nil. If the smaller system is installed, the unused capital can be invested to return 15% before taxes. Assume a 3% per year escalation of capital costs and a 35% tax rate. The after-tax rate of return, r_{AT}, is $(15\%)(1 - 0.35) = 9.75\%$.

Use Equation AT10.1 to find the breakeven point.

$$n_b = \frac{\ln\left(\dfrac{550 - 395}{500}\right)}{\ln\left(\dfrac{1+0.03}{1+0.0975}\right)} = 20 \text{ yrs}$$

Because the extra capacity is needed in about five years — prior to the breakeven point — the higher capacity system should be installed now.

Example AT2: An engineer is sizing a heating, ventilating, and air-conditioning system for a new building. Because potential exists for a major expansion of this building in seven to eight years, the engineer is considering sizing the refrigeration equipment 40% larger than is needed today. If the extra capacity is not installed now, additional parallel refrigeration equipment will be installed when the building is renovated. Thus, the engineer has two options: (a) buying the larger equipment, whose capital cost is $7.4M, or (b) buying the smaller equipment now, and in seven or eight years buying the additional parallel equipment. In today's dollars, the capital cost of the smaller equipment is $5.5M and the additional parallel equipment is $3.0M. Capital costs are expected to escalate at 2.5% per year. The company's capital-investment guidelines require incremental investments such as this to return 12% after-tax.

Because the added capacity will be added via the use of parallel equipment, we will use Equation AT10.2 to find the breakeven point.

$$n_b = \frac{\ln\left(\dfrac{7.4 - 5.5}{3.0}\right)}{\ln\left(\dfrac{1 + 0.025}{1 + 0.12}\right)} = 5.2 \text{ yrs}$$

Because the additional capacity will not be needed in seven or eight years, or after the breakeven point, the smaller unit should be installed now.

REFERENCES

1. Heald, C.C. (Ed.), *Cameron Hydraulic Data, 17th Edition*, Compressed Air, 1992, 3–52 to 3–60.
2. Ulrich, G.D., *Chemical Engineering Process Design and Economics, A Practical Guide, 2nd Edition*, Durham, NH: Process Publishing, 2004.
3. Granger, J.E., Plantsite selection, *Chemical Engineering*, June 15, 1981, 106–109.
4. Granger, J.E., Plantsite selection, *Chemical Engineering*, June 15, 1981, 110–114.
5. Douglas, J.M., *Conceptual Design of Chemical Processes,* Boston: McGraw-Hill, 1988, 417–421.
6. Peters, M.S. and Timmerhaus, K.D., *Plant Design and Economics for Chemical Engineers, 4th Edition*, New York: McGraw-Hill, 1991, 91–95.
7. Jelen, F.C. and Black, J.H., *Cost Optimization Engineering, 2nd Edition*, New York: McGraw-Hill, 1983, 455–456.
8. Schmenner, R.W., *Making Business Location Decisions*, Upper Saddle River, NJ: Prentice-Hall, 1982.
9. Brown, T.R., Easily optimize batch pressure filtration, *Chemical Engineering Progress*, February 1998, 45–50.
10. Brown, T.R., Economic evaluation of future equipment needs, *Chemical Engineering*, January 17, 1977, 125–127.

11 Economic Design Case Studies

The first case study, "Optimal Cooling Water Temperature in a Cooler," explains the rationale and methodology in more detail because it is the first presented. Refer to it to fill in any apparent gaps in logic or problem-solving flow in the other studies.

This chapter uses case studies to illustrate the analysis portion of the Economic Design Model. The case studies show how one sets up and performs an analysis. This involves selecting the independent and dependent variables, estimating the differences in capital and production costs, calculating the NPVs of the options, and selecting the most economic option. The studies are:

- Finding the optimal cooling water outlet temperature in a heat exchanger cooling a hot stream
- Finding the optimal catalyst usage in a reactor/filter system
- Finding the optimal amount of heat recovery in a heat exchanger loop
- Determining whether to build a grass-roots plant or whether to expand an existing plant
- Finding the economically optimal number of plants

11.1 OPTIMAL COOLING WATER TEMPERATURE IN A COOLER

Most every process includes cooling a hot stream with cooling water. The engineer designing the process must select an outlet temperature for the cooling water. Deciding this involves looking at the cost trade-offs. For example, one could pick a higher outlet temperature and reduce cooling water usage and cost but that would increase the size and cost of the heat exchanger.

11.1.1 PROBLEM STATEMENT

A process uses a counter-current flow heat exchanger to cool 50K lb/hr of soybean oil from 155°F to 115°F. The cooler uses 90°F cooling tower water, which cannot return to the cooling tower hotter than 125°F. Find the most economic cooling water exit temperature. Do this for both a fixed-tube sheet and a plate exchanger.

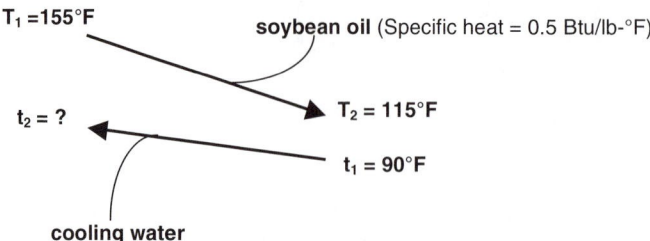

Assumptions/Data:

- Base the calculations upon 2005 costs, a 15% discount rate, a 35% tax rate, and a 10-year project life
- Overall heat transfer coefficients:
  - Fixed-tube sheet: U = 70 Btu/hr-ft²-°F
  - Plate exchanger: U = 120 Btu/hr-ft²-°F
- Both heat exchangers will have a 150 psig rating
- Both carbon steel and stainless steel are acceptable metallurgies
- The exchanger will be typically instrumented for a process plant
- The exchanger will be part of an expansion of an existing plant
- The process is in the U.S.
- Cooling water costs $.07/Kgal
- The process operates 8400 hr/yr

11.1.2 ANALYTICAL METHODOLOGY

- Select the independent and dependent variables. The cooling water outlet temperature should be the independent variable. To find the dependent variables, consider what happens when the outlet temperature changes:
  - The cooling water usage changes and the cooling water cost changes.
  - The log mean temperature difference (LMTD or ΔT_{lm}) changes. This changes the size of the heat exchanger and the cost of the heat exchanger.
  - Investments change.[*]

Capital	Changes	Estimate the Cost of Each Option
Working capital	No change	Ignore
Startup expense	No change	Ignore

---

[*] Chapter 3 suggests estimating startup expenses as a percent of capital, but it is not logical for startup costs to be different for these heat exchanger options.

- Production costs change*

Raw materials	No change	Ignore
Packaging materials	No change	Ignore
Manufacturing		
Operating labor	No change	Ignore
Employee benefits	No change	Ignore
Supervision	No change	Ignore
Laboratory	No change	Ignore
Utilities	*Changes*	*Varies with cooling water usage*
Maintenance	*Changes*	*6% of capital*
Insurance and taxes	*Changes*	*3% of capital*
Operating supplies	*Changes*	*1% of capital*
Plant overhead	*Changes*	*1% of capital*
Depreciation	*Changes*	*Capital/10 years*
Product delivery	No change	Ignore

- Pick a range of outlet temperatures for investigation. For this problem, start with a 5°F interval or temperatures of 95°F, 100°F, 105°F, 110°F, 115°F, 120°F, and 125°F. If the temperature interval does not clearly define the optimum, you would select another interval and run the calculations again.
- Size the heat exchanger for the different outlet temperatures. Estimate the capital and the production cost differences.
- Estimate the investments and the production costs.
- Calculate the NPV (or AC) for each of the temperatures and plot this value versus outlet temperature. The maximum NPV defines the economic outlet temperature.

11.1.3 PROBLEM SOLUTION

The following illustrates the calculations for the fixed-tube exchanger and the 125°F outlet temperature. Table 11.1 and Table 11.2 show all the results for all the calculations.

---

* Note that several of the production costs vary as a percent of capital. You can find the percentages used in Chapter 4, Table 4.8.

11.1.3.1 Size and Price the Exchanger

- Find the heat load based on the oil side:

$$q = WC_p\Delta T = 50000\,\frac{\text{lb}}{\text{hr}}*0.5\,\frac{\text{Btu}}{\text{hr} - {}^{\circ}\text{F}}*(155 - 115{}^{\circ}\text{F}) = 1\times 10^6\,\frac{\text{Btu}}{\text{hr}}$$

- Find the log-mean temperature difference, ΔT_{lm}:

$$\Delta T_{lm} = \frac{\Delta T_2 - \Delta T_1}{\ln\left(\dfrac{\Delta T_2}{\Delta T_1}\right)} = \frac{(155 - 125)-(115 - 90)}{\ln\left(\dfrac{155 - 125}{115 - 90}\right)} = 27.4{}^{\circ}\text{F}$$

- Find the ΔT_{lm} correction factor. Calculate P and R and use the TEMA graphs in *Standards of the Tubular Exchanger Manufacturers Association* or in *Perry's Chemical Engineers' Handbook* to find the correction factor.[1,2] A two-shell pass exchanger is needed, and $F_T = 0.92$.
- Size the exchanger:

$$A = \frac{q}{U\Delta T_{lm}F_T} = \frac{1\times 10^6\,\dfrac{\text{Btu}}{\text{hr}}}{\left(70\,\dfrac{\text{Btu}}{\text{hr} - \text{ft}^2 - {}^{\circ}\text{F}}\right)(27.4{}^{\circ}\text{F})(0.92)} = 566\text{ft}^2$$

- Price the exchanger. Using the Heat Exchanger page in Appendix IV, you find the purchase cost for a 566 ft^2 U-tube exchanger, having carbon steel tubes and shell and rated at 150 psig, is $11K (at a CEPI of 460). Our heat exchanger has two shell passes. Therefore, we either must buy two exchangers, each having an area of 566/2 or 283 ft^2 or buy a single shell unit that is baffled to create two shell passes. The latter is much less expensive, so we will price the unit that way. Because our exchanger is a fixed-tube sheet exchanger, we must adjust the price for this.

$\$_{adjusted}$ = (*Graph price*)(*Factor for one shell/two shell passes*) (*Factor for fixed-tube sheet*)

= ($11K)(1.1)(1.05) = $12.7K

11.1.3.2 Estimate the Capital Cost

Use Equation 3.3.

$Capital\ cost = \Sigma[Equipment\ purchase\ cost * (Hand\ factor * F_m)] * F_i * F_b * F_p$

$Hand\ factor = 3.5$ (Table 3.2)

$F_m = 1$ (Materials are all carbon steel)

$F_i = 1.35$ (Table 3.3)

$F_b = 1.06$

$F_p = 1$ (U.S. construction)

$Capital\ cost = (\$12.7K * 3.5 * 1) * 1.35 * 1.06 * 1 = \$63.6K$

11.1.3.3 Estimate the Production Cost/Expense Differences

- Find the water usage and cost:

$$Water\ usage = \frac{q}{C_p \Delta T}$$

$$= \frac{\left[\dfrac{1 \times 10^6 \dfrac{Btu}{hr}}{\left(1\dfrac{Btu}{lb - °F}\right)(125 - 90°F)}\right]\left(\dfrac{8400\dfrac{hr}{yr}}{8.34\dfrac{lb}{gal}}\right)}{} = 28.8K\ \frac{gal}{yr}$$

Water cost = (28.8K gal/yr)($.07/K gal) = $2.0K/yr

- Find the capital ratioed costs. From above, these total 11% of capital:

$Capital\ ratioed\ costs = (Capital\ cost)(\%\ of\ capital) = (\$63.6K)(0.11)$
$= \$7.0K\ per\ year$

- Find the depreciation cost. Use Equation 2.5:

$$Annual\ deprecation\ writeoff = \frac{Captial\ investment}{Equipment\ life}$$

Annual depreciation write-off = $63.6K/10 yrs = $6.4K per year

- Find the BT annual expenses and the AT cash flow:

  BT expenses = Water cost + Capital related costs + Depreciation

  = 2.0K + 7.0K + 6.4K = $15.4K per year

  AT cash flow (use Equation 2.6):

  $AT_{Cash\,flow}$ = *(Revenues – Expenses)(1 – Tax rate) + Depreciation write-off*

  = (0 – 15.4K/yr)(1 – 0.35) + 6.4K/yr = –$3.6K per year

11.1.3.4 Calculate the NPV

$P_{AT\,cash\,flow}$ = A(P/A, 15%, 10) = (–$3.6K)(5.019) = –$18.1K

NPV = $P_{Capital}$ + $P_{AT\,cash\,flow}$ = –$63.6K + (–$18.1K) = –$81.7K

(Rounding makes these numbers a bit different from the spreadsheet.)

11.1.3.5 Find the Economic Design Point

The economic design point is the water outlet temperature where the NPV is at a maximum. To find this point, plot NPV versus water outlet temperature. Figure 11.1 and Figure 11.2 are the plots for the two types of exchangers. They show that the economic design points are:

- 107°F for the fixed-tube sheet exchanger
- 113°F to 115°F for the plate and frame exchanger

The other conclusion one can make from these calculations is the plate exchanger is a better choice than is the fixed-tube sheet unit because its NPV is the highest at the economic design point.

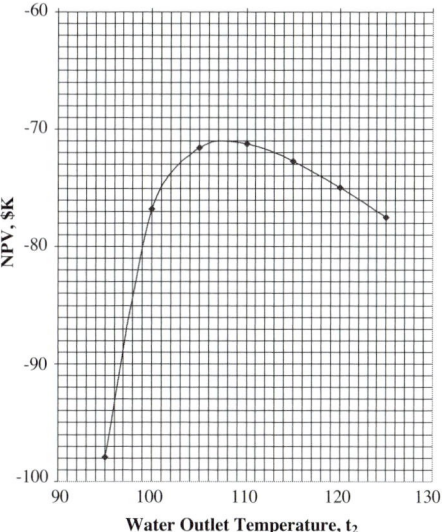

FIGURE 11.1   Water outlet temperature, fixed tube sheet.

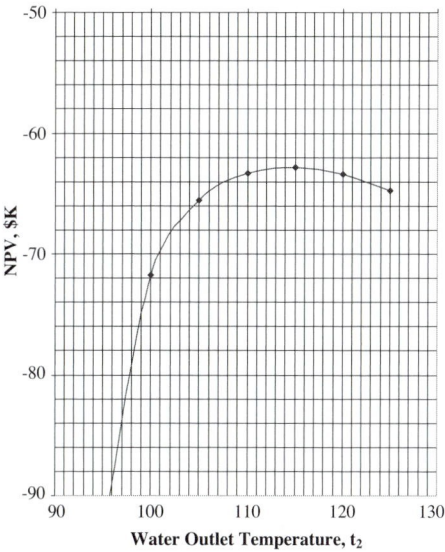

FIGURE 11.2 Water outlet temperature, plate and frame.

TABLE 11.1
Cooler Temperature Approach Optimization, Fixed-Tube Sheet

Data

Water outlet temp, t_2	95	100	105	110	115	120	125
Oil flow, W	50,000	50,000	50,000	50,000	50,000	50,000	50,000
Oil, C_p	0.5	0.5	0.5	0.5	0.5	0.5	0.5
U (B/hr-ft^2-°F)	70	70	70	70	70	70	70
Oil inlet temp, T_1	155	155	155	155	155	155	155
Oil outlet temp, T_2	115	115	115	115	115	115	115
Water inlet temp, t_1	90	90	90	90	90	90	90
CEPI (year installed)	460	460	460	460	460	460	460

Capital ($)

$q = WC_pT$ (Btu/hr)	1,000,000	1,000,000	1,000,000	1,000,000	1,000,000	1,000,000	1,000,000
ΔT_{lm}	40.0	38.0	36.1	34.0	31.9	29.7	27.4
ΔT_{lm} correction factor							
$R = (T_1\ T_2)/(t_2\ t_1)$	8.0	4.0	2.7	2.0	1.6	1.3	1.1
$P = (t_2\ t_1)/(T_1\ t_1)$	0.08	0.15	0.23	0.31	0.38	0.46	0.54
F_T	0.975	0.955	0.92	0.865	0.815	0.945	0.92
Number of shells/shell passes	1/1	1/1	1/1	1/1	1/1	1/2	1/2
Number of shells	1	1	1	1	1	1	1
Area/shell (ft^2)	366	393	431	485	549	509	566
HEX cost							
Figure IV-11, Appendix IV	8.8	9.2	9.6	10.2	10.8	10.4	11.0
Equipment type, fixed-tube sheet	1.05	1.05	1.05	1.05	1.05	1.05	1.05
Materials	1.00	1.00	1.00	1.00	1.00	1.00	1.00
Pressure	1.00	1.00	1.00	1.00	1.00	1.00	1.00
Other, two-shell passes	1.00	1.00	1.00	1.00	1.00	1.10	1.10

Time		1.00	1.00	1.00	1.00	1.00	1.00	1.00
Hand factor		3.5	3.5	3.5	3.5	3.5	3.5	3.5
F_m		1	1	1	1	1	1	1
F_i		1.35	1.35	1.35	1.35	1.35	1.35	1.35
F_b		1.06	1.06	1.06	1.06	1.06	1.06	1.06
	Capital cost	46.5	48.2	50.4	53.5	56.9	60.3	63.6

Production costs/expenses ($/yr)

Water usage, Kgal = (q/C_pT) = (8400hr/yr/8.34lb/gal)		201,439	100,719	67,146	50,360	40,288	33,573	28,777
Water cost ($/Kgal)		0.07	0.07	0.07	0.07	0.07	0.07	0.07
Annual water cost ($K)		14.1	7.1	4.7	3.5	2.8	2.4	2.0
Maintenance, insurance and taxes, operating supplies, plant overhead/miscellaneous: 11% of capital ($K)		5.1	5.3	5.5	5.9	6.3	6.6	7.0
Depreciation ($K)		4.7	4.8	5.0	5.4	5.7	6.0	6.4
Annual expenses, BT ($K)		−23.9	−17.2	−15.3	−14.8	−14.8	−15.0	−15.4
	Annual cash flow, AT ($K)	−10.8	−10.9	−6.3	−4.9	−4.2	−3.9	−3.7

NPV ($)

Capital ($K)		−46.5	−48.2	−50.4	−53.5	−56.9	−60.3	−63.6
Discount rate (%)		15	15	15	15	15	15	15
Project life (yrs)		10	10	10	10	10	10	10
P for annual cash flow		−54.5	−31.8	−24.6	−21.3	−19.6	−18.7	−18.2
	NPV	−101.0	−80.0	−75.0	−74.8	−76.6	−79.0	−81.8

TABLE 11.2
Cooler Temperature Approach Optimization, Plate and Frame Exchanger

Data

Water outlet temp, t_2	95	100	105	110	115	120	125
Oil flow, W	50,000	50,000	50,000	50,000	50,000	50,000	50,000
Oil, C_p	0.5	0.5	0.5	0.5	0.5	0.5	0.5
U (B/hr-ft²-°F)	120	120	120	120	120	120	120
Oil inlet temp, T_1	155	155	155	155	155	155	155
Oil outlet temp, T_2	115	115	115	115	115	115	115
Water inlet temp, t_1	90	90	90	90	90	90	90
CEPI (year installed)	460	460	460	460	460	460	460

Capital, $K

	95	100	105	110	115	120	125
$q = WC_pT$ (Btu/hr)	1,000,000	1,000,000	1,000,000	1,000,000	1,000,000	1,000,000	1,000,000
$T\Delta_{lm}$	40.0	38.0	36.1	34.0	31.9	29.7	27.4
Area (ft²)	208	219	231	245	261	280	304
HEX cost							
Figure IV-11, Appendix IV	8.5	8.7	9.0	9.3	9.6	10.0	10.4
Materials	1.00	1.00	1.00	1.00	1.00	1.00	1.00
Pressure	1.00	1.00	1.00	1.00	1.00	1.00	1.00
Time	1.00	1.00	1.00	1.00	1.00	1.00	1.00
Hand factor	3.5	3.5	3.5	3.5	3.5	3.5	3.5
F_m	1	1	1	1	1	1	1
F_i	1.35	1.35	1.35	1.35	1.35	1.35	1.35
F_b	1.06	1.06	1.06	1.06	1.06	1.06	1.06
Capital cost	40.1	42.5	43.7	45.0	46.4	48.0	49.9

Production costs/expenses ($/yr)

Water usage, Kgal = (q/C_pT) = (8400hr/yr/8.34lb/gal)	201,439	100,719	67,146	50,360	40,288	33,573	28,777
Water cost ($/Kgal)	0.07	0.07	0.07	0.07	0.07	0.07	0.07
Annual water cost ($K)	14.1	7.1	4.7	3.5	2.8	2.4	2.0
Maintenance, insurance and taxes, operating supplies, plant overhead/miscellaneous: 11% of capital ($K)	4.7	4.8	4.9	5.1	5.3	5.5	5.7
Depreciation ($K)	4.3	4.4	4.5	4.6	4.8	5.0	5.2
Annual expenses, BT ($K)	−23	−16	−14	−13	−13	−13	−13
Annual cash flow, AT ($K)	−10.6	−10.7	−6.2	−4.7	−4.0	−3.6	−3.3

NPV ($K)

Capital ($K)	−43	−44	−45	−46	−48	−50	−52
Discount rate (%)	15	15	15	15	15	15	15
Project life (yrs)	10	10	10	10	10	10	10
P for annual cash flow	−54	−31	−24	−20	−18	−17	−16
NPV	−96.3	−74.7	−68.5	−66.4	−66.0	−66.7	−68.2

11.2 OPTIMAL CATALYST USAGE IN A REACTOR/FILTER SYSTEM

This is an example of a problem where the upstream unit operation effects the downstream operation. This opens the possibility of joint optimization. In this study, the amount of catalyst used changes the reaction time, the reactor size, and the size of the downstream filter. Similar problems would include changing reactor conditions to improve or decrease yield. Different yields result in different sized or different types of downstream separation operations. Another example would be changing separator efficiency, which changes the type or size of the downstream environmental equipment.

11.2.1 PROBLEM STATEMENT

The engineer doing the conceptual design for the 70K lb/hr oil hydrogenation process has preliminarily sized a continuous reactor and its batch filter system as shown in Figure 11.3. The process, which will be a new process in the Kansas City plant, will be instrumented typically for a process plant. What is the economic optimal catalyst usage?

Assumptions/Data:
- The process will operate 24 hr/day, 5 days/wk, and 50 wks/yr. Base the calculations upon a 10-year project life, a 10% discount rate, and a tax rate of 32%.

Reactor:
- 3 min of hold/reaction time when using 0.15% catalyst (reactor hold time is inversely proportional to catalyst usage)
- 300 psig pressure rating
- 30 physical mixing stages
- Jacketed
- Stainless steel (SS) construction
- Three vendor quotes were received:
  - 550K for a SS unit having 3 min of hold time
  - $225K for a carbon steel (CS) unit having 3 min of hold time
  - $925K for a SS unit having 6 min of hold time
- Assume a Hand-type factor of 3.4, for multistage, jacketed, and agitated CS reactors.

Filter:
- Filtering area = 760 ft^2 (at a catalyst usage of 0.15%)
- The filter is a totally enclosed, SS, vertical element filter
- The filter has an automated cleaning cycle and requires almost no operator attention
- 300 psig pressure rating
- See Figure 11.4 for filter size versus catalyst usage
- Purchase cost ($\$_{2005}$) = $150K

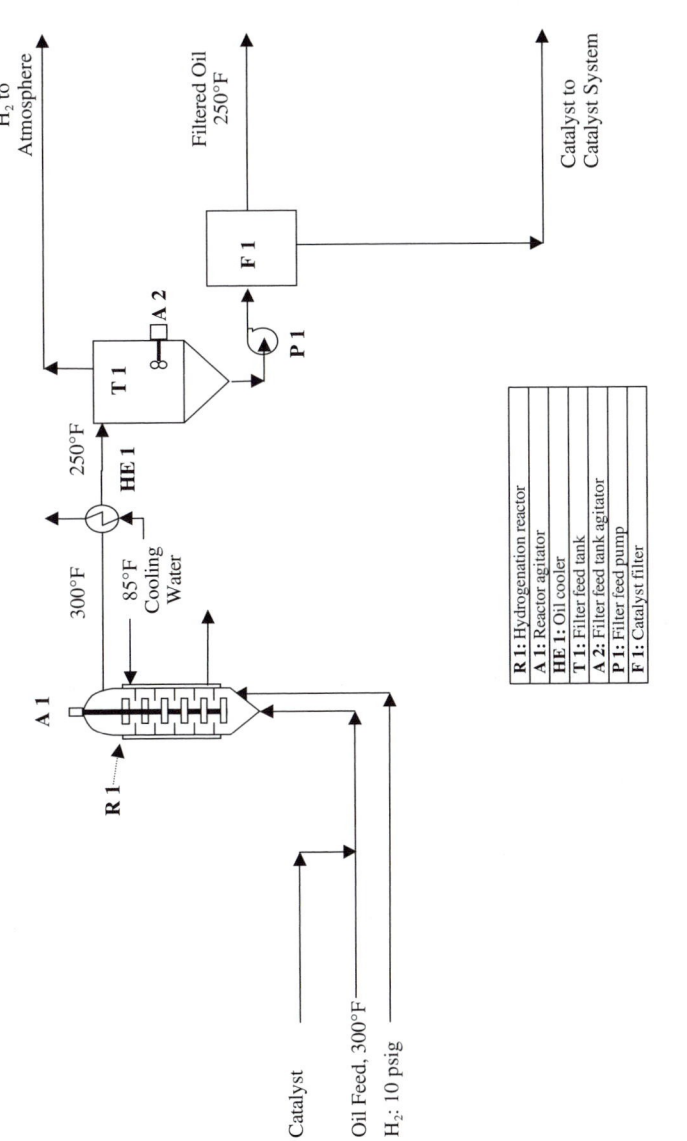

FIGURE 11.3 Oil hydrogenation and filtration.

- The size exponent for the filter is 0.6
- Assume a Hand-type factor of 2.4 for a SS vertical element filter

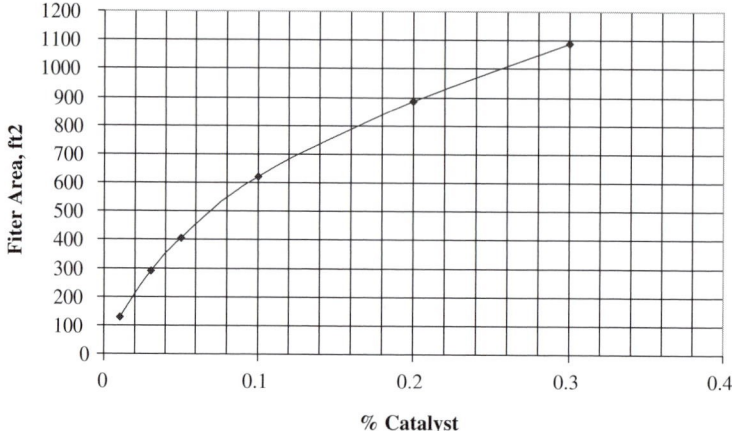

FIGURE 11.4 Filter area vs. % catalyst.

Catalyst:
- The catalyst can be used six times before it must be regenerated
- Catalyst cost = \$18.40/lb (\$_{2005}$)

11.2.2 ANALYTICAL METHODOLOGY

The percent of catalyst is the independent variable. Changes in this value cause the following to occur:

- When the amount of catalyst changes, the reaction time, the reactor size, and the downstream filter size change*
- Investments change†

Capital	Changes	Estimate the Cost of Each Option
Working capital	No change	Ignore
Startup expense	No change	Ignore

---

* As was the case with the cooling water problem, it is not logical for the startup costs to vary with the sizes of the reactor and filter.
† The sizes of the reactor and filter will change but the other equipment will not change or the change will be insignificant.

Production costs change

Raw materials	*Changes*	*Varies with the amount of catalyst used*
Packaging materials	No change	Ignore
Manufacturing		
Operating labor	No change	Ignore
Employee benefits	No change	Ignore
Supervision	No change	Ignore
Laboratory	No change	Ignore
Utilities	No change	Ignore
Maintenance	*Changes*	*6% of capital*
Insurance and taxes	*Changes*	*3% of capital*
Operating supplies	*Changes*	*1% of capital*
Plant overhead	*Changes*	*1% of capital*
Depreciation	*Changes*	*Capital/10 years*
Product delivery	No change	*Ignore*

We will start with a percent catalyst range of 0.05% to 0.2% and an interval of 0.025%. For each percent of catalyst, we will size the reactor and filter, estimate the capital cost and production cost, and calculate the NPV. Note that in the final calculations (Table 11.3), two more catalyst points, 0.02 and 0.03, were added.

We will base the study on 2005 dollars since all of the costs are quoted in that way.

11.2.3 PROBLEM SOLUTION

The following illustrates the calculations at a catalyst concentration of 0.05%. Table 11.3 shows the results for all the calculations.

11.2.3.1 Size and Price the Reactor

- Find the reactor hold time. Hold time is inversely proportional to the catalyst usage:

$$\frac{\text{Hold Time}_2}{\text{Hold Time}_1} = \frac{\text{Catalyst\%}_1}{\text{Catalyst\%}_2}$$

$$\text{Hold Time}_2 = \text{Hold Time}_1 \left(\frac{\text{Catalyst\%}_1}{\text{Catalyst\%}_2} \right)$$

$$= 3 \text{ minutes} \left(\frac{0.15\%}{0.05\%} \right) = 9 \text{ minutes}$$

- Find the size exponent for the reactor. Use hold time or reactor volume as the capacity variable. (Note the quotes for two different reactor sizes.) Use Equation 3.1 and rearrange it:

$$\frac{Cost_{size2}}{Cost_{size1}} = \left(\frac{Capacity_{size2}}{Capacity_{size1}}\right)^n$$

Taking the logarithm of both sides and rearranging:

$$n = \frac{\ln\left(\dfrac{Cost_{size2}}{Cost_{size1}}\right)}{\ln\left(\dfrac{Capacity_{size2}}{Capacity_{size1}}\right)} = \frac{\ln\left(\dfrac{\$925K}{\$550K}\right)}{\ln\left(\dfrac{6\ minutes}{3\ minutes}\right)} = 0.75$$

- Calculate the reactor purchase cost:

$$\$_{9minutes} = \$550K\left(\frac{9\ minutes}{3\ minutes}\right)^{0.75} = \$1254K$$

- Calculate the filter purchase cost. Here we will use the filter area as the capacity factor when size ratioing:

$$\$_2 = \$150K\left(\frac{395ft^2}{760ft^2}\right)^{0.6} = \$101K$$

- Calculate the capital cost. The Hand factor for the reactor is 3.4. Because this is for carbon steel and because our reactor is stainless steel, we will have to use an F_m.

SS/CS ratio = $550K/$225K = 2

Using Figure 3.2, we find $F_m = 0.69$. The Hand factor for the filter is 2.4. Because that is for stainless steel filters, $F_m = 1$. For F_i, use 1.35 because the process is typically instrumented. For F_b, use 1.11 for a fluid process being added to an existing plant:

$$\$_{Capital} = (\$1254K * 3.4 * 0.69 + \$101K * 2.4 * 1) * 1.35 * 1.11 = \$4772K$$

11.2.3.2 Estimate the Production Cost Differences

- Find the annual catalyst usage and cost (annual usage = per pass usage/number of uses):

$$\text{lb/yr} = \frac{\left(70\text{K lb}_{\text{oil}}/\text{yr}\right)\left(0.0005 \text{ lb}_{\text{cat}}/\text{lb}_{\text{oil}}\right)\left(24 \text{ hr/dy}\right)\left(5 \text{ day/wk}\right)\left(50 \text{ wk/yr}\right)}{6 \text{ uses}}$$

$$= 35\text{K lb/yr}$$

$/yr = 35K lb/yr ($18.40/lb) = $644K per year

- Find the capital ratioed costs (maintenance, insurance and taxes, operating supplies, and plant overhead)"

  Capital ratioed costs = $4772 * 0.11 = $525K per year

- Find the depreciation cost:

  Annual depreciation write-off = $4772K/10 years = $477K per year

- Find the BT annual expenses and the AT cash flow

  BT expenses = $644K + 525K + 477K = $1646K

  AT cash flow = *Expenses* (1 − *Tax rate*) + *Depreciation*

  $$= (-1646\text{K})(1 - 0.32) + 477\text{K} = -\$642\text{K}$$

11.2.3.3 Calculate the NPV

$$P_{AT \text{ cash flow}} = A(P/A, 10\%, 10) = (-\$642\text{K})(6.145) = -\$3945\text{K}$$

(This is shown in Table 11.3 as $3948K. The difference is due to rounding in these calculations.)

$$\text{NPV} = P_{Capital} + P_{AT \text{ cash flow}} = -\$4772\text{K} + (-\$3945\text{K}) = -\$8717\text{K}$$

11.2.3.4 Find the Economic Design Point

The economic design is the percent of catalyst usage where the plot of NPV versus catalyst usage is at a maximum. This is 0.06% catalyst (refer to Figure 11.5.):

11.3 OPTIMAL HEAT RECOVERY IN A HEAT EXCHANGER LOOP

Most processes have the opportunity for heat or energy recovery. Usually, one would first perform a pinch analysis to decide which streams would interchange heat with each other. After that, the engineer would perform a more detailed analysis to determine the exact amounts of heat to be transferred between each of the streams.

TABLE 11.3
Optimum Catalyst Use

Catalyst Usage, %	0.02	0.03	0.05	0.075	0.1	0.125	0.15	0.175	0.2
Capital Cost									
Reactor hold time (min)	22.5	15	9	6.00	4.5	3.60	3.00	2.57	2.25
Reactor purchase cost, ($K, exponent = 0.75)	2493	1839	1254	925	745	631	550	490	443
Hand factor	3.4	3.4	3.4	3.4	3.4	3.4	3.4	3.4	3.4
Fm	0.69	0.69	0.69	0.69	0.69	0.69	0.69	0.69	0.69
Reactor installed cost, ($K)	5848	4314	2941	2170	1749	1479	1290	1149	1040
Filter area (ft^2) (from Figure 11.4)	210	290	395	505	600	685	760	820	885
Filter purchase cost, ($K, exponent = 0.6)	69	84	101	117	130	141	150	157	176
Hand factor	2.4	2.4	2.4	2.4	2.4	2.4	2.4	2.4	2.4
F_m	1.00	1.00	1.00	1.00	1.00	1.00	1.00	1.00	1.00
Filter installed cost, ($K)	166	202	243	282	312	338	360	377	394
Reactor + filter installed cost	6014	4516	3184	2452	2061	1818	1650	1526	1434
F_i	1.35	1.35	1.35	1.35	1.35	1.35	1.35	1.35	1.35
F_b	1.11	1.11	1.11	1.11	1.11	1.11	1.11	1.11	1.11
Total capital, ($K)	−9012	−6768	−4772	−3674	−3089	−2724	−2473	−2287	−2149
Production Cost									
Catalyst use, 6 uses (K lb/yr)	14	21	35	52.5	70	87.5	105	122.5	140
Catalyst cost ($K/yr; $18.40/lb)	258	386	644	966	1288	1610	1932	2254	2576
Maintenance, insurance and taxes, operating supplies, plant overhead: 11% of capital ($K/yr)	991	744	525	404	340	300	272	252	236
Depreciation — Capital/10 ($K/yr)	901	677	477	367	309	272	247	229	215
Annual expenses, BT ($/yr)	−2150	−1808	−1646	−1738	−1937	−2182	−2451	−2734	−3027
AT cash flow	−561	−552	−642	−814	−1008	−1211	−1420	−1631	−1844
P for AT cash flow, $K (P/A, 10%, 10 years = 6.145)	−3,447	−3,395	−3,946	−5,003	−6,194	−7,444	−8,723	−10,020	−11,329
NPV									
NPV = NPV$_{Capital}$ + NPV$_{AT\ cash\ flow}$ ($K)	−12,459	−10,162	−8,718	−8,677	−9,283	−10,168	−11,196	−12,307	−13,479

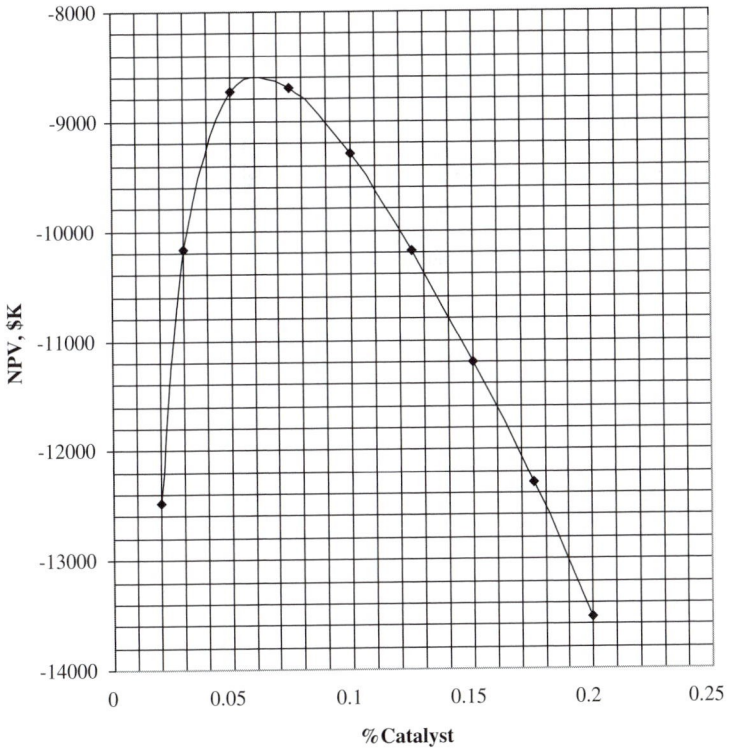

FIGURE 11.5 Catalyst optimization.

In some processes, many interdependent streams may exist, making the analysis quite complex.

The problem below has the interdependencies inherent in this type of problem but is simple enough to more easily show the optimization thought process.

11.3.1 PROBLEM STATEMENT

What is the economic amount of heat recovery for the process in Figure 11.6?
Data:

- Process feed rate = 100K lb/hr
- C_p = 0.5 Btu/lb-°F
- Units for Us are Btu/hr-ft²-°F
- The process operates 8400 hr/yr
- The chilled water leaves the cooler at 60°F
- The cost of 150 psig steam is $5.80/1000 lb (March 2005 dollars)
- The cost of chilled water is $1.35/1000 gal (March 2005 dollars)
- Use March-2005 economics, a 15% discount rate, a 10-year project life, and a 35% tax rate

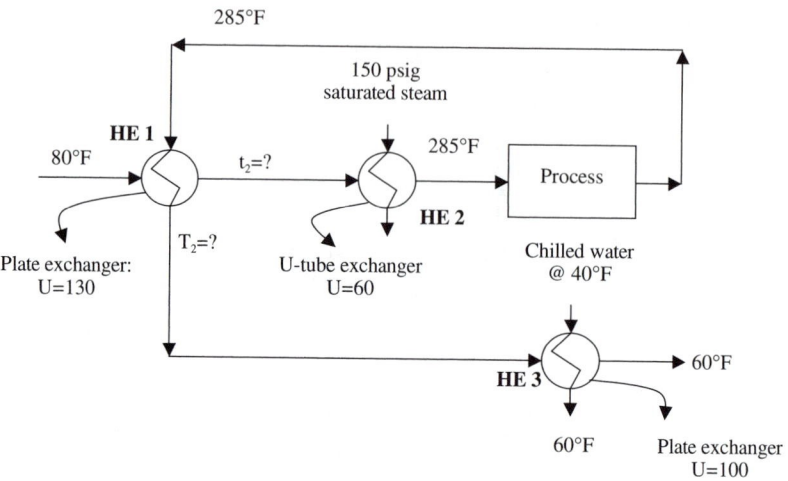

FIGURE 11.6 Heat recovery flowsheet.

Assume:

- C_p for the process fluid is constant
- Ignore ΔT_{lm} correction factors for the heater (assume it has one shell)
- All exchangers will have a 150 psig rating
- Use a Hand factor of 3.5 for SS plate and frame exchangers having nitrile gaskets
- Carbon steel and stainless steel are acceptable metallurgies
- The heat exchanger system will be controlled as in a typical process plant
- This exchanger system will be part of an expansion of an existing U.S. site

11.3.2 ANALYTICAL METHODOLOGY

Several possible choices exist for the dependent variable. One could pick the amount heat transferred in any of the exchangers or either of the two unknown temperatures. The one that makes the analysis easiest to grasp is either of the two temperature unknowns. Of these, I selected the hot stream outlet temperature, T_2.

Selecting a T_2 specifies the inlet temperatures of the other two exchangers in the process. This also sets the heat loads and sizes for each exchanger and the utility usage in the steam heater and the chilled water cooler.

- Investments:[*]

Capital	Changes	Estimate the Cost of Each Option
Working capital	No change	Ignore
Startup expense	No change	Ignore

- Production costs change:

Raw materials	No change	Ignore
Packaging materials	No change	Ignore
Manufacturing		
Operating labor	No change	Ignore
Employee benefits	No change	Ignore
Supervision	No change	Ignore
Laboratory	No change	Ignore
Utilities	*Changes*	*As the amount of heat put into the cold process feed changes, so does the usage of steam and chilled water*
Maintenance	*Changes*	*6% of capital*
Insurance and taxes	*Changes*	*3% of capital*
Operating supplies	*Changes*	*1% of capital*
Plant overhead	*Changes*	*1% of capital*
Depreciation	*Changes*	*Capital/10 years*
Product delivery	No change	Ignore

- We will start with a T_2 range of 90°F to 210°F and an interval of 30°F. The results for these temperatures are shown in Figure 11.7. Because this did not find the optimum T_2, we picked another set of T_2s lower than 90°F.

11.3.3 PROBLEM SOLUTION

The following illustrates some of the calculations for a T_2 of 90°F. Table 11.4 shows the results for all the calculations.

11.3.3.1 Size the Heat Exchangers

- Find the inlet and outlet temperatures for all exchangers:

    HE 1: Hot side temperatures — Inlet = 285°F, Outlet = 90°F

---

[*] As was the case with the cooling water problem, it is not logical for startup costs to vary from option to option.

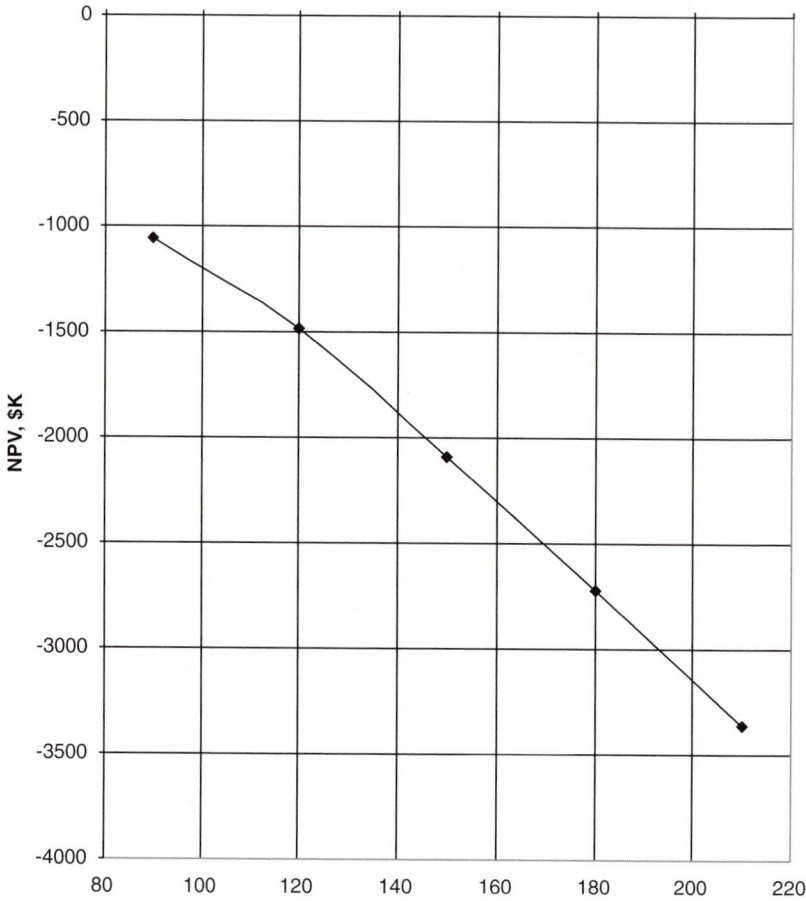

FIGURE 11.7 Heat recovery optimization (first attempt).

Cold side temperatures — knowing the hot side temperatures, one can find the cold side temperatures via an energy balance:

$$q_{cold} = q_{hot} = (WC_p \Delta T)_{cold} = (WC_p \Delta T)_{hot}$$

Because W and C_p are identical for the hot and cold streams,

$$\Delta T_{cold} = \Delta T_{hot} = 285 - 90°F = 195°F.$$

Because the cold side inlet temperature is 80°F, the outlet is 80°F + 195°F = 275°F.

HE 2: Inlet = 275°F, Outlet = 285°F

HE 3: Inlet = 90°F, Outlet = 60°F

- Find the log-mean temperature difference for all exchangers:

$$\Delta T_{lm} = \frac{\Delta T_2 - \Delta T_1}{\ln\left(\dfrac{\Delta T_2}{\Delta T}\right)}$$

HE 1: Because the ΔTs are the same at each end of the exchanger, ΔT_{lm} = 10°F.

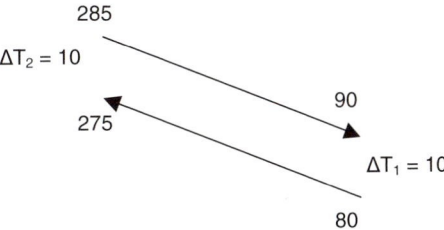

HE 2:

$$\Delta T_{lm} = \frac{91 - 81}{\ln\left(\dfrac{91}{81}\right)} = 85.9$$

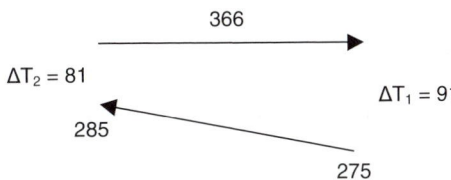

HE 3:

$$\Delta T_{lm} = \frac{30 - 20}{\ln\left(\dfrac{30}{20}\right)} = 24.7$$

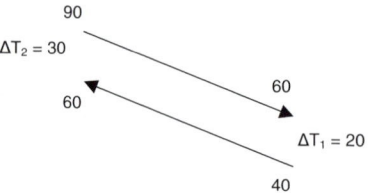

- Size each exchanger:

$$A = \frac{q}{U\Delta T_{lm}}$$

HE 1:

$$A = \frac{q}{U\Delta T_{lm}} = \frac{WC_p\Delta T}{U\Delta T_{lm}} = \frac{(100 Klb/hr)(0.5B/lb - °F)(285 - 90°F)}{(130B/hr - ft^2 - °F)(10°F)} = 7500 ft^2$$

HE 2:

$$A = \frac{(100 Klb/hr)(0.5B/lb - °F)(285 - 275°F)}{(60B/hr - ft^2 - °F)(85.9°F)} = 97 ft^2$$

HE 3:

$$A = \frac{(100 Klb/hr)(0.5B/lb - °F)(90 - 60°F)}{(100B/hr - ft^2 - °F)(24.7°F)} = 607 ft^2$$

11.3.3.2 Estimate the Capital Cost

Use the Heat Exchanger page in Appendix IV to estimate the purchase cost of the three exchangers. For all three, $F_i = 1.35$ because the instrumentation is typical for a process plant, and $F_b = 1.06$ because the process is being added to an existing plant.

Find $\sum(Purchase\ cost * Hand\ factor * F_m)$ for all exchangers:

HE 1: Purchase cost = \$58.8K, a material factor of 1.4 for EPDM gaskets, Hand Factor = 3.5, $F_m = 0.85$ (to adjust for the EPDM gaskets)

HE 2: Purchase cost = \$4.6K, Hand factor = 3.5; $F_m = 1$ (carbon steel construction)

HE 3: Purchase cost = \$15.1K, Hand factor = 3.5; $F_m = 1$

$\sum (Purchase\ cost * Hand\ factor * F_m) = (\$58.8K * 1.4 * 3.5 * 0.85) + (\$4.6K * 3.5) + (\$15.1K * 3.5) = \$313.9K$

$Capital\ cost = \$313.9K * 1.35 * 1.06 = \$449K$

11.3.3.3 Find the Annual Expenses, BT and AT

- Steam usage and cost:

$$lb/yr = \left(\frac{q_{heater}}{\text{heat of condensation, } h_{fg}} \right) (\text{loss factor}) \left(\frac{\text{operating hrs}}{yr} \right)$$

$$= \left(\frac{500K \text{ B/hr}}{857 \text{ B/lb}} \right)(1.075)(8400\text{hr/yr}) = 5270K \text{ lb/yr}$$

$\$/yr = 5270K \text{ lb/yr} * \$5.80 / 1000\text{lb} = \$30.6K/yr$

- Chilled water usage and cost:

$$lb/yr = \left(\frac{q_{heater}}{(C_p \Delta T)_{water}} \right) (\text{Thermal and material losses})$$

$$= \left[\frac{1500K \text{ B/hr}}{(1\text{B/lb} - °F)(60 - 40°F)} \right](1.075)(8400\text{lb/yr}) = 677M \text{ lb/yr}$$

$$gal/yr = \frac{677M \text{ lb/yr}}{8.35\text{lb/gal}} = 81.1M$$

$\$/yr = 81.1M \text{ gal/yr} * \$1.35 / 1000 \text{ gal} = \$109K/yr$

- Find the capital ratioed costs (maintenance, insurance and taxes, operating supplies, and plant overhead):

  Capital ratioed costs = $449K * 0.11 = $49.4K per year

- Find the depreciation cost:

  Annual depreciation write-off = $449K/10 yrs = $44.9 per year

- Find the BT annual expenses and the AT cash flow:

  BT expenses = $30.6 + 109.0 + 49.4 + 44.9 = $234K per year

  AT cash flow = (−234K/yr)(1 − 0.35) + 44.9 = −$107K per year

11.3.3.4 Calculate the NPV

$$P_{AT \text{ cash flow}} = A(P/A, 15\%, 10) = (-\$107K)(5.019) = -\$538$$

$$NPV = P_{Capital} + P_{AT \text{ cash flow}} = -\$449K + (-\$538K) = -\$987K$$

11.3.3.5 Find the Economic Design Point

The economic design point is the T_2 where the NPV is at a maximum. Figure 11.7 shows the results for the first set of calculations (which did not locate the maximum). Figure 11.8 shows the results for the final calculations and the maximum at 89.5°F.

11.4 WHAT TO CHOOSE: A GRASS-ROOTS PLANT OR THE EXPANSION OF AN EXISTING PLANT

The question of where to build a new plant or process routinely comes up in the early part of a design. Usually, some favor building in an existing facility to minimize the total number of plants whereas others wish to build on a new site to get away from some of the problems associated with existing plants. These problems might include high labor costs, poor labor relations, high in-freight costs, and so on. Sorting out the economics of the two options will help the decision-makers decide what to do.

11.4.1 PROBLEM STATEMENT

As a part of the conceptual studies for a fluid process that will make a new product, an engineer is studying whether to build this on a new site (a grass-roots site) or whether to add the process to the existing Dallas plant. Estimates for the capital and production costs at the Dallas plant are $20.6M and $26.14/unit (Table 11.5 shows the details). In Dallas, all supporting systems — steam, cooling water, compressed air, plant facilities (cafeteria, locker room, storeroom, and so on) — have enough capacity to support the new process.

The engineer and a construction contractor have concluded the capital cost to build a grass-roots process will cost 12% more than building in Dallas. This does not include the support systems (steam boiler, cooling water tower, compressed air capacity, and plant facilities) that will have to be built on the grass-roots site.

Compare the economics for each site and recommend which plan should be used and why. Use a 10-year project life, a 35% tax rate, and a 15% discount rate. The cost data below are expressed in March 2005 dollars.

Other data:

- Labor costs: Dallas = $27/hr; grass-roots site = $20/hr. Assume crew sizes will be the same in both plants.
- Employee benefits, supervision, and laboratory: Dallas — 60% of operating labor due to site efficiencies; grass-roots site — assume 70% of operating labor.
- Dallas is closer to the customers and further from the raw and packaging material suppliers; hence, the difference in materials and product delivery costs.
- Startup expenses (spent in Year 1): Dallas = 10% of capital, grass-roots site = 15%.
- Support systems needed at the grass-roots site.
  - Boiler: 125K lb/hr of 400 psig saturated steam, gas-fired.

TABLE 11.4
Optimum Heat Recovery in a Heat Exchanger Loop

	T_2, Hot Stream Outlet Temperature (°F)				
	95	92	90	88	86
Capital Costs					
HE 1: Interchanger (plate and frame)					
$q = WC_pT$ (K Btu/hr)	9500	9650	9750	9850	9950
U	130	130	130	130	130
Temperature in, hot stream	285	285	285	285	285
Temperature out, cold stream	270	273	275	277	279
Temperature in, cold stream	80	80	80	80	80
ΔT_{lm}	15	12	10	8	6
Area (ft²)	4872	6186	7500	9471	12756
HEX cost					
Figure IV-11, Appendix IV	46.6	53.0	58.8	66.7	78.3
Gasket material — EPDM	1.4	1.4	1.4	1.4	1.4
Time	1	1	1	1	1
Hand factor	3.2	3.2	3.2	3.2	3.2
F_m	0.85	0.85	0.85	0.85	0.85
Capital cost at F_i=1.35 and F_b = 1.06	277.5	315.7	350.3	397.4	466.7
HE 2: Heater (U-tube)					
$q = WC_pT$ (K Btu/hr)	750	600	500	400	300
U	60	60	60	60	60
Temperature in and out, steam	366	366	366	366	366
Temperature out, cold stream	285	285	285	285	285
Temperature in, cold stream	270	273	275	277	279
ΔT_{lm}	88.3	86.9	85.9	84.9	84.0

TABLE 11.4
Optimum Heat Recovery in a Heat Exchanger Loop (continued)

	T_2, Hot Stream Outlet Temperature (°F)				
	95	92	90	88	86
Area (ft²)	142	115	97	78	60
HEX cost					
Figure IV-11, Appendix IV	5.5	5.0	4.6	4.1	3.6
Equipment type, U-tube	1	1	1	1	1
Pressure	1	1	1	1	1
Time	1.0	1.0	1.0	1.0	1.0
Hand factor	3.5	3.5	3.5	3.5	3.5
Capital cost at $F_i = 1.35$ and $F_b = 1.06$	27.5	24.8	22.8	20.5	17.9
HE 3: Cooler (plate and frame)					
$q = WC_pT$ (K Btu/hr)	1750	1600	1500	1400	1300
U	100	100	100	100	100
Temperature in, hot stream	95	92	90	88	86
Temperature out, cold stream	60	60	60	60	60
Temperature out, hot stream	60	60	60	60	60
Temperature in, cold stream	40	40	40	40	40
ΔT_{lm}	26.8	25.5	24.7	23.8	22.9
Area (ft²)	653	627	608	589	568
HEX cost					
Figure IV-11, Appendix IV	15.7	15.4	15.1	14.9	14.6
Gasket material — nitrile	1.0	1.0	1.0	1.0	1.0
Time	1.0	1.0	1.0	1.0	1.0
Hand factor	3.2	3.2	3.2	3.2	3.2
Capital cost at $F_i = 1.35$ and $F_b = 1.06$	72.0	70.5	69.3	68.1	66.8
Total capital	383.8	417.6	448.9	492.4	557.7

Annual Expenses

Steam					
M lb/yr (including thermal/material losses of 7.5%)	7.9	6.3	5.3	4.2	3.2
K$/yr at $5.80/1000 lb	46	37	31	24	18
Chilled water					
M gal/yr (including thermal/material losses of 7.5%)	95	87	81	76	70
K$/yr at $1.35/1000 gal	128	117	109	102	95
Maintenance, insurance and taxes, operating supplies, plant overhead/miscellaneous: 11% of capital ($K)	42.2	45.9	49.4	54.2	61.3
Depreciation at capital ($K/life)	38.4	41.8	44.9	49.2	55.8
Annual expenses, BT ($K)	−254	−241	−234	−230	−230
Annual cash flow, AT ($K)	−127	−115	−107	−100	−94

NPV

Capital ($K)	−384	−418	−449	−492	−558
Annual cash flow: $P = (P/A, 15\%, 10) = A \propto 5.019$	−637	−577	−539	−503	−472
NPV	−1020	−995	−988	−996	−1029

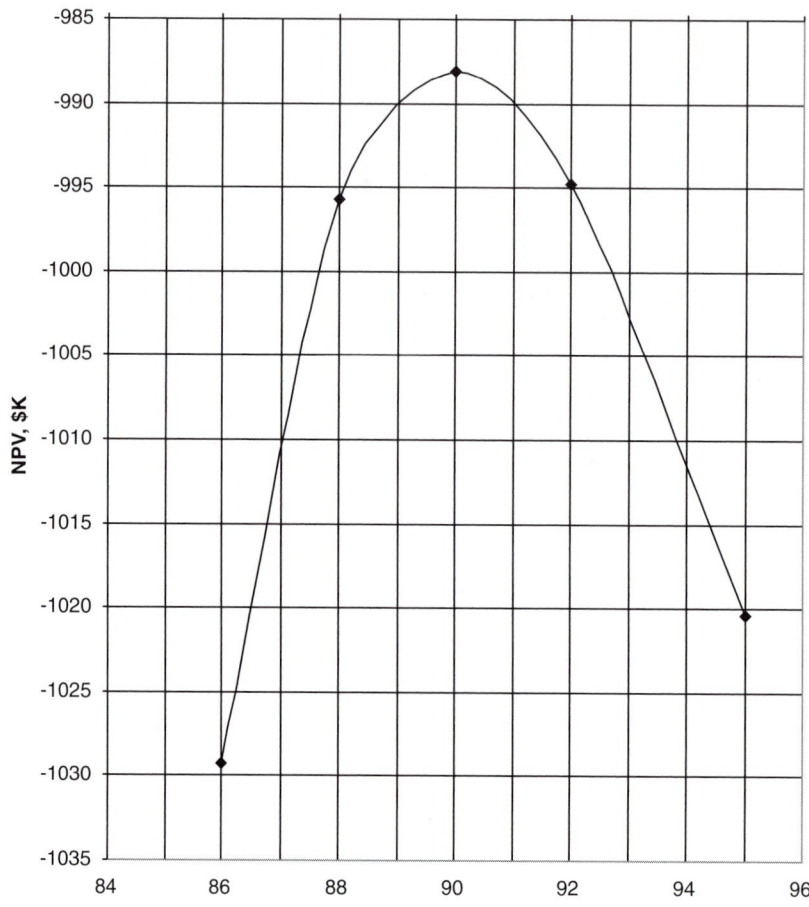

T₂, Hot Stream Outlet Temperature, °F

FIGURE 11.8 Heat recovery optimization.

- Cooling tower: three towers, each having a capacity of 2500 gpm.
- Compressed air: two air compressors, each having a capacity of 1000 cfm.
- Plant facilities: assume they are typical for a fluid processing plant built on a new site.

Controls will be typical for a processing plant.

TABLE 11.5
Dallas Plant Cost Data

	Dallas Plant	Grass-Roots Site
Plant capacity (units/yr)	7.1M	7.1M
Operating hrs/yr	6000	6000
Process capital	$20.6M	
Support system capital	0	See other data
Production cost		
Raw materials	$12.00/unit	$11.40/unit
Packaging materials	6.00	5.70
Manufacturing		
Operating labor	2.60	
Employee benefits	0.99	
Supervision	0.44	
Laboratory	0.13	
Utilities	0.21	0.21
Maintenance	0.17	
Insurance and taxes	0.09	
Operating supplies	0.03	
Plant overhead	0.03	
Depreciation	0.29	
Subtotal	4.98	
Product delivery	3.16	3.46
Total	26.14	

11.4.2 ANALYTICAL METHODOLOGY

In this case study, just two options are to be compared as opposed to the continuum of options found in the proceeding case studies. We will compare the total capital costs and the total production costs for the two options.

11.4.3 PROBLEM SOLUTION

11.4.3.1 Estimate the Capital Cost for the Grass-Roots Site

- Process: $ = (1.12)(20.6M) = $23.1M
- Plant facilities are accounted for by using an F_b of 1.45
- Boiler: Use the Boiler page in Appendix IV. As the large package boiler is less expensive than the field erected boiler, use the package boiler costs. Because the boiler has a 400 psig rating, adjust the cost from the graph for pressure. Interpolating between the 350 psig and 500 psig pressure factors gives a 400 psig factor of 1.08. The purchase cost is $613K * 1.08 = $662K.

$$Capital\ cost = Purchase\ cost * Hand\ factor * F_m * F_i * F_b$$

$$= \$662K * 2.5 * 1 * 1.35 * 1.45 = \$3.24M$$

- Cooling towers: Use the Cooling Tower page in Appendix IV. The purchase cost is ($134K/tower)(3 towers) = $402K.

$$Capital\ cost = Purchase\ cost * Hand\ factor * F_m * F_i * F_b$$

$$= \$402K * 1 * 1.35 * 1.45 = \$0.79M$$

- Air compressors: Use the Compressor page in Appendix IV.

$$Purchase\ cost = (\$118K/compressor)(2\ compressors) = \$236K$$

$$Capital\ cost = Purchase\ cost * Hand\ factor * F_m * F_i * F_b$$

$$= \$236K * 1 * 1.35 * 1.45 = \$0.46M$$

$$Total\ capital\ cost = \$23.1M + 3.24M + 0.79M + 0.46M = \$27.6M$$

11.4.3.2 Estimate the Production Cost for the Grass-Roots Site

- Operating labor (plant capacities and crew sizes will be the same):

$$\$/unit = \left(\frac{(\text{wage rate})_{\text{grass roots}}}{(\text{wage rate})_{\text{Dallas}}}\right)(\$/unit)_{\text{Dallas}} = \left(\frac{\$20/hr}{\$27/hr}\right)(\$2.60/unit) = \$1.93/unit$$

- Labor-related costs (employee benefits, supervision, laboratory):

$$\$/unit = \$1.93 * 0.70 = \$1.35/unit$$

- Capital-related costs (maintenance, insurance and taxes, operating supplies, plant overhead):

$$\$/unit = \left(\frac{0.11(\$27.6M)}{7.1M\ units}\right) = \$0.43/unit$$

- Depreciation:

$$\$/\text{unit} = \frac{\dfrac{\$27.6M}{10 \text{ yrs}}}{7.1M \text{ units}} = \$0.39/\text{unit}$$

- Production Cost Summary:

	Production Cost ($/unit)	Comments
Raw materials	$11.40	From problem data
Packaging materials	5.70	From problem data
Manufacturing		
Operating labor	1.93	
Employee benefits, supervision, laboratory	1.35	
Utilities	0.21	From problem data
Maintenance, insurance and taxes, operating supplies, plant overhead	0.43	
Depreciation	0.39	
Subtotal	4.31	
Product delivery	3.46	From problem data
Total	24.87	

- NPV for Dallas plant:

	P ($M)
Capital	−20.6

Startup expense at 10% of capital
Convert Year 1 (F) expense to Year 0
$\$_{AT} = (0.1)(-20.6M)(1- 0.35) = -\$1.34M$
$P = F(P/F,\ 15\%,\ 1) = (-1.34)(0.87) =$ −1.2
Production cost (these costs are an annuity)
AT cash flow = (Production cost)(1 − *tax rate*) + *Depreciation*
$= (-\$26.14/\text{unit})(7.1M \text{ units})(1 - 0.35) + 20.6M/10 \text{ years}$
$= -\$118.6$
$P = A(P/A,\ 15\%,10) = -\$118.6 * 5.019 =$ −595.3
NPV   −617

- NPV for the grass-roots plant:

	P ($M)
Capital	−27.6
Startup expense at 15% of capital	
Convert Year 1 (F) expense to Year 0	
$\$_{AT} = (0.15)(-27.6M)(1- 0.35) = -\$2.69M$	
$P = F(P/F, 15\%, 1) = (-2.69)(0.87) =$	−2.3
Production cost (these costs are an annuity)	
AT cash flow = (*Production cost*)(1− *tax rate*) + *Depreciation*	
= (−$24.87/unit)(7.1M units)(1 0.35) + 27.6M/10 years	
= $−112.0	
$P = (P/A, 15\%, 10) = -\$112 * 5.019 =$	−562.1
NPV	−592

11.4.4 THE ECONOMIC DESIGN POINT

Because the NPV for the grass-roots plant is the highest, it is the most economic of the two options.

11.5 OPTIMUM NUMBER OF PLANTS

How many plants is it appropriate to build? Quite a few project teams are faced with this question; they can use economic analysis to find the answer. Before continuing, a few words about shipping costs, a key part of the analysis, must be said.

Shipping costs include both in-freight and product shipping. In-freight can be paid directly by the receiving company or it can be included in the price of materials delivered to a plant. It makes no difference who "pays these costs;" they are a part of a company's cost structure. Reducing in-freight costs lowers a company's cost. The lower costs can be either passed on to the customers, improving price competitiveness, or can flow to the profit line. Similar reasoning applies to product shipping costs. Regardless of who pays them — the company or its customer — they are a part of the price a customer pays for a product. Having lower costs either allows a company to lower the price to the customer or increase its profits.

11.5.1 PROBLEM STATEMENT

Company X has authorized the conceptual engineering study for building a new plant or plants for a new liquid product. The customers for the product are located throughout the country. Volume is expected to be 30M cases per year. The next phase in the study is to decide how many plants should be built.

Data/Assumptions:

- Purchasing has decided they will source raw and packaging materials from the same supplier plants regardless of how many plants are built
- The plants will operate 24 hr/day, 5 days/wk, and 50 wks/yr, or 6000 hrs/yr

- Manufacturing has estimated startup expenses at 14% of capital (assume this occurs in Year 1)
- Use a 10-year project life and a 10% discount rate
- Company X's tax rate is 32%

Find the most economic number of plants.

11.5.2 ANALYTICAL METHODOLOGY

- The independent variable is the number of plants. When the number of plants changes, the individual plant capacities change, the crew sizes change, and the distance to suppliers and customers change.
- Investments change:

Capital	Changes	Estimate the Cost of Each Option
Startup expense	Changes	Varies as a percentage of capital
Working capital	No change	Ignore

- Production costs change:*

Raw materials	Changes	In-freight varies depending on the distance from the supplier
Packaging materials	Changes	In-freight varies depending on the distance from the supplier
Manufacturing		
Operating labor and labor-related	Changes	Varies with the size of the plant
Utilities	No change	Ignore
Maintenance	Changes	6% of capital
Insurance and taxes	Changes	3% of capital
Operating supplies	Changes	1% of capital
Plant overhead	Changes	1% of capital
Depreciation	Changes	Capital/10 years
Product delivery	Changes	Varies depending on the distance from the customers

- Once the capital and production costs are estimated for all the options, calculate the NPV for each and select the option having the greatest NPV.

---

* Note that several of the production costs vary as a percent of capital. You can find the percentages used in Chapter 4, Table 4.8.

11.5.3 PROBLEM SOLUTION

The engineer in charge of the siting study has developed the options shown in Table 11.6. To do this, s/he used the methods described in Chapter 10, "Locating Plants Based on Customer and Supplier Locations."

TABLE 11.6
Plant Site Options

Number of Plants	Locations	Plant Capacity (M cases/yr)
1	Cincinnati	30.0
2	Cincinnati	22.5
	San Francisco	7.5
3	New York	15.0
	Kansas City	7.5
	San Francisco	7.5
4	New York	10.2
	Chicago	7.5
	Atlanta	4.8
	San Francisco	7.5
5	New York	10.2
	Chicago	7.5
	Atlanta	2.1
	Dallas, TX	2.7
	San Francisco	7.5
6	New York	10.2
	Chicago	7.5
	Atlanta	2.1
	Dallas, TX	2.7
	Los Angeles	4.8
	San Francisco	2.7

The same engineer has roughly designed two different size plants and has estimated the capital plus the costs for operating labor and labor-related items:

Capacity (M cases/yr)	Capital ($M)	Labor/Labor-Related (hr/case)
7.5	16.9	0.0235
15.0	27.1	0.014

Manufacturing has estimated wage rates for each possible site.

Plant	Wage Rate ($/hr)	Plant	Wage Rate ($/hr)
New York	22	Atlanta	18
Cincinnati	18	Dallas	19
Kansas City	18	Los Angeles	23
Chicago	20	San Francisco	26

Assume utility costs, in $/case, will be the same in all plants. This assumption allows one to ignore the cost in the analysis. (This is a simplifying assumption. Even though the costs will most likely be different, the differences are small enough that the analysis will be unaffected. A rough estimate shows the utility cost for a 7.5M case plant is about $0.07/case.)

The Purchasing and Distribution Departments have estimated in-freight and the product delivery costs:

Number of Plants	In-Freight ($/case)	Delivery ($/case)
1	0.12	2.78
2	0.21	1.51
3	0.25	1.05
4	0.26	0.92
5	0.26	0.85
6	0.25	0.81

The following illustrates the calculations for the four-plant option. Table 11.7 summarizes all the calculations.

11.5.3.1 Estimate the Investments for All Four Plants

- Using Equation 3.1 and the capital costs for the 7.5 and 15M case plants, find the size exponent for capital cost:

$$n = \frac{\ln\left(\dfrac{Cost_{size\ 2}}{Cost_{size\ 1}}\right)}{\ln\left(\dfrac{Capacity_{size\ 2}}{Capacity_{size\ 1}}\right)} = \frac{\ln\left(\dfrac{\$27.1M}{\$16.9M}\right)}{\ln\left(\dfrac{15}{7.5}\right)} = 0.68$$

Using this exponent in Equation 3.1 yields:

Plant	Capacity (M cases/yr)	Capital Cost ($M)
New York	10.2	20.8
Chicago	7.5	16.9
Atlanta	4.8	12.4
San Francisco	7.5	16.9
	Total =	67.0

• Startup expenses:

$M = (14% of capital)(67.0M) = $9.4M

Find $P_{AT\ startup\ expense\ cash\ flow}$:

AT cash flow = (–$9.4M)(1 – 0.32) = –$6.4M

$P = F(P/F, 10\%, 1) = (–\$6.4M)(0.909) = –\$5.8M$

11.5.3.2 Estimate the Items of Production Costs Where Differences are Found

• In-freight and product delivery costs:

$/case = $0.26 + 0.92 = $1.18/case

• Labor and labor-related costs. Use Equation 4.2 and the labor and labor-related costs for the 7.5M and 15M case plant to find the exponent in the Wessell relationship:

$$n = \frac{\ln\left(\dfrac{hr\,/\,case_2}{hr\,/\,case_1}\right)}{\ln\left(\dfrac{Capacity_1}{Capacity_2}\right)} = \frac{\ln\left(\dfrac{0.014}{0.0235}\right)}{\ln\left(\dfrac{7.5}{15}\right)} = 0.75$$

Using this exponent in Equation 4.2 gives:

Plant	Capacity (M cases/yr)	hr/case	Wage Rate ($/hr)	$/case	$M/yr
New York	10.2	0.0186	22	0.409	4.17
Chicago	7.5	0.0235	20	0.470	3.53
Atlanta	4.8	0.0328	18	0.590	2.83
San Francisco	7.5	0.0235	26	0.611	4.58
				Total =	15.1

Average $/case = $15.1M/yr/30M cases/yr = $0.50/case

- Capital-related items (maintenance and so on):

$$\$/case = \frac{(11\% \text{ of capital/yr})(\$67.0M)}{30M \text{ cases/yr}} = \$0.25/case$$

- Depreciation:

$$\$/case = \frac{\$67.0M/10yrs}{30M \text{ cases/yr}} = \$0.22/case$$

- Totals:

	$/case
In-freight and product delivery	1.18
Labor and labor-related	0.50
Capital-related	0.25
Depreciation	0.22
Total	$2.15/case

- Find the AT cash flow:

AT cash flow = Expenses (1 − Tax rate) + Depreciation

$= (-\$2.15/case * 30M \text{ cases/yr})(1 - 0.32) + (\$67.0/10)$

$= -\$37.2M/yr$

11.5.3.3 Calculate the NPV

$$P_{AT\ cash\ flow} = A(P/A,\ 10\%,\ 10) = \$-37.2M\ (6.145) = -\$229M$$

$$NPV = P_{Capital} + P_{AT\ startup\ expenses} + P_{AT\ cash\ flow} = -\$67.0 + (-\$5.8M) + (-\$229M)$$
$$= -\$302M$$

TABLE 11.7
Optimum Number of Plants

	Number of Plants					
	1	2	3	4	5	6
Capital ($M)	43.2	52.5	60.9	67.0	70.1	74.0
Startup expense at14% of capital ($M)	6.1	7.3	8.5	9.4	9.8	10.4
AT: 32% tax rate	4.1	5.0	5.8	6.4	6.7	7.0
$P_{AT\ startup\ expense}$: $(P/F,\ 10\%,\ 1) =$ 0.909	3.7	4.5	5.3	5.8	6.1	6.4
Production cost differences ($/case)						
In-freight and delivery	2.90	1.72	1.30	1.18	1.11	1.06
Labor and labor-related	0.15	0.29	0.41	0.50	0.57	0.66
Capital-related (maintenance, insurance and taxes, etc.)	0.16	0.19	0.22	0.25	0.26	0.27
Depreciation, 10-year life	0.14	0.17	0.20	0.22	0.23	0.25
Total =	3.35	2.38	2.14	2.15	2.17	2.24
AT cash flow, production costs	−64.1	−43.3	−37.5	−37.2	−37.3	−38.2
NPV $						
Capital	−43.2	−52.5	−60.9	−67.0	−70.1	−74.0
Startup expenses	−3.7	−4.5	−5.3	−5.8	−6.1	−6.4
Production costs $(P/A,\ 10\%,\ 10)$ = 6.145	−394	−266	−231	−229	−229	−235
NPV	−441	−323	−297	−302	−306	−315

11.5.3.4 The Economic Design Point

Because the NPV for the three plant option is the greatest, three plants — located in New York, Kansas City, and San Francisco — are economically best.

REFERENCES

1. *Standards of the Tubular Exchanger Manufactures Association (8th Edition)*, Tarrytown, NY: Tubular Exchanger Manufacturers Association, 1999, 106, 111–116.
2. Perry, R.H. and Green, D.W. (Eds.), *Perry's Chemical Engineering Handbook*, New York: McGraw-Hill, 1997, 11–6.

Appendices

Appendix I

Definitions

Annual cost. The sum of the annuitized values of a cash flow series.

Annual percentage rate (APR). This is the annual rate of interest paid or received. The compounding frequency, if other than annual, will be stated, i.e., 6% APR, compounded monthly. This is also called the *nominal interest rate*. When compounding is other than annual, the effective interest rate (on an annual basis) is higher than the APR.

Annuity. A series of uniform payments or withdrawals occurring at equal time intervals.

Breakeven volume. This is the production volume at which the AT expenses and the AT revenues plus depreciation are equal.

By-product credit. The revenues from the sale of byproducts.

Capital. A firm's investment in long-term assets that are not bought or sold in the normal course of business, e.g., plant equipment, buildings, and site upgrades. These assets are depreciated.

Cash flow. The flow of money into or out of a company, a project, a personal account, and so on. Cash flows related to expenses or revenues can be on either a BT or AT basis. To convert a BT cash flow to an AT basis, multiply it by $(1 - \text{Tax rate})$.

Cash flow diagram. A diagram showing all cash flows and the time they occur. Cash flows in are shown by an arrow into the timeline and cash flows out by an arrow away from the timeline.

Chemical engineering plant cost index (CEPI). An index of the costs to design, purchase and install chemical plant equipment. It is maintained by *Chemical Engineering* and includes costs (1) for equipment, machinery and supports (61% of the index weighting); (2) for construction labor (22%); (3) for buildings (7%); and (4) for engineering and supervision (10%). The period 1957 to 1959 is defined as an index of 100.

Controllable activity. Controllable activity accounting identifies costs of specific work activities. Each activity has someone who is responsible for it. The activities are defined so that the responsible person has the power to influence the cost of the activity.

Decision tree. A risk assessment method that evaluates the impact of decisions. It shows all the possible outcomes and their probability of occurring. The result for each possible outcome is probability weighted and combined with other outcomes to determine the most probable overall result.

Depreciation. A deduction from revenues (allowed by the government when calculating income taxes) of a fraction of the capital invested in a plant. This deduction may be considered as a fund to allow eventual replacement of the plant. It is not a cash flow.

Discount rate. The interest rate at which future cash flows are discounted to translate them into present values.

Economic design. A design method for a plant or process that economically balances capital and production costs. The balance point is defined by a company's minimum acceptable rate of return.

Economic life. The life of a project from an economic standpoint. It establishes the value of n in economic calculations. Life is determined by projected product or process obsolescence, projected time before the equipment wears out, or by a company's financial guidelines.

Effective interest rate. The true annualized interest rate when compounding is other than annual.

Engineering economics. The study of how to bring the impact of economics into engineering decision making.

Expense. A firm's costs that are chargeable against sales in a specific period.

Fixed costs. Production costs that do not vary with production volume.

Future worth. This the projected value of a present sum of money when it grows at a specified interest rate for a given number of years.

General expense. Broad corporate level expenses — research and development, marketing, sales, and administrative costs.

Gross savings. Cash flows before the costs associated with new capital are deducted. See also *Net Savings*.

Hurdle rate. The minimum acceptable ROI used to determine whether or not to fund a capital investment. Projects having an ROI below the hurdle rate are not funded.

Inflation. The devaluing of money because the volume of money increases faster than the supply of goods.

Interest. The return from the investment of funds or the money paid for the use of borrowed money.

Internal rate of return. See *Return on investment (ROI)*.

Manufacturing costs. The cost to manufacture a product. It is comprised of operating labor (wages), employee benefits, supervision (wages and benefits), laboratory costs, maintenance costs, utility costs, depreciation, insurance and taxes, operating (consumable) supplies, plant overhead, and contract manufacturing costs.

Net present value (NPV). The sum of the present values of a cash flow series.

Net savings. Cash flows after costs associated with a capital investment (maintenance, operating supplies, insurance and taxes, and plant overhead) are deducted from the gross savings.

Nominal interest rate. This is the annual rate of interest and is the same as the annual percentage rate or APR. See also *Annual percentage rate*.

Outcome analysis. See *Decision tree*.

Present worth. Today's value of a sum of money.

Producer price index (PPI). An index of the selling prices received by domestic producers for their goods and services. The Department of Labor maintains the index. The PPI for Chemical and Allied Products defines December 1984 as an index of 100.

Product cost. The sum of production cost and general expense.

Production cost. The cost to produce a product. It is made up of raw material costs, packaging material costs, manufacturing costs, and delivery costs.

Project life. The years a process or project is expected to operate without major revision. This is determined by the shorter of product or process obsolescence or by depreciable life.

Replacement cost. The capital cost, in today's dollars, to replace an asset (e.g., equipment, a process, a plant) in kind.

Return on investment (ROI). The interest rate at which the net present value of a cash flow series is zero. This is the percent return from an investment.

Risk. The chance that the actual economic results of a project will be different from the estimated results because actual capital spending, production costs, sales volume, selling price, and so on are different from their estimated values.

Salvage value. The estimated value of an asset at disposal. It is expressed in future dollars.

Sensitivity analysis. A risk assessment tool that shows how economic results would vary when key project factors are different from their estimates.

Startup expense. Expenses above normal due to the startup of a new or modified process. They include items such as salaries and benefits for managers and operators hired prior to startup, training costs for managers and operators and other startup-related costs (losses, labor and utilities inefficiencies, reprocessing of off-quality materials, and so on).

Unit cost. Production costs expressed in dollars per unit of production (e.g. $/ton, $/lb, $/case).

Variable costs. Those costs that vary with production volume.

Working capital. A firm's investment in short-term assets. Whereas working capital includes cash on hand and taxes payable, it is essentially made up of:

- Inventories — the raw materials, packaging materials, work in process, and finished product owned by or in the control of a company
- Accounts receivable — the money owed to a company for product sold but not yet paid for
- Accounts payable — the money owed by a company for materials or services received but not yet paid for

Appendix II

Indices

Chemical Engineering Plant Cost Index, Annual Averages (1957–1959 = 100)

Year	CEPI	Year	CEPI	Year	CEPI
1956	93.9	1973	144.1	1990	357.6
1957	98.5	1974	165.4	1991	361.3
1958	99.7	1975	182.4	1992	358.2
1959	101.8	1976	192.1	1993	359.2
1960	102.0	1977	204.1	1994	368.1
1961	101.5	1978	218.8	1995	381.1
1962	102.0	1979	238.7	1996	381.7
1963	102.4	1980	261.2	1997	386.5
1964	103.3	1981	297.0	1998	389.5
1965	104.2	1982	314.0	1999	390.6
1966	107.2	1983	316.9	2000	394.1
1967	109.7	1984	322.7	2001	394.3
1968	113.6	1985	325.3	2002	395.6
1969	119.0	1986	318.4	2003	402.0
1970	125.7	1987	323.8	2004	444.2
1971	132.2	1988	342.5	2005	468.2
1972	137.2	1989	355.4	—	—

Producer Price Index for Chemical and Allied Products, Annual Averages (12/84 = 100)

Year	PPI	Year	PPI
December 1984	100	1996	145.8
1985	100.7	1997	147.1
1986	100.5	1998	148.7
1987	103.6	1999	149.7
1988	113.0	2000	156.7
1989	119.6	2001	158.4
1990	121.0	2002	157.3
1991	124.4	2003	164.6
1992	125.8	2004	172.8
1993	127.2	2005	187.3
1994	130.0		
1995	143.4		

Source: U.S. Department of Labor.

Appendix III
Compound
Interest Tables

Engineering Economics and Economic Design

0.25%

n	F/P (F/P,i%,n)	P/F (P/F,i%,n)	A/F (A/F,i%,n)	A/P (A/P,i%,n)	F/A (F/A,i%,n)	P/A (P/A,i%,n)	n
1	1.00250	0.99751	1.00000	1.00250	1.00000	0.99751	1
2	1.00501	0.99502	0.49938	0.50188	2.00250	1.99252	2
3	1.00752	0.99254	0.33250	0.33500	3.00751	2.98506	3
4	1.01004	0.99006	0.24906	0.25156	4.01503	3.97512	4
5	1.01256	0.98759	0.19900	0.20150	5.02506	4.96272	5
6	1.01509	0.98513	0.16563	0.16813	6.03763	5.94785	6
7	1.01763	0.98267	0.14179	0.14429	7.05272	6.93052	7
8	1.02018	0.98022	0.12391	0.12641	8.07035	7.91074	8
9	1.02273	0.97778	0.11000	0.11250	9.09053	8.88852	9
10	1.02528	0.97534	0.09888	0.10138	10.11325	9.86386	10
11	1.02785	0.97291	0.08978	0.09228	11.13854	10.83677	11
12	1.03042	0.97048	0.08219	0.08469	12.16638	11.80725	12
13	1.03299	0.96806	0.07578	0.07828	13.19680	12.77532	13
14	1.03557	0.96565	0.07028	0.07278	14.22979	13.74096	14
15	1.03816	0.96324	0.06551	0.06801	15.26537	14.70420	15
16	1.04076	0.96084	0.06134	0.06384	16.30353	15.66504	16
17	1.04336	0.95844	0.05766	0.06016	17.34429	16.62348	17
18	1.04597	0.95605	0.05438	0.05688	18.38765	17.57953	18
19	1.04858	0.95367	0.05146	0.05396	19.43362	18.53320	19
20	1.05121	0.95129	0.04882	0.05132	20.48220	19.48449	20
21	1.05383	0.94892	0.04644	0.04894	21.53341	20.43340	21
22	1.05647	0.94655	0.04427	0.04677	22.58724	21.37995	22
23	1.05911	0.94419	0.04229	0.04479	23.64371	22.32414	23
24	1.06176	0.94184	0.04048	0.04298	24.70282	23.26598	24
25	1.06441	0.93949	0.03881	0.04131	25.76457	24.20547	25
26	1.06707	0.93714	0.03727	0.03977	26.82899	25.14261	26
27	1.06974	0.93481	0.03585	0.03835	27.89606	26.07742	27
28	1.07241	0.93248	0.03452	0.03702	28.96580	27.00989	28
29	1.07510	0.93015	0.03329	0.03579	30.03821	27.94004	29
30	1.07778	0.92783	0.03214	0.03464	31.11331	28.86787	30
31	1.08048	0.92552	0.03106	0.03356	32.19109	29.79339	31
32	1.08318	0.92321	0.03006	0.03256	33.27157	30.71660	32
33	1.08589	0.92091	0.02911	0.03161	34.35475	31.63750	33
34	1.08860	0.91861	0.02822	0.03072	35.44064	32.55611	34
35	1.09132	0.91632	0.02738	0.02988	36.52924	33.47243	35
36	1.09405	0.91403	0.02658	0.02908	37.62056	34.38647	36
37	1.09679	0.91175	0.02583	0.02833	38.71461	35.29822	37
38	1.09953	0.90948	0.02512	0.02762	39.81140	36.20770	38
39	1.10228	0.90721	0.02444	0.02694	40.91093	37.11491	39
40	1.10503	0.90495	0.02380	0.02630	42.01320	38.01986	40
41	1.10780	0.90269	0.02319	0.02569	43.11824	38.92256	41
42	1.11057	0.90044	0.02261	0.02511	44.22603	39.82300	42
43	1.11334	0.89820	0.02206	0.02456	45.33660	40.72120	43
44	1.11612	0.89596	0.02153	0.02403	46.44994	41.61715	44
45	1.11892	0.89372	0.02102	0.02352	47.56606	42.51088	45
46	1.12171	0.89149	0.02054	0.02304	48.68498	43.40237	46
47	1.12452	0.88927	0.02008	0.02258	49.80669	44.29164	47
48	1.12733	0.88705	0.01963	0.02213	50.93121	45.17869	48
49	1.13015	0.88484	0.01921	0.02171	52.05854	46.06354	49
50	1.13297	0.88263	0.01880	0.02130	53.18868	46.94617	50

0.50%

n	F/P (F/P,i%,n)	P/F (P/F,i%,n)	A/F (A/F,i%,n)	A/P (A/P,i%,n)	F/A (F/A,i%,n)	P/A (P/A,i%,n)	n
1	1.00500	0.99502	1.00000	1.00500	1.00000	0.99502	1
2	1.01003	0.99007	0.49875	0.50375	2.00500	1.98510	2
3	1.01508	0.98515	0.33167	0.33667	3.01502	2.97025	3
4	1.02015	0.98025	0.24813	0.25313	4.03010	3.95050	4
5	1.02525	0.97537	0.19801	0.20301	5.05025	4.92587	5
6	1.03038	0.97052	0.16460	0.16960	6.07550	5.89638	6
7	1.03553	0.96569	0.14073	0.14573	7.10588	6.86207	7
8	1.04071	0.96089	0.12283	0.12783	8.14141	7.82296	8
9	1.04591	0.95610	0.10891	0.11391	9.18212	8.77906	9
10	1.05114	0.95135	0.09777	0.10277	10.22803	9.73041	10
11	1.05640	0.94661	0.08866	0.09366	11.27917	10.67703	11
12	1.06168	0.94191	0.08107	0.08607	12.33556	11.61893	12
13	1.06699	0.93722	0.07464	0.07964	13.39724	12.55615	13
14	1.07232	0.93256	0.06914	0.07414	14.46423	13.48871	14
15	1.07768	0.92792	0.06436	0.06936	15.53655	14.41662	15
16	1.08307	0.92330	0.06019	0.06519	16.61423	15.33993	16
17	1.08849	0.91871	0.05651	0.06151	17.69730	16.25863	17
18	1.09393	0.91414	0.05323	0.05823	18.78579	17.17277	18
19	1.09940	0.90959	0.05030	0.05530	19.87972	18.08236	19
20	1.10490	0.90506	0.04767	0.05267	20.97912	18.98742	20
21	1.11042	0.90056	0.04528	0.05028	22.08401	19.88798	21
22	1.11597	0.89608	0.04311	0.04811	23.19443	20.78406	22
23	1.12155	0.89162	0.04113	0.04613	24.31040	21.67568	23
24	1.12716	0.88719	0.03932	0.04432	25.43196	22.56287	24
25	1.13280	0.88277	0.03765	0.04265	26.55912	23.44564	25
26	1.13846	0.87838	0.03611	0.04111	27.69191	24.32402	26
27	1.14415	0.87401	0.03469	0.03969	28.83037	25.19803	27
28	1.14987	0.86966	0.03336	0.03836	29.97452	26.06769	28
29	1.15562	0.86533	0.03213	0.03713	31.12439	26.93302	29
30	1.16140	0.86103	0.03098	0.03598	32.28002	27.79405	30
31	1.16721	0.85675	0.02990	0.03490	33.44142	28.65080	31
32	1.17304	0.85248	0.02889	0.03389	34.60862	29.50328	32
33	1.17891	0.84824	0.02795	0.03295	35.78167	30.35153	33
34	1.18480	0.84402	0.02706	0.03206	36.96058	31.19555	34
35	1.19073	0.83982	0.02622	0.03122	38.14538	32.03537	35
36	1.19668	0.83564	0.02542	0.03042	39.33610	32.87102	36
37	1.20266	0.83149	0.02467	0.02967	40.53279	33.70250	37
38	1.20868	0.82735	0.02396	0.02896	41.73545	34.52985	38
39	1.21472	0.82323	0.02329	0.02829	42.94413	35.35309	39
40	1.22079	0.81914	0.02265	0.02765	44.15885	36.17223	40
41	1.22690	0.81506	0.02204	0.02704	45.37964	36.98729	41
42	1.23303	0.81101	0.02146	0.02646	46.60654	37.79830	42
43	1.23920	0.80697	0.02090	0.02590	47.83957	38.60527	43
44	1.24539	0.80296	0.02038	0.02538	49.07877	39.40823	44
45	1.25162	0.79896	0.01987	0.02487	50.32416	40.20720	45
46	1.25788	0.79499	0.01939	0.02439	51.57578	41.00219	46
47	1.26417	0.79103	0.01893	0.02393	52.83366	41.79322	47
48	1.27050	0.78710	0.01849	0.02349	54.09783	42.58032	48
49	1.27684	0.78318	0.01806	0.02306	55.36832	43.36350	49
50	1.28323	0.77929	0.01765	0.02265	56.64516	44.14279	50

0.75%

n	F/P (F/P,i%,n)	P/F (P/F,i%,n)	A/F (A/F,i%,n)	A/P (A/P,i%,n)	F/A (F/A,i%,n)	P/A (P/A,i%,n)	n
1	1.00750	0.99256	1.00000	1.00750	1.00000	0.99256	1
2	1.01506	0.98517	0.49813	0.50563	2.00750	1.97772	2
3	1.02267	0.97783	0.33085	0.33835	3.02256	2.95556	3
4	1.03034	0.97055	0.24721	0.25471	4.04523	3.92611	4
5	1.03807	0.96333	0.19702	0.20452	5.07556	4.88944	5
6	1.04585	0.95616	0.16357	0.17107	6.11363	5.84560	6
7	1.05370	0.94904	0.13967	0.14717	7.15948	6.79464	7
8	1.06160	0.94198	0.12176	0.12926	8.21318	7.73661	8
9	1.06956	0.93496	0.10782	0.11532	9.27478	8.67158	9
10	1.07758	0.92800	0.09667	0.10417	10.34434	9.59958	10
11	1.08566	0.92109	0.08755	0.09505	11.42192	10.52067	11
12	1.09381	0.91424	0.07995	0.08745	12.50759	11.43491	12
13	1.10201	0.90743	0.07352	0.08102	13.60139	12.34235	13
14	1.11028	0.90068	0.06801	0.07551	14.70340	13.24302	14
15	1.11860	0.89397	0.06324	0.07074	15.81368	14.13699	15
16	1.12699	0.88732	0.05906	0.06656	16.93228	15.02431	16
17	1.13544	0.88071	0.05537	0.06287	18.05927	15.90502	17
18	1.14396	0.87416	0.05210	0.05960	19.19472	16.77918	18
19	1.15254	0.86765	0.04917	0.05667	20.33868	17.64683	19
20	1.16118	0.86119	0.04653	0.05403	21.49122	18.50802	20
21	1.16989	0.85478	0.04415	0.05165	22.65240	19.36280	21
22	1.17867	0.84842	0.04198	0.04948	23.82230	20.21121	22
23	1.18751	0.84210	0.04000	0.04750	25.00096	21.05331	23
24	1.19641	0.83583	0.03818	0.04568	26.18847	21.88915	24
25	1.20539	0.82961	0.03652	0.04402	27.38488	22.71876	25
26	1.21443	0.82343	0.03498	0.04248	28.59027	23.54219	26
27	1.22354	0.81730	0.03355	0.04105	29.80470	24.35949	27
28	1.23271	0.81122	0.03223	0.03973	31.02823	25.17071	28
29	1.24196	0.80518	0.03100	0.03850	32.26094	25.97589	29
30	1.25127	0.79919	0.02985	0.03735	33.50290	26.77508	30
31	1.26066	0.79324	0.02877	0.03627	34.75417	27.56832	31
32	1.27011	0.78733	0.02777	0.03527	36.01483	28.35565	32
33	1.27964	0.78147	0.02682	0.03432	37.28494	29.13712	33
34	1.28923	0.77565	0.02593	0.03343	38.56458	29.91278	34
35	1.29890	0.76988	0.02509	0.03259	39.85381	30.68266	35
36	1.30865	0.76415	0.02430	0.03180	41.15272	31.44681	36
37	1.31846	0.75846	0.02355	0.03105	42.46136	32.20527	37
38	1.32835	0.75281	0.02284	0.03034	43.77982	32.95808	38
39	1.33831	0.74721	0.02217	0.02967	45.10817	33.70529	39
40	1.34835	0.74165	0.02153	0.02903	46.44648	34.44694	40
41	1.35846	0.73613	0.02092	0.02842	47.79483	35.18307	41
42	1.36865	0.73065	0.02034	0.02784	49.15329	35.91371	42
43	1.37891	0.72521	0.01979	0.02729	50.52194	36.63892	43
44	1.38926	0.71981	0.01927	0.02677	51.90086	37.35873	44
45	1.39968	0.71445	0.01877	0.02627	53.29011	38.07318	45
46	1.41017	0.70913	0.01828	0.02578	54.68979	38.78231	46
47	1.42075	0.70385	0.01783	0.02533	56.09996	39.48617	47
48	1.43141	0.69861	0.01739	0.02489	57.52071	40.18478	48
49	1.44214	0.69341	0.01696	0.02446	58.95212	40.87820	49
50	1.45296	0.68825	0.01656	0.02406	60.39426	41.56645	50

1.0%

n	F/P (F/P,i%,n)	P/F (P/F,i%,n)	A/F (A/F,i%,n)	A/P (A/P,i%,n)	F/A (F/A,i%,n)	P/A (P/A,i%,n)	n
1	1.01000	0.99010	1.00000	1.01000	1.00000	0.99010	1
2	1.02010	0.98030	0.49751	0.50751	2.01000	1.97040	2
3	1.03030	0.97059	0.33002	0.34002	3.03010	2.94099	3
4	1.04060	0.96098	0.24628	0.25628	4.06040	3.90197	4
5	1.05101	0.95147	0.19604	0.20604	5.10101	4.85343	5
6	1.06152	0.94205	0.16255	0.17255	6.15202	5.79548	6
7	1.07214	0.93272	0.13863	0.14863	7.21354	6.72819	7
8	1.08286	0.92348	0.12069	0.13069	8.28567	7.65168	8
9	1.09369	0.91434	0.10674	0.11674	9.36853	8.56602	9
10	1.10462	0.90529	0.09558	0.10558	10.46221	9.47130	10
11	1.11567	0.89632	0.08645	0.09645	11.56683	10.36763	11
12	1.12683	0.88745	0.07885	0.08885	12.68250	11.25508	12
13	1.13809	0.87866	0.07241	0.08241	13.80933	12.13374	13
14	1.14947	0.86996	0.06690	0.07690	14.94742	13.00370	14
15	1.16097	0.86135	0.06212	0.07212	16.09690	13.86505	15
16	1.17258	0.85282	0.05794	0.06794	17.25786	14.71787	16
17	1.18430	0.84438	0.05426	0.06426	18.43044	15.56225	17
18	1.19615	0.83602	0.05098	0.06098	19.61475	16.39827	18
19	1.20811	0.82774	0.04805	0.05805	20.81090	17.22601	19
20	1.22019	0.81954	0.04542	0.05542	22.01900	18.04555	20
21	1.23239	0.81143	0.04303	0.05303	23.23919	18.85698	21
22	1.24472	0.80340	0.04086	0.05086	24.47159	19.66038	22
23	1.25716	0.79544	0.03889	0.04889	25.71630	20.45582	23
24	1.26973	0.78757	0.03707	0.04707	26.97346	21.24339	24
25	1.28243	0.77977	0.03541	0.04541	28.24320	22.02316	25
26	1.29526	0.77205	0.03387	0.04387	29.52563	22.79520	26
27	1.30821	0.76440	0.03245	0.04245	30.82089	23.55961	27
28	1.32129	0.75684	0.03112	0.04112	32.12910	24.31644	28
29	1.33450	0.74934	0.02990	0.03990	33.45039	25.06579	29
30	1.34785	0.74192	0.02875	0.03875	34.78489	25.80771	30
31	1.36133	0.73458	0.02768	0.03768	36.13274	26.54229	31
32	1.37494	0.72730	0.02667	0.03667	37.49407	27.26959	32
33	1.38869	0.72010	0.02573	0.03573	38.86901	27.98969	33
34	1.40258	0.71297	0.02484	0.03484	40.25770	28.70267	34
35	1.41660	0.70591	0.02400	0.03400	41.66028	29.40858	35
36	1.43077	0.69892	0.02321	0.03321	43.07688	30.10751	36
37	1.44508	0.69200	0.02247	0.03247	44.50765	30.79951	37
38	1.45953	0.68515	0.02176	0.03176	45.95272	31.48466	38
39	1.47412	0.67837	0.02109	0.03109	47.41225	32.16303	39
40	1.48886	0.67165	0.02046	0.03046	48.88637	32.83469	40
41	1.50375	0.66500	0.01985	0.02985	50.37524	33.49969	41
42	1.51879	0.65842	0.01928	0.02928	51.87899	34.15811	42
43	1.53398	0.65190	0.01873	0.02873	53.39778	34.81001	43
44	1.54932	0.64545	0.01820	0.02820	54.93176	35.45545	44
45	1.56481	0.63905	0.01771	0.02771	56.48107	36.09451	45
46	1.58046	0.63273	0.01723	0.02723	58.04589	36.72724	46
47	1.59626	0.62646	0.01677	0.02677	59.62634	37.35370	47
48	1.61223	0.62026	0.01633	0.02633	61.22261	37.97396	48
49	1.62835	0.61412	0.01591	0.02591	62.83483	38.58808	49
50	1.64463	0.60804	0.01551	0.02551	64.46318	39.19612	50

2.0%

n	F/P (F/P,i%,n)	P/F (P/F,i%,n)	A/F (A/F,i%,n)	A/P (A/P,i%,n)	F/A (F/A,i%,n)	P/A (P/A,i%,n)	n
1	1.02000	0.98039	1.00000	1.02000	1.00000	0.98039	1
2	1.04040	0.96117	0.49505	0.51505	2.02000	1.94156	2
3	1.06121	0.94232	0.32675	0.34675	3.06040	2.88388	3
4	1.08243	0.92385	0.24262	0.26262	4.12161	3.80773	4
5	1.10408	0.90573	0.19216	0.21216	5.20404	4.71346	5
6	1.12616	0.88797	0.15853	0.17853	6.30812	5.60143	6
7	1.14869	0.87056	0.13451	0.15451	7.43428	6.47199	7
8	1.17166	0.85349	0.11651	0.13651	8.58297	7.32548	8
9	1.19509	0.83676	0.10252	0.12252	9.75463	8.16224	9
10	1.21899	0.82035	0.09133	0.11133	10.94972	8.98259	10
11	1.24337	0.80426	0.08218	0.10218	12.16872	9.78685	11
12	1.26824	0.78849	0.07456	0.09456	13.41209	10.57534	12
13	1.29361	0.77303	0.06812	0.08812	14.68033	11.34837	13
14	1.31948	0.75788	0.06260	0.08260	15.97394	12.10625	14
15	1.34587	0.74301	0.05783	0.07783	17.29342	12.84926	15
16	1.37279	0.72845	0.05365	0.07365	18.63929	13.57771	16
17	1.40024	0.71416	0.04997	0.06997	20.01207	14.29187	17
18	1.42825	0.70016	0.04670	0.06670	21.41231	14.99203	18
19	1.45681	0.68643	0.04378	0.06378	22.84056	15.67846	19
20	1.48595	0.67297	0.04116	0.06116	24.29737	16.35143	20
21	1.51567	0.65978	0.03878	0.05878	25.78332	17.01121	21
22	1.54598	0.64684	0.03663	0.05663	27.29898	17.65805	22
23	1.57690	0.63416	0.03467	0.05467	28.84496	18.29220	23
24	1.60844	0.62172	0.03287	0.05287	30.42186	18.91393	24
25	1.64061	0.60953	0.03122	0.05122	32.03030	19.52346	25
26	1.67342	0.59758	0.02970	0.04970	33.67091	20.12104	26
27	1.70689	0.58586	0.02829	0.04829	35.34432	20.70690	27
28	1.74102	0.57437	0.02699	0.04699	37.05121	21.28127	28
29	1.77584	0.56311	0.02578	0.04578	38.79223	21.84438	29
30	1.81136	0.55207	0.02465	0.04465	40.56808	22.39646	30
31	1.84759	0.54125	0.02360	0.04360	42.37944	22.93770	31
32	1.88454	0.53063	0.02261	0.04261	44.22703	23.46833	32
33	1.92223	0.52023	0.02169	0.04169	46.11157	23.98856	33
34	1.96068	0.51003	0.02082	0.04082	48.03380	24.49859	34
35	1.99989	0.50003	0.02000	0.04000	49.99448	24.99862	35
36	2.03989	0.49022	0.01923	0.03923	51.99437	25.48884	36
37	2.08069	0.48061	0.01851	0.03851	54.03425	25.96945	37
38	2.12230	0.47119	0.01782	0.03782	56.11494	26.44064	38
39	2.16474	0.46195	0.01717	0.03717	58.23724	26.90259	39
40	2.20804	0.45289	0.01656	0.03656	60.40198	27.35548	40
41	2.25220	0.44401	0.01597	0.03597	62.61002	27.79949	41
42	2.29724	0.43530	0.01542	0.03542	64.86222	28.23479	42
43	2.34319	0.42677	0.01489	0.03489	67.15947	28.66156	43
44	2.39005	0.41840	0.01439	0.03439	69.50266	29.07996	44
45	2.43785	0.41020	0.01391	0.03391	71.89271	29.49016	45
46	2.48661	0.40215	0.01345	0.03345	74.33056	29.89231	46
47	2.53634	0.39427	0.01302	0.03302	76.81718	30.28658	47
48	2.58707	0.38654	0.01260	0.03260	79.35352	30.67312	48
49	2.63881	0.37896	0.01220	0.03220	81.94059	31.05208	49
50	2.69159	0.37153	0.01182	0.03182	84.57940	31.42361	50

2.5%

n	F/P (F/P,i%,n)	P/F (P/F,i%,n)	A/F (A/F,i%,n)	A/P (A/P,i%,n)	F/A (F/A,i%,n)	P/A (P/A,i%,n)	n
1	1.02500	0.97561	1.00000	1.02500	1.00000	0.97561	1
2	1.05063	0.95181	0.49383	0.51883	2.02500	1.92742	2
3	1.07689	0.92860	0.32514	0.35014	3.07563	2.85602	3
4	1.10381	0.90595	0.24082	0.26582	4.15252	3.76197	4
5	1.13141	0.88385	0.19025	0.21525	5.25633	4.64583	5
6	1.15969	0.86230	0.15655	0.18155	6.38774	5.50813	6
7	1.18869	0.84127	0.13250	0.15750	7.54743	6.34939	7
8	1.21840	0.82075	0.11447	0.13947	8.73612	7.17014	8
9	1.24886	0.80073	0.10046	0.12546	9.95452	7.97087	9
10	1.28008	0.78120	0.08926	0.11426	11.20338	8.75206	10
11	1.31209	0.76214	0.08011	0.10511	12.48347	9.51421	11
12	1.34489	0.74356	0.07249	0.09749	13.79555	10.25776	12
13	1.37851	0.72542	0.06605	0.09105	15.14044	10.98318	13
14	1.41297	0.70773	0.06054	0.08554	16.51895	11.69091	14
15	1.44830	0.69047	0.05577	0.08077	17.93193	12.38138	15
16	1.48451	0.67362	0.05160	0.07660	19.38022	13.05500	16
17	1.52162	0.65720	0.04793	0.07293	20.86473	13.71220	17
18	1.55966	0.64117	0.04467	0.06967	22.38635	14.35336	18
19	1.59865	0.62553	0.04176	0.06676	23.94601	14.97889	19
20	1.63862	0.61027	0.03915	0.06415	25.54466	15.58916	20
21	1.67958	0.59539	0.03679	0.06179	27.18327	16.18455	21
22	1.72157	0.58086	0.03465	0.05965	28.86286	16.76541	22
23	1.76461	0.56670	0.03270	0.05770	30.58443	17.33211	23
24	1.80873	0.55288	0.03091	0.05591	32.34904	17.88499	24
25	1.85394	0.53939	0.02928	0.05428	34.15776	18.42438	25
26	1.90029	0.52623	0.02777	0.05277	36.01171	18.95061	26
27	1.94780	0.51340	0.02638	0.05138	37.91200	19.46401	27
28	1.99650	0.50088	0.02509	0.05009	39.85980	19.96489	28
29	2.04641	0.48866	0.02389	0.04889	41.85630	20.45355	29
30	2.09757	0.47674	0.02278	0.04778	43.90270	20.93029	30
31	2.15001	0.46511	0.02174	0.04674	46.00027	21.39541	31
32	2.20376	0.45377	0.02077	0.04577	48.15028	21.84918	32
33	2.25885	0.44270	0.01986	0.04486	50.35403	22.29188	33
34	2.31532	0.43191	0.01901	0.04401	52.61289	22.72379	34
35	2.37321	0.42137	0.01821	0.04321	54.92821	23.14516	35
36	2.43254	0.41109	0.01745	0.04245	57.30141	23.55625	36
37	2.49335	0.40107	0.01674	0.04174	59.73395	23.95732	37
38	2.55568	0.39128	0.01607	0.04107	62.22730	24.34860	38
39	2.61957	0.38174	0.01544	0.04044	64.78298	24.73034	39
40	2.68506	0.37243	0.01484	0.03984	67.40255	25.10278	40
41	2.75219	0.36335	0.01427	0.03927	70.08762	25.46612	41
42	2.82100	0.35448	0.01373	0.03873	72.83981	25.82061	42
43	2.89152	0.34584	0.01322	0.03822	75.66080	26.16645	43
44	2.96381	0.33740	0.01273	0.03773	78.55232	26.50385	44
45	3.03790	0.32917	0.01227	0.03727	81.51613	26.83302	45
46	3.11385	0.32115	0.01183	0.03683	84.55403	27.15417	46
47	3.19170	0.31331	0.01141	0.03641	87.66789	27.46748	47
48	3.27149	0.30567	0.01101	0.03601	90.85958	27.77315	48
49	3.35328	0.29822	0.01062	0.03562	94.13107	28.07137	49
50	3.43711	0.29094	0.01026	0.03526	97.48435	28.36231	50

Engineering Economics and Economic Design

3.0%

n	F/P (F/P,i%,n)	P/F (P/F,i%,n)	A/F (A/F,i%,n)	A/P (A/P,i%,n)	F/A (F/A,i%,n)	P/A (P/A,i%,n)	n
1	1.03000	0.97087	1.00000	1.03000	1.00000	0.97087	1
2	1.06090	0.94260	0.49261	0.52261	2.03000	1.91347	2
3	1.09273	0.91514	0.32353	0.35353	3.09090	2.82861	3
4	1.12551	0.88849	0.23903	0.26903	4.18363	3.71710	4
5	1.15927	0.86261	0.18835	0.21835	5.30914	4.57971	5
6	1.19405	0.83748	0.15460	0.18460	6.46841	5.41719	6
7	1.22987	0.81309	0.13051	0.16051	7.66246	6.23028	7
8	1.26677	0.78941	0.11246	0.14246	8.89234	7.01969	8
9	1.30477	0.76642	0.09843	0.12843	10.15911	7.78611	9
10	1.34392	0.74409	0.08723	0.11723	11.46388	8.53020	10
11	1.38423	0.72242	0.07808	0.10808	12.80780	9.25262	11
12	1.42576	0.70138	0.07046	0.10046	14.19203	9.95400	12
13	1.46853	0.68095	0.06403	0.09403	15.61779	10.63496	13
14	1.51259	0.66112	0.05853	0.08853	17.08632	11.29607	14
15	1.55797	0.64186	0.05377	0.08377	18.59891	11.93794	15
16	1.60471	0.62317	0.04961	0.07961	20.15688	12.56110	16
17	1.65285	0.60502	0.04595	0.07595	21.76159	13.16612	17
18	1.70243	0.58739	0.04271	0.07271	23.41444	13.75351	18
19	1.75351	0.57029	0.03981	0.06981	25.11687	14.32380	19
20	1.80611	0.55368	0.03722	0.06722	26.87037	14.87747	20
21	1.86029	0.53755	0.03487	0.06487	28.67649	15.41502	21
22	1.91610	0.52189	0.03275	0.06275	30.53678	15.93692	22
23	1.97359	0.50669	0.03081	0.06081	32.45288	16.44361	23
24	2.03279	0.49193	0.02905	0.05905	34.42647	16.93554	24
25	2.09378	0.47761	0.02743	0.05743	36.45926	17.41315	25
26	2.15659	0.46369	0.02594	0.05594	38.55304	17.87684	26
27	2.22129	0.45019	0.02456	0.05456	40.70963	18.32703	27
28	2.28793	0.43708	0.02329	0.05329	42.93092	18.76411	28
29	2.35657	0.42435	0.02211	0.05211	45.21885	19.18845	29
30	2.42726	0.41199	0.02102	0.05102	47.57542	19.60044	30
31	2.50008	0.39999	0.02000	0.05000	50.00268	20.00043	31
32	2.57508	0.38834	0.01905	0.04905	52.50276	20.38877	32
33	2.65234	0.37703	0.01816	0.04816	55.07784	20.76579	33
34	2.73191	0.36604	0.01732	0.04732	57.73018	21.13184	34
35	2.81386	0.35538	0.01654	0.04654	60.46208	21.48722	35
36	2.89828	0.34503	0.01580	0.04580	63.27594	21.83225	36
37	2.98523	0.33498	0.01511	0.04511	66.17422	22.16724	37
38	3.07478	0.32523	0.01446	0.04446	69.15945	22.49246	38
39	3.16703	0.31575	0.01384	0.04384	72.23423	22.80822	39
40	3.26204	0.30656	0.01326	0.04326	75.40126	23.11477	40
41	3.35990	0.29763	0.01271	0.04271	78.66330	23.41240	41
42	3.46070	0.28896	0.01219	0.04219	82.02320	23.70136	42
43	3.56452	0.28054	0.01170	0.04170	85.48389	23.98190	43
44	3.67145	0.27237	0.01123	0.04123	89.04841	24.25427	44
45	3.78160	0.26444	0.01079	0.04079	92.71986	24.51871	45
46	3.89504	0.25674	0.01036	0.04036	96.50146	24.77545	46
47	4.01190	0.24926	0.00996	0.03996	100.39650	25.02471	47
48	4.13225	0.24200	0.00958	0.03958	104.40840	25.26671	48
49	4.25622	0.23495	0.00921	0.03921	108.54065	25.50166	49
50	4.38391	0.22811	0.00887	0.03887	112.79687	25.72976	50

3.5%

n	F/P (F/P,i%,n)	P/F (P/F,i%,n)	A/F (A/F,i%,n)	A/P (A/P,i%,n)	F/A (F/A,i%,n)	P/A (P/A,i%,n)	n
1	1.03500	0.96618	1.00000	1.03500	1.00000	0.96618	1
2	1.07123	0.93351	0.49140	0.52640	2.03500	1.89969	2
3	1.10872	0.90194	0.32193	0.35693	3.10622	2.80164	3
4	1.14752	0.87144	0.23725	0.27225	4.21494	3.67308	4
5	1.18769	0.84197	0.18648	0.22148	5.36247	4.51505	5
6	1.22926	0.81350	0.15267	0.18767	6.55015	5.32855	6
7	1.27228	0.78599	0.12854	0.16354	7.77941	6.11454	7
8	1.31681	0.75941	0.11048	0.14548	9.05169	6.87396	8
9	1.36290	0.73373	0.09645	0.13145	10.36850	7.60769	9
10	1.41060	0.70892	0.08524	0.12024	11.73139	8.31661	10
11	1.45997	0.68495	0.07609	0.11109	13.14199	9.00155	11
12	1.51107	0.66178	0.06848	0.10348	14.60196	9.66333	12
13	1.56396	0.63940	0.06206	0.09706	16.11303	10.30274	13
14	1.61869	0.61778	0.05657	0.09157	17.67699	10.92052	14
15	1.67535	0.59689	0.05183	0.08683	19.29568	11.51741	15
16	1.73399	0.57671	0.04768	0.08268	20.97103	12.09412	16
17	1.79468	0.55720	0.04404	0.07904	22.70502	12.65132	17
18	1.85749	0.53836	0.04082	0.07582	24.49969	13.18968	18
19	1.92250	0.52016	0.03794	0.07294	26.35718	13.70984	19
20	1.98979	0.50257	0.03536	0.07036	28.27968	14.21240	20
21	2.05943	0.48557	0.03304	0.06804	30.26947	14.69797	21
22	2.13151	0.46915	0.03093	0.06593	32.32890	15.16712	22
23	2.20611	0.45329	0.02902	0.06402	34.46041	15.62041	23
24	2.28333	0.43796	0.02727	0.06227	36.66653	16.05837	24
25	2.36324	0.42315	0.02567	0.06067	38.94986	16.48151	25
26	2.44596	0.40884	0.02421	0.05921	41.31310	16.89035	26
27	2.53157	0.39501	0.02285	0.05785	43.75906	17.28536	27
28	2.62017	0.38165	0.02160	0.05660	46.29063	17.66702	28
29	2.71188	0.36875	0.02045	0.05545	48.91080	18.03577	29
30	2.80679	0.35628	0.01937	0.05437	51.62268	18.39205	30
31	2.90503	0.34423	0.01837	0.05337	54.42947	18.73628	31
32	3.00671	0.33259	0.01744	0.05244	57.33450	19.06887	32
33	3.11194	0.32134	0.01657	0.05157	60.34121	19.39021	33
34	3.22086	0.31048	0.01576	0.05076	63.45315	19.70068	34
35	3.33359	0.29998	0.01500	0.05000	66.67401	20.00066	35
36	3.45027	0.28983	0.01428	0.04928	70.00760	20.29049	36
37	3.57103	0.28003	0.01361	0.04861	73.45787	20.57053	37
38	3.69601	0.27056	0.01298	0.04798	77.02889	20.84109	38
39	3.82537	0.26141	0.01239	0.04739	80.72491	21.10250	39
40	3.95926	0.25257	0.01183	0.04683	84.55028	21.35507	40
41	4.09783	0.24403	0.01130	0.04630	88.50954	21.59910	41
42	4.24126	0.23578	0.01080	0.04580	92.60737	21.83488	42
43	4.38970	0.22781	0.01033	0.04533	96.84863	22.06269	43
44	4.54334	0.22010	0.00988	0.04488	101.238331	22.28279	44
45	4.70236	0.21266	0.00945	0.04445	105.781673	22.49545	45
46	4.86694	0.20547	0.00905	0.04405	110.484031	22.70092	46
47	5.03728	0.19852	0.00867	0.04367	115.350973	22.89944	47
48	5.21359	0.19181	0.00831	0.04331	120.388257	23.09124	48
49	5.39606	0.18532	0.00796	0.04296	125.601846	23.27656	49
50	5.58493	0.17905	0.00763	0.04263	130.997910	23.45562	50

4.0%

n	F/P (F/P,i%,n)	P/F (P/F,i%,n)	A/F (A/F,i%,n)	A/P (A/P,i%,n)	F/A (F/A,i%,n)	P/A (P/A,i%,n)	n
1	1.04000	0.96154	1.00000	1.04000	1.00000	0.96154	1
2	1.08160	0.92456	0.49020	0.53020	2.04000	1.88609	2
3	1.12486	0.88900	0.32035	0.36035	3.12160	2.77509	3
4	1.16986	0.85480	0.23549	0.27549	4.24646	3.62990	4
5	1.21665	0.82193	0.18463	0.22463	5.41632	4.45182	5
6	1.26532	0.79031	0.15076	0.19076	6.63298	5.24214	6
7	1.31593	0.75992	0.12661	0.16661	7.89829	6.00205	7
8	1.36857	0.73069	0.10853	0.14853	9.21423	6.73274	8
9	1.42331	0.70259	0.09449	0.13449	10.58280	7.43533	9
10	1.48024	0.67556	0.08329	0.12329	12.00611	8.11090	10
11	1.53945	0.64958	0.07415	0.11415	13.48635	8.76048	11
12	1.60103	0.62460	0.06655	0.10655	15.02581	9.38507	12
13	1.66507	0.60057	0.06014	0.10014	16.62684	9.98565	13
14	1.73168	0.57748	0.05467	0.09467	18.29191	10.56312	14
15	1.80094	0.55526	0.04994	0.08994	20.02359	11.11839	15
16	1.87298	0.53391	0.04582	0.08582	21.82453	11.65230	16
17	1.94790	0.51337	0.04220	0.08220	23.69751	12.16567	17
18	2.02582	0.49363	0.03899	0.07899	25.64541	12.65930	18
19	2.10685	0.47464	0.03614	0.07614	27.67123	13.13394	19
20	2.19112	0.45639	0.03358	0.07358	29.77808	13.59033	20
21	2.27877	0.43883	0.03128	0.07128	31.96920	14.02916	21
22	2.36992	0.42196	0.02920	0.06920	34.24797	14.45112	22
23	2.46472	0.40573	0.02731	0.06731	36.61789	14.85684	23
24	2.56330	0.39012	0.02559	0.06559	39.08260	15.24696	24
25	2.66584	0.37512	0.02401	0.06401	41.64591	15.62208	25
26	2.77247	0.36069	0.02257	0.06257	44.31174	15.98277	26
27	2.88337	0.34682	0.02124	0.06124	47.08421	16.32959	27
28	2.99870	0.33348	0.02001	0.06001	49.96758	16.66306	28
29	3.11865	0.32065	0.01888	0.05888	52.96629	16.98371	29
30	3.24340	0.30832	0.01783	0.05783	56.08494	17.29203	30
31	3.37313	0.29646	0.01686	0.05686	59.32834	17.58849	31
32	3.50806	0.28506	0.01595	0.05595	62.70147	17.87355	32
33	3.64838	0.27409	0.01510	0.05510	66.20953	18.14765	33
34	3.79432	0.26355	0.01431	0.05431	69.85791	18.41120	34
35	3.94609	0.25342	0.01358	0.05358	73.65222	18.66461	35
36	4.10393	0.24367	0.01289	0.05289	77.59831	18.90828	36
37	4.26809	0.23430	0.01224	0.05224	81.70225	19.14258	37
38	4.43881	0.22529	0.01163	0.05163	85.97034	19.36786	38
39	4.61637	0.21662	0.01106	0.05106	90.40915	19.58448	39
40	4.80102	0.20829	0.01052	0.05052	95.02552	19.79277	40
41	4.99306	0.20028	0.01002	0.05002	99.82654	19.99305	41
42	5.19278	0.19257	0.00954	0.04954	104.81960	20.18563	42
43	5.40050	0.18517	0.00909	0.04909	110.01238	20.37079	43
44	5.61652	0.17805	0.00866	0.04866	115.41288	20.54884	44
45	5.84118	0.17120	0.00826	0.04826	121.02939	20.72004	45
46	6.07482	0.16461	0.00788	0.04788	126.87057	20.88465	46
47	6.31782	0.15828	0.00752	0.04752	132.94539	21.04294	47
48	6.57053	0.15219	0.00718	0.04718	139.26321	21.19513	48
49	6.83335	0.14634	0.00686	0.04686	145.83373	21.34147	49
50	7.10668	0.14071	0.00655	0.04655	152.66708	21.48218	50

4.5%

n	F/P (F/P,i%,n)	P/F (P/F,i%,n)	A/F (A/F,i%,n)	A/P (A/P,i%,n)	F/A (F/A,i%,n)	P/A (P/A,i%,n)	n
1	1.04500	0.95694	1.00000	1.04500	1.00000	0.95694	1
2	1.09203	0.91573	0.48900	0.53400	2.04500	1.87267	2
3	1.14117	0.87630	0.31877	0.36377	3.13703	2.74896	3
4	1.19252	0.83856	0.23374	0.27874	4.27819	3.58753	4
5	1.24618	0.80245	0.18279	0.22779	5.47071	4.38998	5
6	1.30226	0.76790	0.14888	0.19388	6.71689	5.15787	6
7	1.36086	0.73483	0.12470	0.16970	8.01915	5.89270	7
8	1.42210	0.70319	0.10661	0.15161	9.38001	6.59589	8
9	1.48610	0.67290	0.09257	0.13757	10.80211	7.26879	9
10	1.55297	0.64393	0.08138	0.12638	12.28821	7.91272	10
11	1.62285	0.61620	0.07225	0.11725	13.84118	8.52892	11
12	1.69588	0.58966	0.06467	0.10967	15.46403	9.11858	12
13	1.77220	0.56427	0.05828	0.10328	17.15991	9.68285	13
14	1.85194	0.53997	0.05282	0.09782	18.93211	10.22283	14
15	1.93528	0.51672	0.04811	0.09311	20.78405	10.73955	15
16	2.02237	0.49447	0.04402	0.08902	22.71934	11.23402	16
17	2.11338	0.47318	0.04042	0.08542	24.74171	11.70719	17
18	2.20848	0.45280	0.03724	0.08224	26.85508	12.15999	18
19	2.30786	0.43330	0.03441	0.07941	29.06356	12.59329	19
20	2.41171	0.41464	0.03188	0.07688	31.37142	13.00794	20
21	2.52024	0.39679	0.02960	0.07460	33.78314	13.40472	21
22	2.63365	0.37970	0.02755	0.07255	36.30338	13.78442	22
23	2.75217	0.36335	0.02568	0.07068	38.93703	14.14777	23
24	2.87601	0.34770	0.02399	0.06899	41.68920	14.49548	24
25	3.00543	0.33273	0.02244	0.06744	44.56521	14.82821	25
26	3.14068	0.31840	0.02102	0.06602	47.57064	15.14661	26
27	3.28201	0.30469	0.01972	0.06472	50.71132	15.45130	27
28	3.42970	0.29157	0.01852	0.06352	53.99333	15.74287	28
29	3.58404	0.27902	0.01741	0.06241	57.42303	16.02189	29
30	3.74532	0.26700	0.01639	0.06139	61.00707	16.28889	30
31	3.91386	0.25550	0.01544	0.06044	64.75239	16.54439	31
32	4.08998	0.24450	0.01456	0.05956	68.66625	16.78889	32
33	4.27403	0.23397	0.01374	0.05874	72.75623	17.02286	33
34	4.46636	0.22390	0.01298	0.05798	77.03026	17.24676	34
35	4.66735	0.21425	0.01227	0.05727	81.49662	17.46101	35
36	4.87738	0.20503	0.01161	0.05661	86.16397	17.66604	36
37	5.09686	0.19620	0.01098	0.05598	91.04134	17.86224	37
38	5.32622	0.18775	0.01040	0.05540	96.13820	18.04999	38
39	5.56590	0.17967	0.00986	0.05486	101.46442	18.22966	39
40	5.81636	0.17193	0.00934	0.05434	107.03032	18.40158	40
41	6.07810	0.16453	0.00886	0.05386	112.84669	18.56611	41
42	6.35162	0.15744	0.00841	0.05341	118.92479	18.72355	42
43	6.63744	0.15066	0.00798	0.05298	125.27640	18.87421	43
44	6.93612	0.14417	0.00758	0.05258	131.91384	19.01838	44
45	7.24825	0.13796	0.00720	0.05220	138.84997	19.15635	45
46	7.57442	0.13202	0.00684	0.05184	146.09821	19.28837	46
47	7.91527	0.12634	0.00651	0.05151	153.67263	19.41471	47
48	8.27146	0.12090	0.00619	0.05119	161.58790	19.53561	48
49	8.64367	0.11569	0.00589	0.05089	169.85936	19.65130	49
50	9.03264	0.11071	0.00560	0.05060	178.50303	19.76201	50

5.0%

n	F/P (F/P,i%,n)	P/F (P/F,i%,n)	A/F (A/F,i%,n)	A/P (A/P,i%,n)	F/A (F/A,i%,n)	P/A (P/A,i%,n)	n
1	1.05000	0.95238	1.00000	1.05000	1.00000	0.95238	1
2	1.10250	0.90703	0.48780	0.53780	2.05000	1.85941	2
3	1.15763	0.86384	0.31721	0.36721	3.15250	2.72325	3
4	1.21551	0.82270	0.23201	0.28201	4.31013	3.54595	4
5	1.27628	0.78353	0.18097	0.23097	5.52563	4.32948	5
6	1.34010	0.74622	0.14702	0.19702	6.80191	5.07569	6
7	1.40710	0.71068	0.12282	0.17282	8.14201	5.78637	7
8	1.47746	0.67684	0.10472	0.15472	9.54911	6.46321	8
9	1.55133	0.64461	0.09069	0.14069	11.02656	7.10782	9
10	1.62889	0.61391	0.07950	0.12950	12.57789	7.72173	10
11	1.71034	0.58468	0.07039	0.12039	14.20679	8.30641	11
12	1.79586	0.55684	0.06283	0.11283	15.91713	8.86325	12
13	1.88565	0.53032	0.05646	0.10646	17.71298	9.39357	13
14	1.97993	0.50507	0.05102	0.10102	19.59863	9.89864	14
15	2.07893	0.48102	0.04634	0.09634	21.57856	10.37966	15
16	2.18287	0.45811	0.04227	0.09227	23.65749	10.83777	16
17	2.29202	0.43630	0.03870	0.08870	25.84037	11.27407	17
18	2.40662	0.41552	0.03555	0.08555	28.13238	11.68959	18
19	2.52695	0.39573	0.03275	0.08275	30.53900	12.08532	19
20	2.65330	0.37689	0.03024	0.08024	33.06595	12.46221	20
21	2.78596	0.35894	0.02800	0.07800	35.71925	12.82115	21
22	2.92526	0.34185	0.02597	0.07597	38.50521	13.16300	22
23	3.07152	0.32557	0.02414	0.07414	41.43048	13.48857	23
24	3.22510	0.31007	0.02247	0.07247	44.50200	13.79864	24
25	3.38635	0.29530	0.02095	0.07095	47.72710	14.09394	25
26	3.55567	0.28124	0.01956	0.06956	51.11345	14.37519	26
27	3.73346	0.26785	0.01829	0.06829	54.66913	14.64303	27
28	3.92013	0.25509	0.01712	0.06712	58.40258	14.89813	28
29	4.11614	0.24295	0.01605	0.06605	62.32271	15.14107	29
30	4.32194	0.23138	0.01505	0.06505	66.43885	15.37245	30
31	4.53804	0.22036	0.01413	0.06413	70.76079	15.59281	31
32	4.76494	0.20987	0.01328	0.06328	75.29883	15.80268	32
33	5.00319	0.19987	0.01249	0.06249	80.06377	16.00255	33
34	5.25335	0.19035	0.01176	0.06176	85.06696	16.19290	34
35	5.51602	0.18129	0.01107	0.06107	90.32031	16.37419	35
36	5.79182	0.17266	0.01043	0.06043	95.83632	16.54685	36
37	6.08141	0.16444	0.00984	0.05984	101.62814	16.71129	37
38	6.38548	0.15661	0.00928	0.05928	107.70955	16.86789	38
39	6.70475	0.14915	0.00876	0.05876	114.09502	17.01704	39
40	7.03999	0.14205	0.00828	0.05828	120.79977	17.15909	40
41	7.39199	0.13528	0.00782	0.05782	127.83976	17.29437	41
42	7.76159	0.12884	0.00739	0.05739	135.23175	17.42321	42
43	8.14967	0.12270	0.00699	0.05699	142.99334	17.54591	43
44	8.55715	0.11686	0.00662	0.05662	151.14301	17.66277	44
45	8.98501	0.11130	0.00626	0.05626	159.70016	17.77407	45
46	9.43426	0.10600	0.00593	0.05593	168.68516	17.88007	46
47	9.90597	0.10095	0.00561	0.05561	178.11942	17.98102	47
48	10.40127	0.09614	0.00532	0.05532	188.02539	18.07716	48
49	10.92133	0.09156	0.00504	0.05504	198.42666	18.16872	49
50	11.46740	0.08720	0.00478	0.05478	209.34800	18.25593	50

6.0%

n	F/P (F/P,i%,n)	P/F (P/F,i%,n)	A/F (A/F,i%,n)	A/P (A/P,i%,n)	F/A (F/A,i%,n)	P/A (P/A,i%,n)	n
1	1.06000	0.94340	1.00000	1.06000	1.00000	0.94340	1
2	1.12360	0.89000	0.48544	0.54544	2.06000	1.83339	2
3	1.19102	0.83962	0.31411	0.37411	3.18360	2.67301	3
4	1.26248	0.79209	0.22859	0.28859	4.37462	3.46511	4
5	1.33823	0.74726	0.17740	0.23740	5.63709	4.21236	5
6	1.41852	0.70496	0.14336	0.20336	6.97532	4.91732	6
7	1.50363	0.66506	0.11914	0.17914	8.39384	5.58238	7
8	1.59385	0.62741	0.10104	0.16104	9.89747	6.20979	8
9	1.68948	0.59190	0.08702	0.14702	11.49132	6.80169	9
10	1.79085	0.55839	0.07587	0.13587	13.18079	7.36009	10
11	1.89830	0.52679	0.06679	0.12679	14.97164	7.88687	11
12	2.01220	0.49697	0.05928	0.11928	16.86994	8.38384	12
13	2.13293	0.46884	0.05296	0.11296	18.88214	8.85268	13
14	2.26090	0.44230	0.04758	0.10758	21.01507	9.29498	14
15	2.39656	0.41727	0.04296	0.10296	23.27597	9.71225	15
16	2.54035	0.39365	0.03895	0.09895	25.67253	10.10590	16
17	2.69277	0.37136	0.03544	0.09544	28.21288	10.47726	17
18	2.85434	0.35034	0.03236	0.09236	30.90565	10.82760	18
19	3.02560	0.33051	0.02962	0.08962	33.75999	11.15812	19
20	3.20714	0.31180	0.02718	0.08718	36.78559	11.46992	20
21	3.39956	0.29416	0.02500	0.08500	39.99273	11.76408	21
22	3.60354	0.27751	0.02305	0.08305	43.39229	12.04158	22
23	3.81975	0.26180	0.02128	0.08128	46.99583	12.30338	23
24	4.04893	0.24698	0.01968	0.07968	50.81558	12.55036	24
25	4.29187	0.23300	0.01823	0.07823	54.86451	12.78336	25
26	4.54938	0.21981	0.01690	0.07690	59.15638	13.00317	26
27	4.82235	0.20737	0.01570	0.07570	63.70577	13.21053	27
28	5.11169	0.19563	0.01459	0.07459	68.52811	13.40616	28
29	5.41839	0.18456	0.01358	0.07358	73.63980	13.59072	29
30	5.74349	0.17411	0.01265	0.07265	79.05819	13.76483	30
31	6.08810	0.16425	0.01179	0.07179	84.80168	13.92909	31
32	6.45339	0.15496	0.01100	0.07100	90.88978	14.08404	32
33	6.84059	0.14619	0.01027	0.07027	97.34316	14.23023	33
34	7.25103	0.13791	0.00960	0.06960	104.18375	14.36814	34
35	7.68609	0.13011	0.00897	0.06897	111.43478	14.49825	35
36	8.14725	0.12274	0.00839	0.06839	119.12087	14.62099	36
37	8.63609	0.11579	0.00786	0.06786	127.26812	14.73678	37
38	9.15425	0.10924	0.00736	0.06736	135.90421	14.84602	38
39	9.70351	0.10306	0.00689	0.06689	145.05846	14.94907	39
40	10.28572	0.09722	0.00646	0.06646	154.76197	15.04630	40
41	10.90286	0.09172	0.00606	0.06606	165.04768	15.13802	41
42	11.55703	0.08653	0.00568	0.06568	175.95054	15.22454	42
43	12.25045	0.08163	0.00533	0.06533	187.50758	15.30617	43
44	12.98548	0.07701	0.00501	0.06501	199.75803	15.38318	44
45	13.76461	0.07265	0.00470	0.06470	212.74351	15.45583	45
46	14.59049	0.06854	0.00441	0.06441	226.50812	15.52437	46
47	15.46592	0.06466	0.00415	0.06415	241.09861	15.58903	47
48	16.39387	0.06100	0.00390	0.06390	256.56453	15.65003	48
49	17.37750	0.05755	0.00366	0.06366	272.95840	15.70757	49
50	18.42015	0.05429	0.00344	0.06344	290.33590	15.76186	50

7.0%

n	F/P (F/P,i%,n)	P/F (P/F,i%,n)	A/F (A/F,i%,n)	A/P (A/P,i%,n)	F/A (F/A,i%,n)	P/A (P/A,i%,n)	n
1	1.07000	0.93458	1.00000	1.07000	1.00000	0.93458	1
2	1.14490	0.87344	0.48309	0.55309	2.07000	1.80802	2
3	1.22504	0.81630	0.31105	0.38105	3.21490	2.62432	3
4	1.31080	0.76290	0.22523	0.29523	4.43994	3.38721	4
5	1.40255	0.71299	0.17389	0.24389	5.75074	4.10020	5
6	1.50073	0.66634	0.13980	0.20980	7.15329	4.76654	6
7	1.60578	0.62275	0.11555	0.18555	8.65402	5.38929	7
8	1.71819	0.58201	0.09747	0.16747	10.25980	5.97130	8
9	1.83846	0.54393	0.08349	0.15349	11.97799	6.51523	9
10	1.96715	0.50835	0.07238	0.14238	13.81645	7.02358	10
11	2.10485	0.47509	0.06336	0.13336	15.78360	7.49867	11
12	2.25219	0.44401	0.05590	0.12590	17.88845	7.94269	12
13	2.40985	0.41496	0.04965	0.11965	20.14064	8.35765	13
14	2.57853	0.38782	0.04434	0.11434	22.55049	8.74547	14
15	2.75903	0.36245	0.03979	0.10979	25.12902	9.10791	15
16	2.95216	0.33873	0.03586	0.10586	27.88805	9.44665	16
17	3.15882	0.31657	0.03243	0.10243	30.84022	9.76322	17
18	3.37993	0.29586	0.02941	0.09941	33.99903	10.05909	18
19	3.61653	0.27651	0.02675	0.09675	37.37896	10.33560	19
20	3.86968	0.25842	0.02439	0.09439	40.99549	10.59401	20
21	4.14056	0.24151	0.02229	0.09229	44.86518	10.83553	21
22	4.43040	0.22571	0.02041	0.09041	49.00574	11.06124	22
23	4.74053	0.21095	0.01871	0.08871	53.43614	11.27219	23
24	5.07237	0.19715	0.01719	0.08719	58.17667	11.46933	24
25	5.42743	0.18425	0.01581	0.08581	63.24904	11.65358	25
26	5.80735	0.17220	0.01456	0.08456	68.67647	11.82578	26
27	6.21387	0.16093	0.01343	0.08343	74.48382	11.98671	27
28	6.64884	0.15040	0.01239	0.08239	80.69769	12.13711	28
29	7.11426	0.14056	0.01145	0.08145	87.34653	12.27767	29
30	7.61226	0.13137	0.01059	0.08059	94.46079	12.40904	30
31	8.14511	0.12277	0.00980	0.07980	102.07304	12.53181	31
32	8.71527	0.11474	0.00907	0.07907	110.21815	12.64656	32
33	9.32534	0.10723	0.00841	0.07841	118.93343	12.75379	33
34	9.97811	0.10022	0.00780	0.07780	128.25876	12.85401	34
35	10.67658	0.09366	0.00723	0.07723	138.23688	12.94767	35
36	11.42394	0.08754	0.00672	0.07672	148.91346	13.03521	36
37	12.22362	0.08181	0.00624	0.07624	160.33740	13.11702	37
38	13.07927	0.07646	0.00580	0.07580	172.56102	13.19347	38
39	13.99482	0.07146	0.00539	0.07539	185.64029	13.26493	39
40	14.97446	0.06678	0.00501	0.07501	199.63511	13.33171	40
41	16.02267	0.06241	0.00466	0.07466	214.60957	13.39412	41
42	17.14426	0.05833	0.00434	0.07434	230.63224	13.45245	42
43	18.34435	0.05451	0.00404	0.07404	247.77650	13.50696	43
44	19.62846	0.05095	0.00376	0.07376	266.12085	13.55791	44
45	21.00245	0.04761	0.00350	0.07350	285.74931	13.60552	45
46	22.47262	0.04450	0.00326	0.07326	306.75176	13.65002	46
47	24.04571	0.04159	0.00304	0.07304	329.22439	13.69161	47
48	25.72891	0.03887	0.00283	0.07283	353.27009	13.73047	48
49	27.52993	0.03632	0.00264	0.07264	378.99900	13.76680	49
50	29.45703	0.03395	0.00246	0.07246	406.52893	13.80075	50

8.0%

n	F/P (F/P,i%,n)	P/F (P/F,i%,n)	A/F (A/F,i%,n)	A/P (A/P,i%,n)	F/A (F/A,i%,n)	P/A (P/A,i%,n)	n
1	1.08000	0.92593	1.00000	1.08000	1.00000	0.92593	1
2	1.16640	0.85734	0.48077	0.56077	2.08000	1.78326	2
3	1.25971	0.79383	0.30803	0.38803	3.24640	2.57710	3
4	1.36049	0.73503	0.22192	0.30192	4.50611	3.31213	4
5	1.46933	0.68058	0.17046	0.25046	5.86660	3.99271	5
6	1.58687	0.63017	0.13632	0.21632	7.33593	4.62288	6
7	1.71382	0.58349	0.11207	0.19207	8.92280	5.20637	7
8	1.85093	0.54027	0.09401	0.17401	10.63663	5.74664	8
9	1.99900	0.50025	0.08008	0.16008	12.48756	6.24689	9
10	2.15892	0.46319	0.06903	0.14903	14.48656	6.71008	10
11	2.33164	0.42888	0.06008	0.14008	16.64549	7.13896	11
12	2.51817	0.39711	0.05270	0.13270	18.97713	7.53608	12
13	2.71962	0.36770	0.04652	0.12652	21.49530	7.90378	13
14	2.93719	0.34046	0.04130	0.12130	24.21492	8.24424	14
15	3.17217	0.31524	0.03683	0.11683	27.15211	8.55948	15
16	3.42594	0.29189	0.03298	0.11298	30.32428	8.85137	16
17	3.70002	0.27027	0.02963	0.10963	33.75023	9.12164	17
18	3.99602	0.25025	0.02670	0.10670	37.45024	9.37189	18
19	4.31570	0.23171	0.02413	0.10413	41.44626	9.60360	19
20	4.66096	0.21455	0.02185	0.10185	45.76196	9.81815	20
21	5.03383	0.19866	0.01983	0.09983	50.42292	10.01680	21
22	5.43654	0.18394	0.01803	0.09803	55.45676	10.20074	22
23	5.87146	0.17032	0.01642	0.09642	60.89330	10.37106	23
24	6.34118	0.15770	0.01498	0.09498	66.76476	10.52876	24
25	6.84848	0.14602	0.01368	0.09368	73.10594	10.67478	25
26	7.39635	0.13520	0.01251	0.09251	79.95442	10.80998	26
27	7.98806	0.12519	0.01145	0.09145	87.35077	10.93516	27
28	8.62711	0.11591	0.01049	0.09049	95.33883	11.05108	28
29	9.31727	0.10733	0.00962	0.08962	103.96594	11.15841	29
30	10.06266	0.09938	0.00883	0.08883	113.28321	11.25778	30
31	10.86767	0.09202	0.00811	0.08811	123.34587	11.34980	31
32	11.73708	0.08520	0.00745	0.08745	134.21354	11.43500	32
33	12.67605	0.07889	0.00685	0.08685	145.95062	11.51389	33
34	13.69013	0.07305	0.00630	0.08630	158.62667	11.58693	34
35	14.78534	0.06763	0.00580	0.08580	172.31680	11.65457	35
36	15.96817	0.06262	0.00534	0.08534	187.10215	11.71719	36
37	17.24563	0.05799	0.00492	0.08492	203.07032	11.77518	37
38	18.62528	0.05369	0.00454	0.08454	220.31595	11.82887	38
39	20.11530	0.04971	0.00419	0.08419	238.94122	11.87858	39
40	21.72452	0.04603	0.00386	0.08386	259.05652	11.92461	40
41	23.46248	0.04262	0.00356	0.08356	280.78104	11.96723	41
42	25.33948	0.03946	0.00329	0.08329	304.24352	12.00670	42
43	27.36664	0.03654	0.00303	0.08303	329.58301	12.04324	43
44	29.55597	0.03383	0.00280	0.08280	356.94965	12.07707	44
45	31.92045	0.03133	0.00259	0.08259	386.50562	12.10840	45
46	34.47409	0.02901	0.00239	0.08239	418.42607	12.13741	46
47	37.23201	0.02686	0.00221	0.08221	452.90015	12.16427	47
48	40.21057	0.02487	0.00204	0.08204	490.13216	12.18914	48
49	43.42742	0.02303	0.00189	0.08189	530.34274	12.21216	49
50	46.90161	0.02132	0.00174	0.08174	573.77016	12.23348	50

9.0%

n	F/P (F/P,i%,n)	P/F (P/F,i%,n)	A/F (A/F,i%,n)	A/P (A/P,i%,n)	F/A (F/A,i%,n)	P/A (P/A,i%,n)	n
1	1.09000	0.91743	1.00000	1.09000	1.00000	0.91743	1
2	1.18810	0.84168	0.47847	0.56847	2.09000	1.75911	2
3	1.29503	0.77218	0.30505	0.39505	3.27810	2.53129	3
4	1.41158	0.70843	0.21867	0.30867	4.57313	3.23972	4
5	1.53862	0.64993	0.16709	0.25709	5.98471	3.88965	5
6	1.67710	0.59627	0.13292	0.22292	7.52333	4.48592	6
7	1.82804	0.54703	0.10869	0.19869	9.20043	5.03295	7
8	1.99256	0.50187	0.09067	0.18067	11.02847	5.53482	8
9	2.17189	0.46043	0.07680	0.16680	13.02104	5.99525	9
10	2.36736	0.42241	0.06582	0.15582	15.19293	6.41766	10
11	2.58043	0.38753	0.05695	0.14695	17.56029	6.80519	11
12	2.81266	0.35553	0.04965	0.13965	20.14072	7.16073	12
13	3.06580	0.32618	0.04357	0.13357	22.95338	7.48690	13
14	3.34173	0.29925	0.03843	0.12843	26.01919	7.78615	14
15	3.64248	0.27454	0.03406	0.12406	29.36092	8.06069	15
16	3.97031	0.25187	0.03030	0.12030	33.00340	8.31256	16
17	4.32763	0.23107	0.02705	0.11705	36.97370	8.54363	17
18	4.71712	0.21199	0.02421	0.11421	41.30134	8.75563	18
19	5.14166	0.19449	0.02173	0.11173	46.01846	8.95011	19
20	5.60441	0.17843	0.01955	0.10955	51.16012	9.12855	20
21	6.10881	0.16370	0.01762	0.10762	56.76453	9.29224	21
22	6.65860	0.15018	0.01590	0.10590	62.87334	9.44243	22
23	7.25787	0.13778	0.01438	0.10438	69.53194	9.58021	23
24	7.91108	0.12640	0.01302	0.10302	76.78981	9.70661	24
25	8.62308	0.11597	0.01181	0.10181	84.70090	9.82258	25
26	9.39916	0.10639	0.01072	0.10072	93.32398	9.92897	26
27	10.24508	0.09761	0.00973	0.09973	102.72313	10.02658	27
28	11.16714	0.08955	0.00885	0.09885	112.96822	10.11613	28
29	12.17218	0.08215	0.00806	0.09806	124.13536	10.19828	29
30	13.26768	0.07537	0.00734	0.09734	136.30754	10.27365	30
31	14.46177	0.06915	0.00669	0.09669	149.57522	10.34280	31
32	15.76333	0.06344	0.00610	0.09610	164.03699	10.40624	32
33	17.18203	0.05820	0.00556	0.09556	179.80032	10.46444	33
34	18.72841	0.05339	0.00508	0.09508	196.98234	10.51784	34
35	20.41397	0.04899	0.00464	0.09464	215.71075	10.56682	35
36	22.25123	0.04494	0.00424	0.09424	236.12472	10.61176	36
37	24.25384	0.04123	0.00387	0.09387	258.37595	10.65299	37
38	26.43668	0.03783	0.00354	0.09354	282.62978	10.69082	38
39	28.81598	0.03470	0.00324	0.09324	309.06646	10.72552	39
40	31.40942	0.03184	0.00296	0.09296	337.88245	10.75736	40
41	34.23627	0.02921	0.00271	0.09271	369.29187	10.78657	41
42	37.31753	0.02680	0.00248	0.09248	403.52813	10.81337	42
43	40.67611	0.02458	0.00227	0.09227	440.84566	10.83795	43
44	44.33696	0.02255	0.00208	0.09208	481.52177	10.86051	44
45	48.32729	0.02069	0.00190	0.09190	525.85873	10.88120	45
46	52.67674	0.01898	0.00174	0.09174	574.18602	10.90018	46
47	57.41765	0.01742	0.00160	0.09160	626.86276	10.91760	47
48	62.58524	0.01598	0.00146	0.09146	684.28041	10.93358	48
49	68.21791	0.01466	0.00134	0.09134	746.86565	10.94823	49
50	74.35752	0.01345	0.00123	0.09123	815.08356	10.96168	50

10.0%

n	F/P (F/P,i%,n)	P/F (P/F,i%,n)	A/F (A/F,i%,n)	A/P (A/P,i%,n)	F/A (F/A,i%,n)	P/A (P/A,i%,n)	n
1	1.10000	0.90909	1.00000	1.10000	1.00000	0.90909	1
2	1.21000	0.82645	0.47619	0.57619	2.10000	1.73554	2
3	1.33100	0.75131	0.30211	0.40211	3.31000	2.48685	3
4	1.46410	0.68301	0.21547	0.31547	4.64100	3.16987	4
5	1.61051	0.62092	0.16380	0.26380	6.10510	3.79079	5
6	1.77156	0.56447	0.12961	0.22961	7.71561	4.35526	6
7	1.94872	0.51316	0.10541	0.20541	9.48717	4.86842	7
8	2.14359	0.46651	0.08744	0.18744	11.43589	5.33493	8
9	2.35795	0.42410	0.07364	0.17364	13.57948	5.75902	9
10	2.59374	0.38554	0.06275	0.16275	15.93742	6.14457	10
11	2.85312	0.35049	0.05396	0.15396	18.53117	6.49506	11
12	3.13843	0.31863	0.04676	0.14676	21.38428	6.81369	12
13	3.45227	0.28966	0.04078	0.14078	24.52271	7.10336	13
14	3.79750	0.26333	0.03575	0.13575	27.97498	7.36669	14
15	4.17725	0.23939	0.03147	0.13147	31.77248	7.60608	15
16	4.59497	0.21763	0.02782	0.12782	35.94973	7.82371	16
17	5.05447	0.19784	0.02466	0.12466	40.54470	8.02155	17
18	5.55992	0.17986	0.02193	0.12193	45.59917	8.20141	18
19	6.11591	0.16351	0.01955	0.11955	51.15909	8.36492	19
20	6.72750	0.14864	0.01746	0.11746	57.27500	8.51356	20
21	7.40025	0.13513	0.01562	0.11562	64.00250	8.64869	21
22	8.14027	0.12285	0.01401	0.11401	71.40275	8.77154	22
23	8.95430	0.11168	0.01257	0.11257	79.54302	8.88322	23
24	9.84973	0.10153	0.01130	0.11130	88.49733	8.98474	24
25	10.83471	0.09230	0.01017	0.11017	98.34706	9.07704	25
26	11.91818	0.08391	0.00916	0.10916	109.181765	9.16095	26
27	13.10999	0.07628	0.00826	0.10826	121.099942	9.23722	27
28	14.42099	0.06934	0.00745	0.10745	134.209936	9.30657	28
29	15.86309	0.06304	0.00673	0.10673	148.630930	9.36961	29
30	17.44940	0.05731	0.00608	0.10608	164.494023	9.42691	30
31	19.19434	0.05210	0.00550	0.10550	181.943425	9.47901	31
32	21.11378	0.04736	0.00497	0.10497	201.137767	9.52638	32
33	23.22515	0.04306	0.00450	0.10450	222.251544	9.56943	33
34	25.54767	0.03914	0.00407	0.10407	245.476699	9.60857	34
35	28.10244	0.03558	0.00369	0.10369	271.024368	9.64416	35
36	30.91268	0.03235	0.00334	0.10334	299.126805	9.67651	36
37	34.00395	0.02941	0.00303	0.10303	330.039486	9.70592	37
38	37.40434	0.02673	0.00275	0.10275	364.043434	9.73265	38
39	41.14478	0.02430	0.00249	0.10249	401.447778	9.75696	39
40	45.25926	0.02209	0.00226	0.10226	442.592556	9.77905	40
41	49.78518	0.02009	0.00205	0.10205	487.851811	9.79914	41
42	54.76370	0.01826	0.00186	0.10186	537.636992	9.81740	42
43	60.24007	0.01660	0.00169	0.10169	592.400692	9.83400	43
44	66.26408	0.01509	0.00153	0.10153	652.640761	9.84909	44
45	72.89048	0.01372	0.00139	0.10139	718.904837	9.86281	45
46	80.17953	0.01247	0.00126	0.10126	791.795321	9.87528	46
47	88.19749	0.01134	0.00115	0.10115	871.974853	9.88662	47
48	97.01723	0.01031	0.00104	0.10104	960.172338	9.89693	48
49	106.71896	0.00937	0.00095	0.10095	1057.189572	9.90630	49
50	117.39085	0.00852	0.00086	0.10086	1163.908529	9.91481	50

11.0%

n	F/P (F/P,i%,n)	P/F (P/F,i%,n)	A/F (A/F,i%,n)	A/P (A/P,i%,n)	F/A (F/A,i%,n)	P/A (P/A,i%,n)	n
1	1.11000	0.90090	1.00000	1.11000	1.00000	0.90090	1
2	1.23210	0.81162	0.47393	0.58393	2.11000	1.71252	2
3	1.36763	0.73119	0.29921	0.40921	3.34210	2.44371	3
4	1.51807	0.65873	0.21233	0.32233	4.70973	3.10245	4
5	1.68506	0.59345	0.16057	0.27057	6.22780	3.69590	5
6	1.87041	0.53464	0.12638	0.23638	7.91286	4.23054	6
7	2.07616	0.48166	0.10222	0.21222	9.78327	4.71220	7
8	2.30454	0.43393	0.08432	0.19432	11.85943	5.14612	8
9	2.55804	0.39092	0.07060	0.18060	14.16397	5.53705	9
10	2.83942	0.35218	0.05980	0.16980	16.72201	5.88923	10
11	3.15176	0.31728	0.05112	0.16112	19.56143	6.20652	11
12	3.49845	0.28584	0.04403	0.15403	22.71319	6.49236	12
13	3.88328	0.25751	0.03815	0.14815	26.21164	6.74987	13
14	4.31044	0.23199	0.03323	0.14323	30.09492	6.98187	14
15	4.78459	0.20900	0.02907	0.13907	34.40536	7.19087	15
16	5.31089	0.18829	0.02552	0.13552	39.18995	7.37916	16
17	5.89509	0.16963	0.02247	0.13247	44.50084	7.54879	17
18	6.54355	0.15282	0.01984	0.12984	50.39594	7.70162	18
19	7.26334	0.13768	0.01756	0.12756	56.93949	7.83929	19
20	8.06231	0.12403	0.01558	0.12558	64.20283	7.96333	20
21	8.94917	0.11174	0.01384	0.12384	72.26514	8.07507	21
22	9.93357	0.10067	0.01231	0.12231	81.21431	8.17574	22
23	11.02627	0.09069	0.01097	0.12097	91.14788	8.26643	23
24	12.23916	0.08170	0.00979	0.11979	102.17415	8.34814	24
25	13.58546	0.07361	0.00874	0.11874	114.41331	8.42174	25
26	15.07986	0.06631	0.00781	0.11781	127.99877	8.48806	26
27	16.73865	0.05974	0.00699	0.11699	143.07864	8.54780	27
28	18.57990	0.05382	0.00626	0.11626	159.81729	8.60162	28
29	20.62369	0.04849	0.00561	0.11561	178.39719	8.65011	29
30	22.89230	0.04368	0.00502	0.11502	199.02088	8.69379	30
31	25.41045	0.03935	0.00451	0.11451	221.91317	8.73315	31
32	28.20560	0.03545	0.00404	0.11404	247.32362	8.76860	32
33	31.30821	0.03194	0.00363	0.11363	275.52922	8.80054	33
34	34.75212	0.02878	0.00326	0.11326	306.83744	8.82932	34
35	38.57485	0.02592	0.00293	0.11293	341.58955	8.85524	35
36	42.81808	0.02335	0.00263	0.11263	380.16441	8.87859	36
37	47.52807	0.02104	0.00236	0.11236	422.98249	8.89963	37
38	52.75616	0.01896	0.00213	0.11213	470.51056	8.91859	38
39	58.55934	0.01708	0.00191	0.11191	523.26673	8.93567	39
40	65.00087	0.01538	0.00172	0.11172	581.82607	8.95105	40
41	72.15096	0.01386	0.00155	0.11155	646.82693	8.96491	41
42	80.08757	0.01249	0.00139	0.11139	718.97790	8.97740	42
43	88.89720	0.01125	0.00125	0.11125	799.06547	8.98865	43
44	98.67589	0.01013	0.00113	0.11113	887.96267	8.99878	44
45	109.53024	0.00913	0.00101	0.11101	986.63856	9.00791	45
46	121.57857	0.00823	0.00091	0.11091	1096.16880	9.01614	46
47	134.95221	0.00741	0.00082	0.11082	1217.74737	9.02355	47
48	149.79695	0.00668	0.00074	0.11074	1352.69958	9.03022	48
49	166.27462	0.00601	0.00067	0.11067	1502.49653	9.03624	49
50	184.56483	0.00542	0.00060	0.11060	1668.77115	9.04165	50

12.0%

n	F/P (F/P,i%,n)	P/F (P/F,i%,n)	A/F (A/F,i%,n)	A/P (A/P,i%,n)	F/A (F/A,i%,n)	P/A (P/A,i%,n)	n
1	1.12000	0.89286	1.00000	1.12000	1.00000	0.89286	1
2	1.25440	0.79719	0.47170	0.59170	2.12000	1.69005	2
3	1.40493	0.71178	0.29635	0.41635	3.37440	2.40183	3
4	1.57352	0.63552	0.20923	0.32923	4.77933	3.03735	4
5	1.76234	0.56743	0.15741	0.27741	6.35285	3.60478	5
6	1.97382	0.50663	0.12323	0.24323	8.11519	4.11141	6
7	2.21068	0.45235	0.09912	0.21912	10.08901	4.56376	7
8	2.47596	0.40388	0.08130	0.20130	12.29969	4.96764	8
9	2.77308	0.36061	0.06768	0.18768	14.77566	5.32825	9
10	3.10585	0.32197	0.05698	0.17698	17.54874	5.65022	10
11	3.47855	0.28748	0.04842	0.16842	20.65458	5.93770	11
12	3.89598	0.25668	0.04144	0.16144	24.13313	6.19437	12
13	4.36349	0.22917	0.03568	0.15568	28.02911	6.42355	13
14	4.88711	0.20462	0.03087	0.15087	32.39260	6.62817	14
15	5.47357	0.18270	0.02682	0.14682	37.27971	6.81086	15
16	6.13039	0.16312	0.02339	0.14339	42.75328	6.97399	16
17	6.86604	0.14564	0.02046	0.14046	48.88367	7.11963	17
18	7.68997	0.13004	0.01794	0.13794	55.74971	7.24967	18
19	8.61276	0.11611	0.01576	0.13576	63.43968	7.36578	19
20	9.64629	0.10367	0.01388	0.13388	72.05244	7.46944	20
21	10.80385	0.09256	0.01224	0.13224	81.69874	7.56200	21
22	12.10031	0.08264	0.01081	0.13081	92.50258	7.64465	22
23	13.55235	0.07379	0.00956	0.12956	104.60289	7.71843	23
24	15.17863	0.06588	0.00846	0.12846	118.15524	7.78432	24
25	17.00006	0.05882	0.00750	0.12750	133.33387	7.84314	25
26	19.04007	0.05252	0.00665	0.12665	150.33393	7.89566	26
27	21.32488	0.04689	0.00590	0.12590	169.37401	7.94255	27
28	23.88387	0.04187	0.00524	0.12524	190.69889	7.98442	28
29	26.74993	0.03738	0.00466	0.12466	214.58275	8.02181	29
30	29.95992	0.03338	0.00414	0.12414	241.33268	8.05518	30
31	33.55511	0.02980	0.00369	0.12369	271.29261	8.08499	31
32	37.58173	0.02661	0.00328	0.12328	304.84772	8.11159	32
33	42.09153	0.02376	0.00292	0.12292	342.42945	8.13535	33
34	47.14252	0.02121	0.00260	0.12260	384.52098	8.15656	34
35	52.79962	0.01894	0.00232	0.12232	431.66350	8.17550	35
36	59.13557	0.01691	0.00206	0.12206	484.46312	8.19241	36
37	66.23184	0.01510	0.00184	0.12184	543.59869	8.20751	37
38	74.17966	0.01348	0.00164	0.12164	609.83053	8.22099	38
39	83.08122	0.01204	0.00146	0.12146	684.01020	8.23303	39
40	93.05097	0.01075	0.00130	0.12130	767.09142	8.24378	40
41	104.21709	0.00960	0.00116	0.12116	860.14239	8.25337	41
42	116.72314	0.00857	0.00104	0.12104	964.35948	8.26194	42
43	130.72991	0.00765	0.00092	0.12092	1081.08262	8.26959	43
44	146.41750	0.00683	0.00083	0.12083	1211.81253	8.27642	44
45	163.98760	0.00610	0.00074	0.12074	1358.23003	8.28252	45
46	183.66612	0.00544	0.00066	0.12066	1522.21764	8.28796	46
47	205.70605	0.00486	0.00059	0.12059	1705.88375	8.29282	47
48	230.39078	0.00434	0.00052	0.12052	1911.58980	8.29716	48
49	258.03767	0.00388	0.00047	0.12047	2141.98058	8.30104	49
50	289.00219	0.00346	0.00042	0.12042	2400.01825	8.30450	50

13.0%

n	F/P (F/P,i%,n)	P/F (P/F,i%,n)	A/F (A/F,i%,n)	A/P (A/P,i%,n)	F/A (F/A,i%,n)	P/A (P/A,i%,n)	n
1	1.13000	0.88496	1.00000	1.13000	1.00000	0.88496	1
2	1.27690	0.78315	0.46948	0.59948	2.13000	1.66810	2
3	1.44290	0.69305	0.29352	0.42352	3.40690	2.36115	3
4	1.63047	0.61332	0.20619	0.33619	4.84980	2.97447	4
5	1.84244	0.54276	0.15431	0.28431	6.48027	3.51723	5
6	2.08195	0.48032	0.12015	0.25015	8.32271	3.99755	6
7	2.35261	0.42506	0.09611	0.22611	10.40466	4.42261	7
8	2.65844	0.37616	0.07839	0.20839	12.75726	4.79877	8
9	3.00404	0.33288	0.06487	0.19487	15.41571	5.13166	9
10	3.39457	0.29459	0.05429	0.18429	18.41975	5.42624	10
11	3.83586	0.26070	0.04584	0.17584	21.81432	5.68694	11
12	4.33452	0.23071	0.03899	0.16899	25.65018	5.91765	12
13	4.89801	0.20416	0.03335	0.16335	29.98470	6.12181	13
14	5.53475	0.18068	0.02867	0.15867	34.88271	6.30249	14
15	6.25427	0.15989	0.02474	0.15474	40.41746	6.46238	15
16	7.06733	0.14150	0.02143	0.15143	46.67173	6.60388	16
17	7.98608	0.12522	0.01861	0.14861	53.73906	6.72909	17
18	9.02427	0.11081	0.01620	0.14620	61.72514	6.83991	18
19	10.19742	0.09806	0.01413	0.14413	70.74941	6.93797	19
20	11.52309	0.08678	0.01235	0.14235	80.94683	7.02475	20
21	13.02109	0.07680	0.01081	0.14081	92.46992	7.10155	21
22	14.71383	0.06796	0.00948	0.13948	105.49101	7.16951	22
23	16.62663	0.06014	0.00832	0.13832	120.20484	7.22966	23
24	18.78809	0.05323	0.00731	0.13731	136.83147	7.28288	24
25	21.23054	0.04710	0.00643	0.13643	155.61956	7.32998	25
26	23.99051	0.04168	0.00565	0.13565	176.85010	7.37167	26
27	27.10928	0.03689	0.00498	0.13498	200.84061	7.40856	27
28	30.63349	0.03264	0.00439	0.13439	227.94989	7.44120	28
29	34.61584	0.02889	0.00387	0.13387	258.58338	7.47009	29
30	39.11590	0.02557	0.00341	0.13341	293.19922	7.49565	30
31	44.20096	0.02262	0.00301	0.13301	332.31511	7.51828	31
32	49.94709	0.02002	0.00266	0.13266	376.51608	7.53830	32
33	56.44021	0.01772	0.00234	0.13234	426.46317	7.55602	33
34	63.77744	0.01568	0.00207	0.13207	482.90338	7.57170	34
35	72.06851	0.01388	0.00183	0.13183	546.68082	7.58557	35
36	81.43741	0.01228	0.00162	0.13162	618.74933	7.59785	36
37	92.02428	0.01087	0.00143	0.13143	700.18674	7.60872	37
38	103.98743	0.00962	0.00126	0.13126	792.21101	7.61833	38
39	117.50580	0.00851	0.00112	0.13112	896.19845	7.62684	39
40	132.78155	0.00753	0.00099	0.13099	1013.70424	7.63438	40
41	150.04315	0.00666	0.00087	0.13087	1146.48579	7.64104	41
42	169.54876	0.00590	0.00077	0.13077	1296.52895	7.64694	42
43	191.59010	0.00522	0.00068	0.13068	1466.07771	7.65216	43
44	216.49682	0.00462	0.00060	0.13060	1657.66781	7.65678	44
45	244.64140	0.00409	0.00053	0.13053	1874.16463	7.66086	45
46	276.44478	0.00362	0.00047	0.13047	2118.80603	7.66448	46
47	312.38261	0.00320	0.00042	0.13042	2395.25082	7.66768	47
48	352.99234	0.00283	0.00037	0.13037	2707.63342	7.67052	48
49	398.88135	0.00251	0.00033	0.13033	3060.62577	7.67302	49
50	450.73593	0.00222	0.00029	0.13029	3459.50712	7.67524	50

15.0%

n	F/P (F/P,i%,n)	P/F (P/F,i%,n)	A/F (A/F,i%,n)	A/P (A/P,i%,n)	F/A (F/A,i%,n)	P/A (P/A,i%,n)	n
1	1.15000	0.86957	1.00000	1.15000	1.00000	0.86957	1
2	1.32250	0.75614	0.46512	0.61512	2.15000	1.62571	2
3	1.52088	0.65752	0.28798	0.43798	3.47250	2.28323	3
4	1.74901	0.57175	0.20027	0.35027	4.99338	2.85498	4
5	2.01136	0.49718	0.14832	0.29832	6.74238	3.35216	5
6	2.31306	0.43233	0.11424	0.26424	8.75374	3.78448	6
7	2.66002	0.37594	0.09036	0.24036	11.06680	4.16042	7
8	3.05902	0.32690	0.07285	0.22285	13.72682	4.48732	8
9	3.51788	0.28426	0.05957	0.20957	16.78584	4.77158	9
10	4.04556	0.24718	0.04925	0.19925	20.30372	5.01877	10
11	4.65239	0.21494	0.04107	0.19107	24.34928	5.23371	11
12	5.35025	0.18691	0.03448	0.18448	29.00167	5.42062	12
13	6.15279	0.16253	0.02911	0.17911	34.35192	5.58315	13
14	7.07571	0.14133	0.02469	0.17469	40.50471	5.72448	14
15	8.13706	0.12289	0.02102	0.17102	47.58041	5.84737	15
16	9.35762	0.10686	0.01795	0.16795	55.71747	5.95423	16
17	10.76126	0.09293	0.01537	0.16537	65.07509	6.04716	17
18	12.37545	0.08081	0.01319	0.16319	75.83636	6.12797	18
19	14.23177	0.07027	0.01134	0.16134	88.21181	6.19823	19
20	16.36654	0.06110	0.00976	0.15976	102.44358	6.25933	20
21	18.82152	0.05313	0.00842	0.15842	118.81012	6.31246	21
22	21.64475	0.04620	0.00727	0.15727	137.63164	6.35866	22
23	24.89146	0.04017	0.00628	0.15628	159.27638	6.39884	23
24	28.62518	0.03493	0.00543	0.15543	184.16784	6.43377	24
25	32.91895	0.03038	0.00470	0.15470	212.79302	6.46415	25
26	37.85680	0.02642	0.00407	0.15407	245.71197	6.49056	26
27	43.53531	0.02297	0.00353	0.15353	283.56877	6.51353	27
28	50.06561	0.01997	0.00306	0.15306	327.10408	6.53351	28
29	57.57545	0.01737	0.00265	0.15265	377.16969	6.55088	29
30	66.21177	0.01510	0.00230	0.15230	434.74515	6.56598	30
31	76.14354	0.01313	0.00200	0.15200	500.95692	6.57911	31
32	87.56507	0.01142	0.00173	0.15173	577.10046	6.59053	32
33	100.69983	0.00993	0.00150	0.15150	664.66552	6.60046	33
34	115.80480	0.00864	0.00131	0.15131	765.36535	6.60910	34
35	133.17552	0.00751	0.00113	0.15113	881.17016	6.61661	35
36	153.15185	0.00653	0.00099	0.15099	1014.34568	6.62314	36
37	176.12463	0.00568	0.00086	0.15086	1167.49753	6.62881	37
38	202.54332	0.00494	0.00074	0.15074	1343.62216	6.63375	38
39	232.92482	0.00429	0.00065	0.15065	1546.16549	6.63805	39
40	267.86355	0.00373	0.00056	0.15056	1779.09031	6.64178	40
41	308.04308	0.00325	0.00049	0.15049	2046.95385	6.64502	41
42	354.24954	0.00282	0.00042	0.15042	2354.99693	6.64785	42
43	407.38697	0.00245	0.00037	0.15037	2709.24647	6.65030	43
44	468.49502	0.00213	0.00032	0.15032	3116.63344	6.65244	44
45	538.76927	0.00186	0.00028	0.15028	3585.12846	6.65429	45
46	619.58466	0.00161	0.00024	0.15024	4123.89773	6.65591	46
47	712.52236	0.00140	0.00021	0.15021	4743.48239	6.65731	47
48	819.40071	0.00122	0.00018	0.15018	5456.00475	6.65853	48
49	942.31082	0.00106	0.00016	0.15016	6275.40546	6.65959	49
50	1083.6574	0.00092	0.00014	0.15014	7217.71628	6.66051	50

17.0%

n	F/P (F/P,i%,n)	P/F (P/F,i%,n)	A/F (A/F,i%,n)	A/P (A/P,i%,n)	F/A (F/A,i%,n)	P/A (P/A,i%,n)	n
1	1.17000	0.85470	1.00000	1.17000	1.00000	0.85470	1
2	1.36890	0.73051	0.46083	0.63083	2.17000	1.58521	2
3	1.60161	0.62437	0.28257	0.45257	3.53890	2.20958	3
4	1.87389	0.53365	0.19453	0.36453	5.14051	2.74324	4
5	2.19245	0.45611	0.14256	0.31256	7.01440	3.19935	5
6	2.56516	0.38984	0.10861	0.27861	9.20685	3.58918	6
7	3.00124	0.33320	0.08495	0.25495	11.77201	3.92238	7
8	3.51145	0.28478	0.06769	0.23769	14.77325	4.20716	8
9	4.10840	0.24340	0.05469	0.22469	18.28471	4.45057	9
10	4.80683	0.20804	0.04466	0.21466	22.39311	4.65860	10
11	5.62399	0.17781	0.03676	0.20676	27.19994	4.83641	11
12	6.58007	0.15197	0.03047	0.20047	32.82393	4.98839	12
13	7.69868	0.12989	0.02538	0.19538	39.40399	5.11828	13
14	9.00745	0.11102	0.02123	0.19123	47.10267	5.22930	14
15	10.53872	0.09489	0.01782	0.18782	56.11013	5.32419	15
16	12.33030	0.08110	0.01500	0.18500	66.64885	5.40529	16
17	14.42646	0.06932	0.01266	0.18266	78.97915	5.47461	17
18	16.87895	0.05925	0.01071	0.18071	93.40561	5.53385	18
19	19.74838	0.05064	0.00907	0.17907	110.28456	5.58449	19
20	23.10560	0.04328	0.00769	0.17769	130.03294	5.62777	20
21	27.03355	0.03699	0.00653	0.17653	153.13854	5.66476	21
22	31.62925	0.03162	0.00555	0.17555	180.17209	5.69637	22
23	37.00623	0.02702	0.00472	0.17472	211.80134	5.72340	23
24	43.29729	0.02310	0.00402	0.17402	248.80757	5.74649	24
25	50.65783	0.01974	0.00342	0.17342	292.10486	5.76623	25
26	59.26966	0.01687	0.00292	0.17292	342.76268	5.78311	26
27	69.34550	0.01442	0.00249	0.17249	402.03234	5.79753	27
28	81.13423	0.01233	0.00212	0.17212	471.37783	5.80985	28
29	94.92705	0.01053	0.00181	0.17181	552.51207	5.82039	29
30	111.06465	0.00900	0.00154	0.17154	647.43912	5.82939	30
31	129.94564	0.00770	0.00132	0.17132	758.50377	5.83709	31
32	152.03640	0.00658	0.00113	0.17113	888.44941	5.84366	32
33	177.88259	0.00562	0.00096	0.17096	1040.48581	5.84928	33
34	208.12263	0.00480	0.00082	0.17082	1218.36839	5.85409	34
35	243.50347	0.00411	0.00070	0.17070	1426.49102	5.85820	35
36	284.89906	0.00351	0.00060	0.17060	1669.99450	5.86171	36
37	333.33191	0.00300	0.00051	0.17051	1954.89356	5.86471	37
38	389.99833	0.00256	0.00044	0.17044	2288.22547	5.86727	38
39	456.29805	0.00219	0.00037	0.17037	2678.22379	5.86946	39
40	533.86871	0.00187	0.00032	0.17032	3134.52184	5.87133	40
41	624.62639	0.00160	0.00027	0.17027	3668.39055	5.87294	41
42	730.81288	0.00137	0.00023	0.17023	4293.01695	5.87430	42
43	855.05107	0.00117	0.00020	0.17020	5023.82983	5.87547	43
44	1000.40975	0.00100	0.00017	0.17017	5878.88090	5.87647	44
45	1170.47941	0.00085	0.00015	0.17015	6879.29065	5.87733	45
46	1369.46091	0.00073	0.00012	0.17012	8050	5.87806	46
47	1602.26927	0.00062	0.00011	0.17011	9419	5.87868	47
48	1874.65504	0.00053	0.00009	0.17009	11022	5.87922	48
49	2193.34640	0.00046	0.00008	0.17008	12896	5.87967	49
50	2566.21528	0.00039	0.00007	0.17007	15090	5.88006	50

20.0%

n	F/P (F/P,i%,n)	P/F (P/F,i%,n)	A/F (A/F,i%,n)	A/P (A/P,i%,n)	F/A (F/A,i%,n)	P/A (P/A,i%,n)	n
1	1.20000	0.83333	1.00000	1.20000	1.00000	0.83333	1
2	1.44000	0.69444	0.45455	0.65455	2.20000	1.52778	2
3	1.72800	0.57870	0.27473	0.47473	3.64000	2.10648	3
4	2.07360	0.48225	0.18629	0.38629	5.36800	2.58873	4
5	2.48832	0.40188	0.13438	0.33438	7.44160	2.99061	5
6	2.98598	0.33490	0.10071	0.30071	9.92992	3.32551	6
7	3.58318	0.27908	0.07742	0.27742	12.91590	3.60459	7
8	4.29982	0.23257	0.06061	0.26061	16.49908	3.83716	8
9	5.15978	0.19381	0.04808	0.24808	20.79890	4.03097	9
10	6.19174	0.16151	0.03852	0.23852	25.95868	4.19247	10
11	7.43008	0.13459	0.03110	0.23110	32.15042	4.32706	11
12	8.91610	0.11216	0.02526	0.22526	39.58050	4.43922	12
13	10.69932	0.09346	0.02062	0.22062	48.49660	4.53268	13
14	12.83918	0.07789	0.01689	0.21689	59.19592	4.61057	14
15	15.40702	0.06491	0.01388	0.21388	72.03511	4.67547	15
16	18.48843	0.05409	0.01144	0.21144	87.44213	4.72956	16
17	22.18611	0.04507	0.00944	0.20944	105.93056	4.77463	17
18	26.62333	0.03756	0.00781	0.20781	128.11667	4.81219	18
19	31.94800	0.03130	0.00646	0.20646	154.74000	4.84350	19
20	38.33760	0.02608	0.00536	0.20536	186.68800	4.86958	20
21	46.00512	0.02174	0.00444	0.20444	225.02560	4.89132	21
22	55.20614	0.01811	0.00369	0.20369	271.03072	4.90943	22
23	66.24737	0.01509	0.00307	0.20307	326.23686	4.92453	23
24	79.49685	0.01258	0.00255	0.20255	392.48424	4.93710	24
25	95.39622	0.01048	0.00212	0.20212	471.98108	4.94759	25
26	114.47546	0.00874	0.00176	0.20176	567.37730	4.95632	26
27	137.37055	0.00728	0.00147	0.20147	681.85276	4.96360	27
28	164.84466	0.00607	0.00122	0.20122	819.22331	4.96967	28
29	197.81359	0.00506	0.00102	0.20102	984.06797	4.97472	29
30	237.37631	0.00421	0.00085	0.20085	1181.88157	4.97894	30
31	284.85158	0.00351	0.00070	0.20070	1419.25788	4.98245	31
32	341.82189	0.00293	0.00059	0.20059	1704.10946	4.98537	32
33	410.18627	0.00244	0.00049	0.20049	2045.93135	4.98781	33
34	492.22352	0.00203	0.00041	0.20041	2456.11762	4.98984	34
35	590.66823	0.00169	0.00034	0.20034	2948.34115	4.99154	35
36	708.80187	0.00141	0.00028	0.20028	3539.00937	4.99295	36
37	850.56225	0.00118	0.00024	0.20024	4247.81125	4.99412	37
38	1020.67470	0.00098	0.00020	0.20020	5098.37350	4.99510	38
39	1224.80964	0.00082	0.00016	0.20016	6119.04820	4.99592	39
40	1469.77157	0.00068	0.00014	0.20014	7343.85784	4.99660	40
41	1764	0.00057	0.00011	0.20011	8814	4.99717	41
42	2116	0.00047	0.00009	0.20009	10577	4.99764	42
43	2540	0.00039	0.00008	0.20008	12694	4.99803	43
44	3048	0.00033	0.00007	0.20007	15234	4.99836	44
45	3657	0.00027	0.00005	0.20005	18281	4.99863	45
46	4389	0.00023	0.00005	0.20005	21939	4.99886	46
47	5266	0.00019	0.00004	0.20004	26327	4.99905	47
48	6320	0.00016	0.00003	0.20003	31594	4.99921	48
49	7584	0.00013	0.00003	0.20003	37913	4.99934	49
50	9100	0.00011	0.00002	0.20002	45497	4.99945	50

25.0%

n	F/P (F/P,i%,n)	P/F (P/F,i%,n)	A/F (A/F,i%,n)	A/P (A/P,i%,n)	F/A (F/A,i%,n)	P/A (P/A,i%,n)	n
1	1.25000	0.80000	1.00000	1.25000	1.00000	0.80000	1
2	1.56250	0.64000	0.44444	0.69444	2.25000	1.44000	2
3	1.95313	0.51200	0.26230	0.51230	3.81250	1.95200	3
4	2.44141	0.40960	0.17344	0.42344	5.76563	2.36160	4
5	3.05176	0.32768	0.12185	0.37185	8.20703	2.68928	5
6	3.81470	0.26214	0.08882	0.33882	11.25879	2.95142	6
7	4.76837	0.20972	0.06634	0.31634	15.07349	3.16114	7
8	5.96046	0.16777	0.05040	0.30040	19.84186	3.32891	8
9	7.45058	0.13422	0.03876	0.28876	25.80232	3.46313	9
10	9.31323	0.10737	0.03007	0.28007	33.25290	3.57050	10
11	11.64153	0.08590	0.02349	0.27349	42.56613	3.65640	11
12	14.55192	0.06872	0.01845	0.26845	54.20766	3.72512	12
13	18.18989	0.05498	0.01454	0.26454	68.75958	3.78010	13
14	22.73737	0.04398	0.01150	0.26150	86.94947	3.82408	14
15	28.42171	0.03518	0.00912	0.25912	109.68684	3.85926	15
16	35.52714	0.02815	0.00724	0.25724	138.10855	3.88741	16
17	44.40892	0.02252	0.00576	0.25576	173.63568	3.90993	17
18	55.51115	0.01801	0.00459	0.25459	218.04460	3.92794	18
19	69.38894	0.01441	0.00366	0.25366	273.55576	3.94235	19
20	86.73617	0.01153	0.00292	0.25292	342.94470	3.95388	20
21	108.42022	0.00922	0.00233	0.25233	429.68087	3.96311	21
22	135.52527	0.00738	0.00186	0.25186	538.10109	3.97049	22
23	169.40659	0.00590	0.00148	0.25148	673.62636	3.97639	23
24	211.75824	0.00472	0.00119	0.25119	843.03295	3.98111	24
25	264.69780	0.00378	0.00095	0.25095	1054.79118	3.98489	25
26	330.87225	0.00302	0.00076	0.25076	1319.48898	3.98791	26
27	413.59031	0.00242	0.00061	0.25061	1650.36123	3.99033	27
28	516.98788	0.00193	0.00048	0.25048	2063.95153	3.99226	28
29	646.23485	0.00155	0.00039	0.25039	2580.93941	3.99381	29
30	807.79357	0.00124	0.00031	0.25031	3227.17427	3.99505	30
31	1009.74196	0.00099	0.00025	0.25025	4034.96783	3.99604	31
32	1262.17745	0.00079	0.00020	0.25020	5044.70979	3.99683	32
33	1577.72181	0.00063	0.00016	0.25016	6306.88724	3.99746	33
34	1972.15226	0.00051	0.00013	0.25013	7884.60905	3.99797	34
35	2465.19033	0.00041	0.00010	0.25010	9856.76132	3.99838	35
36	3081	0.0003245	0.0000812	0.2500812	12322	3.99870	36
37	3852	0.0002596	0.0000649	0.2500649	15403	3.99896	37
38	4815	0.0002077	0.0000519	0.2500519	19255	3.99917	38
39	6019	0.0001662	0.0000415	0.2500415	24070	3.99934	39
40	7523	0.0001329	0.0000332	0.2500332	30089	3.99947	40
41	9404	0.0001063	0.0000266	0.2500266	37612	3.99957	41
42	11755	0.0000851	0.0000213	0.2500213	47016	3.99966	42
43	14694	0.0000681	0.0000170	0.2500170	58771	3.99973	43
44	18367	0.0000544	0.0000136	0.2500136	73464	3.99978	44
45	22959	0.0000436	0.0000109	0.2500109	91831	3.99983	45
46	28699	0.0000348	0.0000087	0.2500087	114790	3.99986	46
47	35873	0.0000279	0.0000070	0.2500070	143489	3.99989	47
48	44842	0.0000223	0.0000056	0.2500056	179362	3.99991	48
49	56052	0.0000178	0.0000045	0.2500045	224204	3.99993	49
50	70065	0.0000143	0.0000036	0.2500036	280256	3.99994	50

30.0%

n	F/P (F/P,i%,n)	P/F (P/F,i%,n)	A/F (A/F,i%,n)	A/P (A/P,i%,n)	F/A (F/A,i%,n)	P/A (P/A,i%,n)	n
1	1.30000	0.76923	1.00000	1.30000	1.00000	0.76923	1
2	1.69000	0.59172	0.43478	0.73478	2.30000	1.36095	2
3	2.19700	0.45517	0.25063	0.55063	3.99000	1.81611	3
4	2.85610	0.35013	0.16163	0.46163	6.18700	2.16624	4
5	3.71293	0.26933	0.11058	0.41058	9.04310	2.43557	5
6	4.82681	0.20718	0.07839	0.37839	12.75603	2.64275	6
7	6.27485	0.15937	0.05687	0.35687	17.58284	2.80211	7
8	8.15731	0.12259	0.04192	0.34192	23.85769	2.92470	8
9	10.60450	0.09430	0.03124	0.33124	32.01500	3.01900	9
10	13.78585	0.07254	0.02346	0.32346	42.61950	3.09154	10
11	17.92160	0.05580	0.01773	0.31773	56.40535	3.14734	11
12	23.29809	0.04292	0.01345	0.31345	74.32695	3.19026	12
13	30.28751	0.03302	0.01024	0.31024	97.62504	3.22328	13
14	39.37376	0.02540	0.00782	0.30782	127.91255	3.24867	14
15	51.18589	0.01954	0.00598	0.30598	167.28631	3.26821	15
16	66.54166	0.01503	0.00458	0.30458	218.47220	3.28324	16
17	86.50416	0.01156	0.00351	0.30351	285.01386	3.29480	17
18	112.45541	0.00889	0.00269	0.30269	371.51802	3.30369	18
19	146.19203	0.00684	0.00207	0.30207	483.97343	3.31053	19
20	190.04964	0.00526	0.00159	0.30159	630.16546	3.31579	20
21	247.06453	0.00405	0.00122	0.30122	820.21510	3.31984	21
22	321.18389	0.00311	0.00094	0.30094	1067.27963	3.32296	22
23	417.53905	0.00239	0.00072	0.30072	1388.46351	3.32535	23
24	542.80077	0.00184	0.00055	0.30055	1806.00257	3.32719	24
25	705.64100	0.00142	0.00043	0.30043	2348.80334	3.32861	25
26	917.33330	0.00109	0.00033	0.30033	3054.44434	3.32970	26
27	1192.53329	0.00084	0.00025	0.30025	3971.77764	3.33054	27
28	1550.29328	0.00065	0.00019	0.30019	5164.31093	3.33118	28
29	2015.38126	0.00050	0.00015	0.30015	6714.60421	3.33168	29
30	2619.99564	0.00038	0.00011	0.30011	8729.98548	3.33206	30
31	3406	0.000294	0.000088	0.300088	11350	3.332355	31
32	4428	0.000226	0.000068	0.300068	14756	3.332581	32
33	5756	0.000174	0.000052	0.300052	19184	3.332754	33
34	7483	0.000134	0.000040	0.300040	24940	3.332888	34
35	9728	0.000103	0.000031	0.300031	32423	3.332991	35
36	12646	0.000079	0.000024	0.300024	42151	3.333070	36
37	16440	0.000061	0.000018	0.300018	54797	3.333131	37
38	21372	0.000047	0.000014	0.300014	71237	3.333177	38
39	27784	0.000036	0.000011	0.300011	92609	3.333213	39
40	36119	0.000028	0.000008	0.300008	120393	3.333241	40
41	46955	0.000021	0.000006	0.300006	156512	3.333262	41
42	61041	0.000016	0.000005	0.300005	203466	3.333279	42
43	79353	0.000013	0.000004	0.300004	264507	3.333291	43
44	103159	0.000010	0.000003	0.300003	343860	3.333301	44
45	134107	0.000007	0.000002	0.300002	447019	3.333308	45
46	174339	0.000006	0.000002	0.300002	581126	3.333314	46
47	226641	0.000004	0.000001	0.300001	755465	3.333319	47
48	294633	0.000003	0.000001	0.300001	982106	3.333322	48
49	383022	0.000003	0.000001	0.300001	1276738	3.333325	49
50	497929	0.000002	0.000001	0.300001	1659761	3.333327	50

Engineering Economics and Economic Design

35.0%

n	F/P (F/P,i%,n)	P/F (P/F,i%,n)	A/F (A/F,i%,n)	A/P (A/P,i%,n)	F/A (F/A,i%,n)	P/A (P/A,i%,n)	n
1	1.35000	0.74074	1.00000	1.35000	1.00000	0.74074	1
2	1.82250	0.54870	0.42553	0.77553	2.35000	1.28944	2
3	2.46038	0.40644	0.23966	0.58966	4.17250	1.69588	3
4	3.32151	0.30107	0.15076	0.50076	6.63288	1.99695	4
5	4.48403	0.22301	0.10046	0.45046	9.95438	2.21996	5
6	6.05345	0.16520	0.06926	0.41926	14.43841	2.38516	6
7	8.17215	0.12237	0.04880	0.39880	20.49186	2.50752	7
8	11.03240	0.09064	0.03489	0.38489	28.66401	2.59817	8
9	14.89375	0.06714	0.02519	0.37519	39.69641	2.66531	9
10	20.10656	0.04974	0.01832	0.36832	54.59016	2.71504	10
11	27.14385	0.03684	0.01339	0.36339	74.69672	2.75188	11
12	36.64420	0.02729	0.00982	0.35982	101.84057	2.77917	12
13	49.46967	0.02021	0.00722	0.35722	138.48476	2.79939	13
14	66.78405	0.01497	0.00532	0.35532	187.95443	2.81436	14
15	90.15847	0.01109	0.00393	0.35393	254.73848	2.82545	15
16	121.71393	0.00822	0.00290	0.35290	344.89695	2.83367	16
17	164.31381	0.00609	0.00214	0.35214	466.61088	2.83975	17
18	221.82364	0.00451	0.00158	0.35158	630.92469	2.84426	18
19	299.46192	0.00334	0.00117	0.35117	852.74834	2.84760	19
20	404.27359	0.00247	0.00087	0.35087	1152.21025	2.85008	20
21	545.76935	0.00183	0.00064	0.35064	1556.48384	2.85191	21
22	736.78862	0.00136	0.00048	0.35048	2102.25319	2.85327	22
23	994.66463	0.00101	0.00035	0.35035	2839.04180	2.85427	23
24	1342.79725	0.00074	0.00026	0.35026	3833.70643	2.85502	24
25	1812.77629	0.00055	0.00019	0.35019	5176.50369	2.85557	25
26	2447.24799	0.0004086	0.0001431	0.3501431	6989	2.8559754	26
27	3303.78479	0.0003027	0.0001060	0.3501060	9437	2.8562780	27
28	4460.10947	0.0002242	0.0000785	0.3500785	12740	2.8565023	28
29	6021.14778	0.0001661	0.0000581	0.3500581	17200	2.8566683	29
30	8128.54950	0.0001230	0.0000431	0.3500431	23222	2.8567914	30
31	10974	0.0000911	0.0000319	0.3500319	31350	2.8568825	31
32	14814	0.0000675	0.0000236	0.3500236	42324	2.8569500	32
33	19999	0.0000500	0.0000175	0.3500175	57138	2.8570000	33
34	26999	0.0000370	0.0000130	0.3500130	77137	2.8570370	34
35	36449	0.0000274	0.0000096	0.3500096	104136	2.8570645	35
36	49206	0.0000203	0.0000071	0.3500071	140585	2.8570848	36
37	66428	0.0000151	0.0000053	0.3500053	189791	2.8570998	37
38	89677	0.0000112	0.0000039	0.3500039	256218	2.8571110	38
39	121065	0.0000083	0.0000029	0.3500029	345896	2.8571193	39
40	163437	0.0000061	0.0000021	0.3500021	466960	2.8571254	40
41	220640	0.0000045	0.0000016	0.3500016	630398	2.8571299	41
42	297864	0.0000034	0.0000012	0.3500012	851038	2.8571333	42
43	402117	0.0000025	0.0000009	0.3500009	1148902	2.8571358	43
44	542857	0.0000018	0.0000006	0.3500006	1551018	2.8571376	44
45	732858	0.0000014	0.0000005	0.3500005	2093876	2.8571390	45
46	989358	0.0000010	0.0000004	0.3500004	2826734	2.8571400	46
47	1335633	0.0000007	0.0000003	0.3500003	3816091	2.8571407	47
48	1803104	0.0000006	0.0000002	0.3500002	5151724	2.8571413	48
49	2434191	0.0000004	0.0000001	0.3500001	6954829	2.8571417	49
50	3286158	0.0000003	0.0000001	0.3500001	9389020	2.8571420	50

40.0%

n	F/P (F/P,i%,n)	P/F (P/F,i%,n)	A/F (A/F,i%,n)	A/P (A/P,i%,n)	F/A (F/A,i%,n)	P/A (P/A,i%,n)	n
1	1.40000	0.71429	1.00000	1.40000	1.00000	0.71429	1
2	1.96000	0.51020	0.41667	0.81667	2.40000	1.22449	2
3	2.74400	0.36443	0.22936	0.62936	4.36000	1.58892	3
4	3.84160	0.26031	0.14077	0.54077	7.10400	1.84923	4
5	5.37824	0.18593	0.09136	0.49136	10.94560	2.03516	5
6	7.52954	0.13281	0.06126	0.46126	16.32384	2.16797	6
7	10.54135	0.09486	0.04192	0.44192	23.85338	2.26284	7
8	14.75789	0.06776	0.02907	0.42907	34.39473	2.33060	8
9	20.66105	0.04840	0.02034	0.42034	49.15262	2.37900	9
10	28.92547	0.03457	0.01432	0.41432	69.81366	2.41357	10
11	40.49565	0.02469	0.01013	0.41013	98.73913	2.43826	11
12	56.69391	0.01764	0.00718	0.40718	139.23478	2.45590	12
13	79.37148	0.01260	0.00510	0.40510	195.92869	2.46850	13
14	111.12007	0.00900	0.00363	0.40363	275.30017	2.47750	14
15	155.56810	0.00643	0.00259	0.40259	386.42024	2.48393	15
16	217.79533	0.00459	0.00185	0.40185	541.98833	2.48852	16
17	304.91347	0.00328	0.00132	0.40132	759.78367	2.49180	17
18	426.87885	0.00234	0.00094	0.40094	1064.69714	2.49414	18
19	597.63040	0.00167	0.00067	0.40067	1491.57599	2.49582	19
20	836.68255	0.00120	0.00048	0.40048	2089.20639	2.49701	20
21	1171.35558	0.00085	0.00034	0.40034	2925.88894	2.49787	21
22	1639.89781	0.00061	0.00024	0.40024	4097.24452	2.49848	22
23	2295.85693	0.00044	0.00017	0.40017	5737.14232	2.49891	23
24	3214.19970	0.00031	0.00012	0.40012	8032.99925	2.49922	24
25	4499.87958	0.00022	0.00009	0.40009	11247.19895	2.49944	25
26	6300	0.00015873	0.00006350	0.40006350	15747	2.49960	26
27	8820	0.00011338	0.00004536	0.40004536	22047	2.49972	27
28	12348	0.00008099	0.00003240	0.40003240	30867	2.49980	28
29	17287	0.00005785	0.00002314	0.40002314	43214	2.49986	29
30	24201	0.00004132	0.00001653	0.40001653	60501	2.49990	30
31	33882	0.00002951	0.00001181	0.40001181	84703	2.4999262	31
32	47435	0.00002108	0.00000843	0.40000843	118585	2.4999473	32
33	66409	0.00001506	0.00000602	0.40000602	166019	2.4999624	33
34	92972	0.00001076	0.00000430	0.40000430	232428	2.4999731	34
35	130161	0.00000768	0.00000307	0.40000307	325400	2.4999808	35
36	182226	0.00000549	0.00000220	0.40000220	455561	2.4999863	36
37	255116	0.00000392	0.00000157	0.40000157	637787	2.4999902	37
38	357162	0.00000280	0.00000112	0.40000112	892903	2.4999930	38
39	500027	0.00000200	0.00000080	0.40000080	1250065	2.4999950	39
40	700038	0.00000143	0.00000057	0.40000057	1750092	2.4999964	40
41	980053	0.00000102	0.00000041	0.40000041	2450129	2.4999974	41
42	1372074	0.00000073	0.00000029	0.40000029	3430182	2.4999982	42
43	1920903	0.00000052	0.00000021	0.40000021	4802256	2.4999987	43
44	2689265	0.00000037	0.00000015	0.40000015	6723160	2.4999991	44
45	3764971	0.00000027	0.00000011	0.40000011	9412424	2.4999993	45
46	5270959	0.00000019	0.00000008	0.40000008	13177395	2.4999995	46
47	7379343	0.00000014	0.00000005	0.40000005	18448354	2.4999997	47
48	10331080	0.00000010	0.00000004	0.40000004	25827697	2.4999998	48
49	14463512	0.00000007	0.00000003	0.40000003	36158776	2.4999998	49
50	20248916	0.00000005	0.00000002	0.40000002	50622288	2.4999999	50

50.0%

n	F/P (F/P,i%,n)	P/F (P/F,i%,n)	A/F (A/F,i%,n)	A/P (A/P,i%,n)	F/A (F/A,i%,n)	P/A (P/A,i%,n)	n
1	1.50000	0.66667	1.00000	1.50000	1.00000	0.66667	1
2	2.25000	0.44444	0.40000	0.90000	2.50000	1.11111	2
3	3.37500	0.29630	0.21053	0.71053	4.75000	1.40741	3
4	5.06250	0.19753	0.12308	0.62308	8.12500	1.60494	4
5	7.59375	0.13169	0.07583	0.57583	13.18750	1.73663	5
6	11.39063	0.08779	0.04812	0.54812	20.78125	1.82442	6
7	17.08594	0.05853	0.03108	0.53108	32.17188	1.88294	7
8	25.62891	0.03902	0.02030	0.52030	49.25781	1.92196	8
9	38.44336	0.02601	0.01335	0.51335	74.88672	1.94798	9
10	57.66504	0.01734	0.00882	0.50882	113.33008	1.96532	10
11	86.49756	0.01156	0.00585	0.50585	170.99512	1.97688	11
12	129.74634	0.00771	0.00388	0.50388	257.49268	1.98459	12
13	194.61951	0.00514	0.00258	0.50258	387.23901	1.98972	13
14	291.92926	0.00343	0.00172	0.50172	581.85852	1.99315	14
15	437.89389	0.00228	0.00114	0.50114	873.78778	1.99543	15
16	656.84084	0.00152	0.00076	0.50076	1311.7	1.99696	16
17	985.26125	0.00101	0.00051	0.50051	1968.5	1.99797	17
18	1477.89188	0.00068	0.00034	0.50034	2953.8	1.99865	18
19	2216.83782	0.00045	0.00023	0.50023	4431.7	1.99910	19
20	3325.25673	0.00030	0.00015	0.50015	6648.5	1.99940	20
21	4988	0.000200486	0.000100263	0.500100263	9973.8	1.99959903	21
22	7482	0.000133657	0.000066838	0.500066838	14961.7	1.99973269	22
23	11223	0.000089105	0.000044556	0.500044556	22443.5	1.99982179	23
24	16834	0.000059403	0.000029703	0.500029703	33666.2	1.99988119	24
25	25251	0.000039602	0.000019802	0.500019802	50500.3	1.99992080	25
26	37877	0.000026401	0.000013201	0.500013201	75752	1.99994720	26
27	56815	0.000017601	0.000008801	0.500008801	113628	1.99996480	27
28	85223	0.000011734	0.000005867	0.500005867	170443	1.99997653	28
29	127834	0.000007823	0.000003911	0.500003911	255666	1.99998435	29
30	191751	0.000005215	0.000002608	0.500002608	383500	1.99998957	30
31	287627	0.000003477	0.000001738	0.500001738	575251	1.99999305	31
32	431440	0.000002318	0.000001159	0.500001159	862878	1.99999536	32
33	647160	0.000001545	0.000000773	0.500000773	1294318	1.99999691	33
34	970740	0.000001030	0.000000515	0.500000515	1941477	1.99999794	34
35	1456110	0.000000687	0.000000343	0.500000343	2912217	1.99999863	35
36	2184164	0.000000458	0.000000229	0.500000229	4368327	1.99999908	36
37	3276247	0.000000305	0.000000153	0.500000153	6552491	1.99999939	37
38	4914370	0.000000203	0.000000102	0.500000102	9828738	1.99999959	38
39	7371555	0.000000136	0.000000068	0.500000068	14743108	1.99999973	39
40	11057332	0.000000090	0.000000045	0.500000045	22114663	1.99999982	40
41	16585998	0.000000060	0.000000030	0.500000030	33171995	1.99999988	41
42	24878998	0.000000040	0.000000020	0.500000020	49757993	1.99999992	42
43	37318497	0.000000027	0.000000013	0.500000013	74636991	1.99999995	43
44	55977745	0.000000018	0.000000009	0.500000009	111955488	1.99999996	44
45	83966617	0.000000012	0.000000006	0.500000006	167933233	1.99999998	45
46	125949926	0.000000008	0.000000004	0.500000004	251899850	1.99999998	46
47	188924889	0.000000005	0.000000003	0.500000003	377849776	1.99999999	47
48	283387333	0.000000004	0.000000002	0.500000002	566774665	1.99999999	48
49	425081000	0.000000002	0.000000001	0.500000001	850161998	2.00000000	49
50	637621500	0.000000002	0.000000001	0.500000001	1275242998	2.00000000	50

Appendix IV

Equipment Pricing Data

This appendix provides up-to-date equipment purchase cost data for 15 different types of process equipment. Included in the data are purchase prices, price versus capacity equations, size exponents, factors for different materials of construction, factors for different design pressures, and factors for different types of equipment (for example: floating head versus U-tube heat exchangers). The data can be used for a variety of purposes:

- Preliminary estimating of the cost of a piece of equipment. Recall that cost data such as this is the least accurate method for estimating purchase costs. See the "Equipment Purchase Cost Estimating" section of Chapter 3 for a review of this topic.
- The creation of mathematical models for cost evaluation and option analysis.
- The adjusting of price data from other sources such as budget quotes, personal, or company data. Adjustments could be for equipment size, materials of construction, design pressure, or equipment type.

All the data in the following pages are based upon a CEPI of 460, which roughly corresponds to a date of March 2005.

The data is based upon actual purchase cost data, budget quotes, and previously published data. Each of the cost charts blends data from several sources. The sources used are:

- Actual purchase costs from several companies
- Budget quotes from Alpha-Laval, APV Baker, Chemineer, Hamilton Tanks, Jaeger Products, Manning & Lewis Engineering Co., and Mueller Industries
- Publications:
  - *Chemical Engineering, Process Design and Economics, 2nd Edition*, Ulrich, G.D. and Vasudeven, P.T., Durham, NH: Process Publishing, 2004
  - *Plant Design and Economics for Chemical Engineers, 5th Edition*, Peters, M.S., Timmerhaus, K.D., and West, R.E., New York: McGraw-Hill, 2003

- *Manual of Process Economic Evaluation, 2nd Edition*, Chauvel, A. et al., Paris: Editions TECHNIP, 2001
- *Planning, Estimating, and Control of Chemical Construction Projects, 2nd Edition*, Navarrete, P.F. and Cole, W.C., New York: Marcel Dekker, 2001
- "Shedding New Light on Titanium in CPI Construction," Grauman, J.S. and Wiley, B., *Chemical Engineering*, August 1998, 106–111
- *Perry's Chemical Engineers' Handbook, 7th Edition*, Perry, R.H. and Green, D.W. (Eds.), New York: McGraw-Hill, 1997
- "A Potpourri of Equipment Prices," Vatavuk, W.M., *Chemical Engineering*, August 1995, 68–73
- "Piping Systems: How Installation Costs Stack Up," Lindley, N.L. and Floyd, J.C., *Chemical Engineering*, January 1993, 94–100
- *Chemical Process Equipment, Selection and Design*, Walas, S.W., Boston: Butterworth-Heinemann, 1990
- *Chemical Engineering Economics*, Garrett, D.E., New York: Von Nostrand Reinhold, 1989
- "Estimating Process Equipment Costs," Hall, R.S., Vatavuk, W.M., and Matley, J., *Chemical Engineering*, November 21, 1988, 66–75

AGITATORS

Side-entering propeller: 316 SS wetted parts, Single propeller, Stuffing box, Motor
Top-entering turbine: SS wetted parts, Single blade, Stuffing box, Gear reducer, Motor. Shaft speed: 30–45 rpm

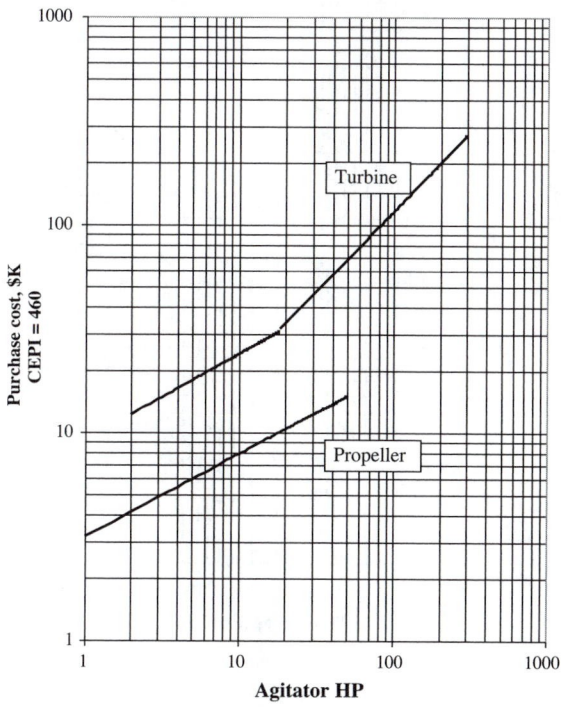

Equations		Material Factors	
Propeller:		Carbon steel	0.8
$\$K = 3.14\ HP^{0.40}$		304 Stainless steel	0.9
Turbine:		316 Stainless steel	1.0
$\leq 18\ HP,\ \$K = 9.13\ HP^{0.42}$		Hastelloy C276	1.9
$> 18\ HP,\ \$K = 3.33\ HP^{0.77}$		Monel	1.2

Size Exponents		Other Factors	
Propeller	0.40	Turbine speed, > 45 rpm	$1.08\ HP^{0.095}$
Turbine, ≤ 15 HP	0.42	Two turbine blades	1.1
Turbine, > 15 HP	0.77	Mechanical seal	1.3

BLOWERS

Cast iron with motor

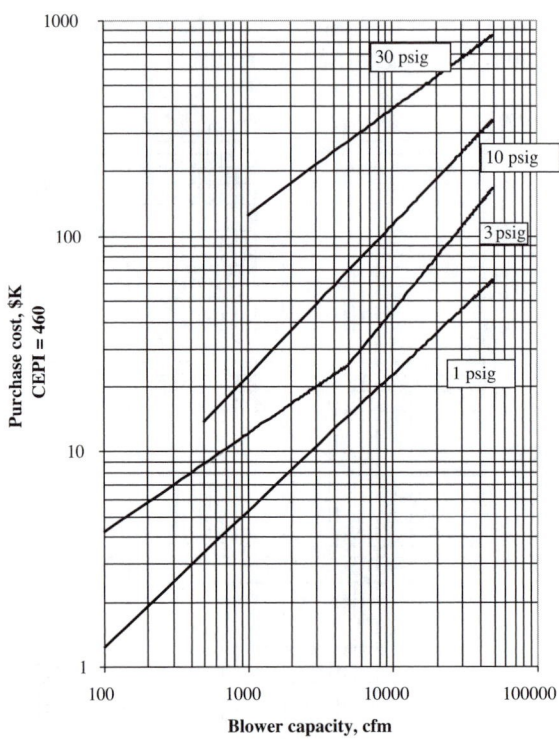

Blower capacity, cfm

Equations

1 psig	$K = 0.066 \text{ cfm}^{0.63}$
3 psig, < 5000 cfm	$K = 0.516 \text{ cfm}^{0.46}$
3 psig, > 5000 cfm	$K = 0.023 \text{ cfm}^{0.82}$
10 psig	$K = 0.174 \text{ cfm}^{0.70}$
30 psig	$K = 4.09 \text{ cfm}^{0.49}$

Size Exponents

1 psig	0.63
3 psig, < 5000 cfm	0.46
3 psig, > 5000 cfm	0.82
10 psig	0.70
30 psig	0.49

BOILERS

Oil and gas fired, No superheat
Small package boilers, 10–1000 HP, 150 psig
Large package boilers, 10K–300K lb/hr, 250 psig
Field-erected boilers, 100K–1000K lb/hr, 400 psig

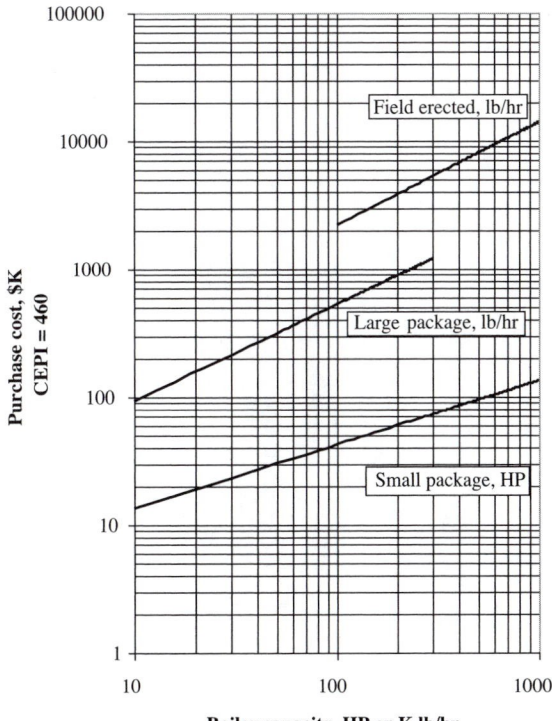

Boiler capacity, HP or K lb/hr

	Equations
Small package	$\$K = 4.24\ HP^{0.50}$
Large package	$\$K = 16.4 \left(\dfrac{lb/hr}{1000}\right)^{0.75}$
Field-erected	$\$K = 53.5 \left(\dfrac{lb/hr}{1000}\right)^{0.81}$

Size Exponents

Small package	0.50
Large package	0.75
Field-erected	0.81

Pressure Factors

Small Package

15 psig	0.85
150	1.00
200	1.05
250	1.11

Large Package

150 psig	0.93
250	1.00
350	1.05
500	1.15
600	1.25

Field-Erected

400 psig	1.00
500	1.04
600	1.08
700	1.14
1000	1.32

CENTRIFUGES

Filtering centrifuges: Solid-bowl, Screen-bowl, and Pusher,
316 Stainless steel, including motor
Sedimenting centrifuges: Disc and sedimenting-scroll,
316 Stainless steel, including motor

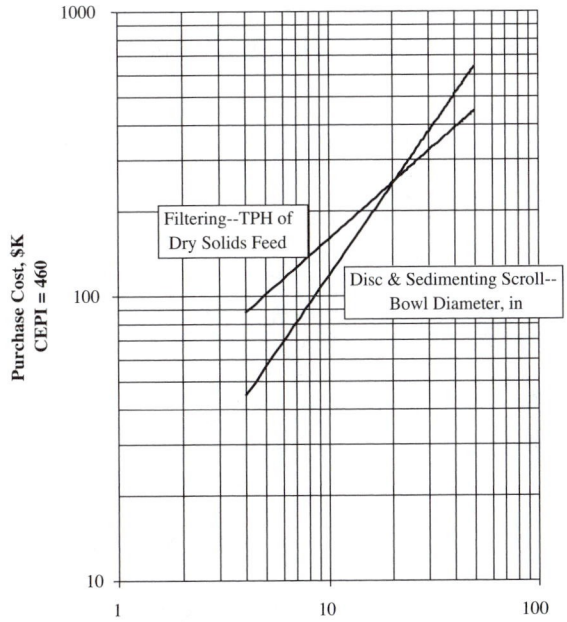

Capacity, TPH of Dry Solids Feed or Bowl Diameter, in

	Equations		**Material Factors**	
Filtering	$\$K = 35.5 \ (TPH)^{0.65}$		Carbon steel	0.7
Sedimenting	$\$K = 10.2 \ (Bowl \ Diameter, \ in)^{1.06}$		Stainless steel	1.0
			Monel	1.4
	Size Exponents		Nickel	1.7
Filtering	0.65		Hasteloy C	2.5
Sedimenting	1.06			

Figure labels: Filtering--TPH of Dry Solids Feed; Disc & Sedimenting Scroll-- Bowl Diameter, in

Y-axis: Purchase Cost, $K   CEPI = 460

COLUMNS

Distillation, Stripping, Absorption — Tray or Packed

To estimate the cost of a column, first find the price of the vessel using the Pressure Vessel page. Next, adjust that cost using the appropriate factor from the graph below. If the metallurgy is other than carbon steel or the design pressure is different from 150 psig, adjust the pressure vessel cost using factors on the pressure vessel page.

Estimate the cost of column internals, trays, and packing using the graph on the next page.

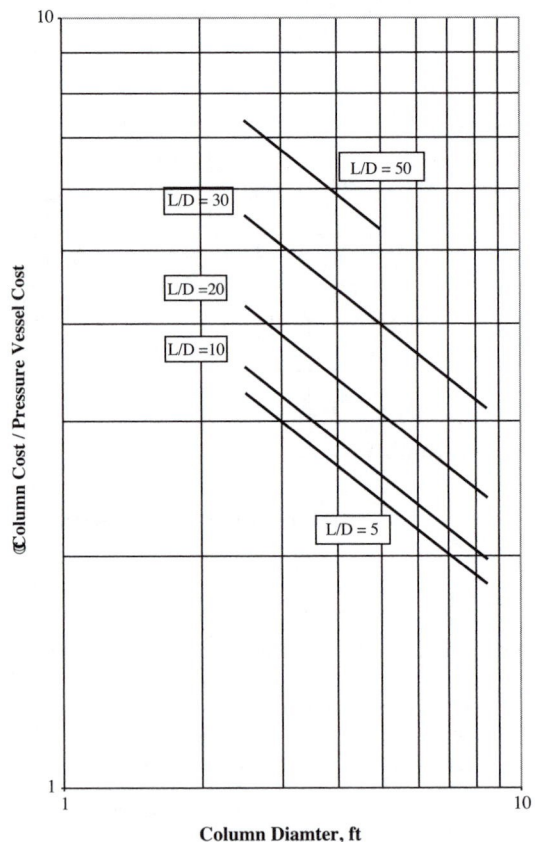

Column Diamter, ft

Equations

L/D = 5	$F = 5.02 * (\text{Diameter, ft})^{-0.47}$
L/D = 10	$F = 5.42 * (\text{Diameter, ft})^{-0.47}$
L/D = 20	$F = 6.51 * (\text{Diameter, ft})^{-0.47}$
L/D = 30	$F = 8.51 * (\text{Diameter, ft})^{-0.47}$
L/D = 50	$F = 11.3 * (\text{Diameter, ft})^{-0.47}$

COLUMN INTERNALS

Trays: Carbon steel Bubble cap, Sieve, and Valve
Column packing: Pall rings and Intalox saddles

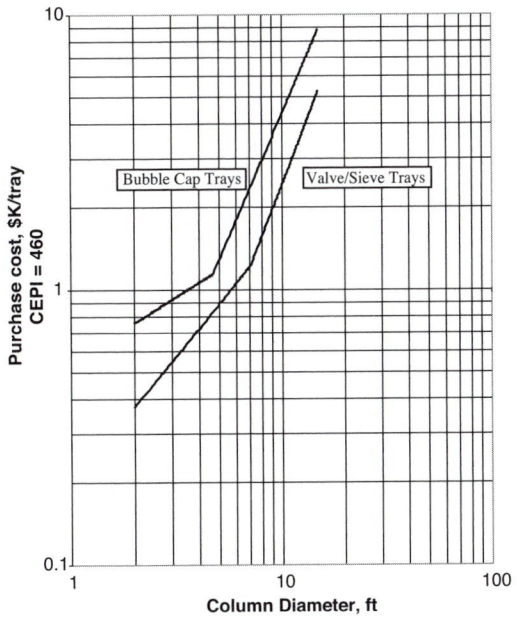

	Equations	**Tray Material Factors**	
Bubble cap ≤ 4.75 ft	$K/tray = 0.0632 * (Diameter, ft)^{0.71}$	Carbon steel	1.0
Bubble cap ≥ 4.75 ft	$K/tray = 0.0731 * (Diameter, ft)^{1.8}$	Stainless steel	1.9
Valve/Sieve ≤ 7 ft	$K/tray = 0.167 * (Diameter, ft)^{0.41}$	Inconel 600	3.6
Valve/Sieve ≥ 7 ft	$K/tray = 0.026 * (Diameter, ft)^{2}$		

Quantity Factor

Size Exponents		**Number of Trays**	**Factor**
Bubble cap ≤ 4.75 ft	0.71	1	3.0
Bubble cap ≥ 4.75 ft	1.8	5	2.0
Valve/Sieve ≤ 7 ft	0.41	10	1.5
Valve/Sieve ≥ 7 ft	2	15	1.3
		20+	1.0

Column Packing, \$/ft^3, CEPI = 460

Packing Size (in)	Pall Rings		Intalox Saddles	
	Stainless Steel	Plastic	Ceramic	Plastic
$^1/_2$	216	—	35	—
5/8	—	120	—	—
1	146	34	29	35
2	98	18	19	23
3	81	13	16	13
$3^1/_2$	—	11	—	—

COMPRESSORS

Air compressors: Reciprocating, Centrifugal, or Lubricated screw, Carbon steel, 150 psig. Price includes: Motor and auxiliaries (coolers, separators, tank, and so on)

Gas compressors: Reciprocating and Centrifugal, Carbon steel, 150 psig. Price includes: Motor and auxiliaries (coolers, separators, tank, and so on)

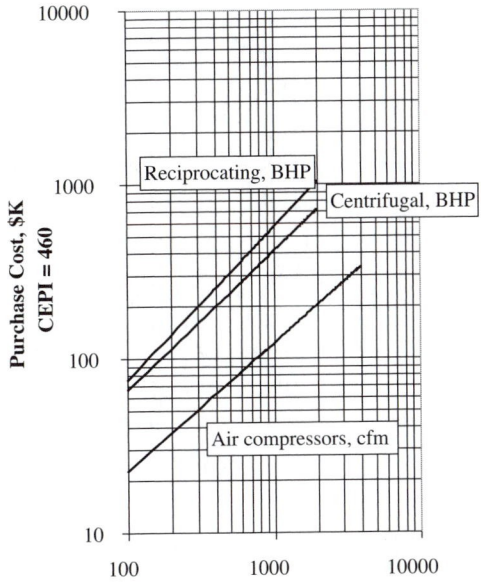

Compressor capacity, cfm or HP

Equations		Material Factors	
Air compressors	$\$K = 0.76\,(cfm)^{0.73}$	Carbon steel	1.0
Reciprocating, gas	$\$K = 1.31\,(HP)^{0.88}$	Stainless steel	2.5
Centrifugal, gas	$\$K = 1.64\,(HP)^{0.80}$		

Size Exponents		Driver Factors	
Air compressors	0.73	Electric motor	1.0
Reciprocating gas	0.88	Turbine	1.2
Centrifugal gas	0.80	Gas engine	1.4

COOLING TOWERS

10–20°F range (water temperature change)

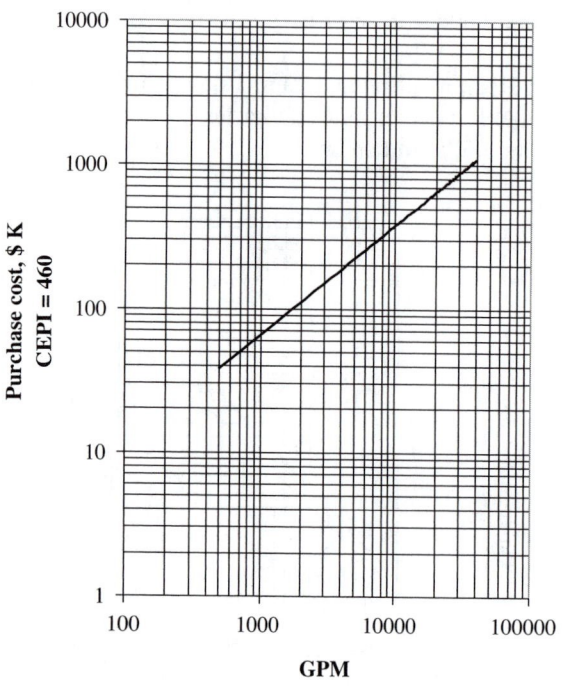

Equation	$\$K = 0.324\ (\text{gpm})^{0.77}$
Size Exponent	0.77

DUST COLLECTORS

Bag filters: Carbon steel bodyCyclones:
Carbon steel Electrostatic precipitators: Carbon steel,
Dry operation, Low-efficiency Venturi scrubbers: Carbon steel

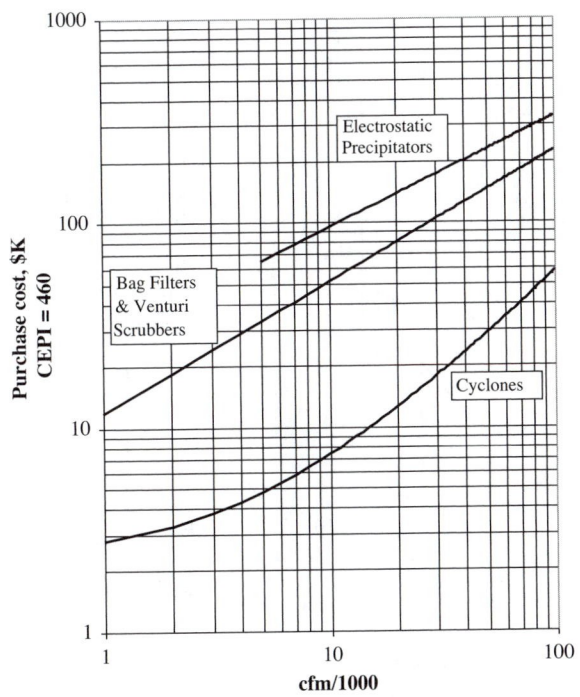

cfm/1000

Equations

Cyclones

$$\$K = 0.0006 \left(\frac{cfm}{1000}\right)^2 + 0.498 \left(\frac{cfm}{1000}\right) + 2.28$$

Bag filters and Venturi scrubbers

$$\$K = 11.6 \left(\frac{cfm}{1000}\right)^{0.64}$$

Electrostatic precipitators

$$\$K = 26.9 \left(\frac{cfm}{1000}\right)^{0.54}$$

Material Factors

Cyclones, Electrostatic Filters, and Venturi Scrubbers	
Carbon steel	1.0
Stainless steel	1.9

Size Exponents

Bag filters and Venturi scrubbers	0.64
Electrostatic precipitators	0.54

FANS

Carbon steel with Motor

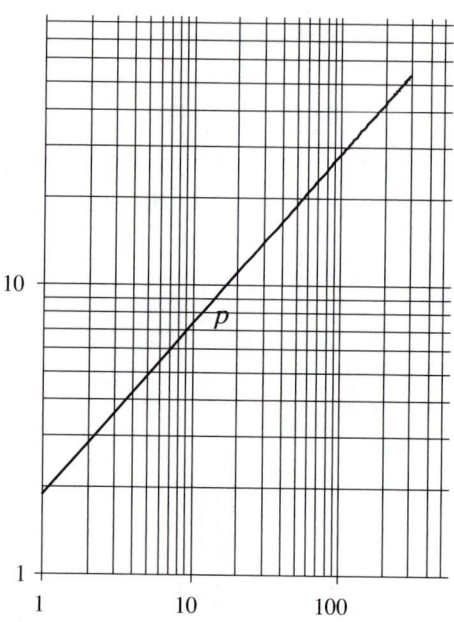

Equation $$\$K = 1.88 \left(\frac{\text{cfm}}{1000}\right)^{0.59}$$

Size Exponent   0.59

HEAT EXCHANGERS

Shell and Tube: U-tube, Carbon steel shell and tubes, 150 psig
Plate and Frame: Stainless steel plates, Nitrile (NBR) gaskets, 150 psig
Spiral: Stainless steel, 150 psig

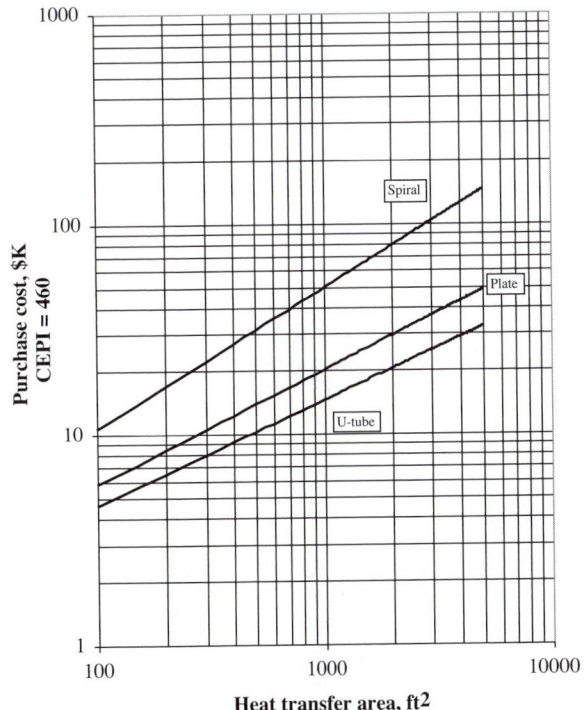

Equations

Shell and Tube	$\$K = 0.462 \ (Area, \ ft^2)^{0.5}$
Plate and Frame	$\$K = 0.475 \ (Area, \ ft^2)^{0.54}$
Spiral	$\$K = 0.494 \ (Area, \ ft^2)^{0.67}$

Size Exponents

Shell and Tube	0.5
Plate and Frame	0.54
Spiral	0.67

Shell and Tube Type Factors

U-tube	1.0
Fixed tube sheet	1.05
Floating head	1.3
Kettle reboiler	1.35
1 shell/2 shell passes	1.1

Pressure Factors

U-tube, Fixed Tube Sheet and Floating Head

200 psig	1.06
400	1.16
600	1.26
800	1.34
1000	1.44

Kettle Reboiler

200 psig	1.14
400	1.25
600	1.36
800	1.45
1000	1.56

Plate and Frame

235 psig	1.23
370	1.35

Material Factors

Shell and Tube

CS shell/CS tube	1.0
CS/SS	1.7
SS/SS	2.0
CS/Monel	3.3
Monel/Monel	4.0
CS/Titanium (Ti)	2.6
Ti/Ti	3.0
CS/Hastelloy C (HC)	7.8
HC/HC	9.0

Plate and Frame

Stainless steel	1.0
Ti	1.6
HC	3.5

Plate Gaskets

Nitrile (NBR)	1.0
EPDM	1.4
Viton	2.6

Spiral

CS	0.5
SS	1.0
Ti	2.5
HC	1.2

PIPING

Installed cost of schedule 40 carbon steel pipe

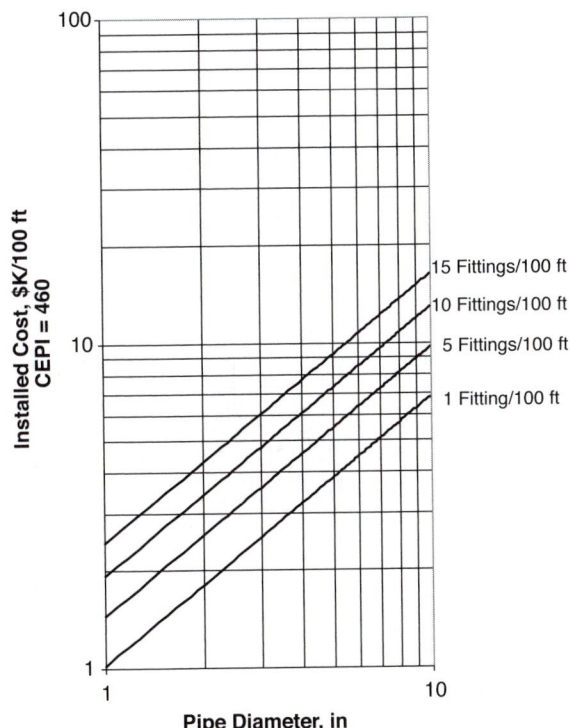

Equation	Material Factors		
$\$ = (0.1F + 0.924)D^{0.83}$	PVC (schedule 80)		0.6
where:	304L SS (schedule 10)		1.2
$\$$ = Installed cost, $K/100ft	316L, SS (schedule 10)		1.4
F = Fittings/100 ft of pipe	Monel (schedule 10)	2 in	3.0
D = Nominal pipe diameter (in)		4 in	4.5
		6 in	7.5
		8 in	9.5
		10 in	12.0
	Hastelloy (schedule 10)	2 in	4.5
		4 in	5.5
		6 in	7.5
		8 in	9.0
Size exponent: 0.83		10 in	10.5

PRESSURE VESSELS

Carbon steel, 150 psig design pressure

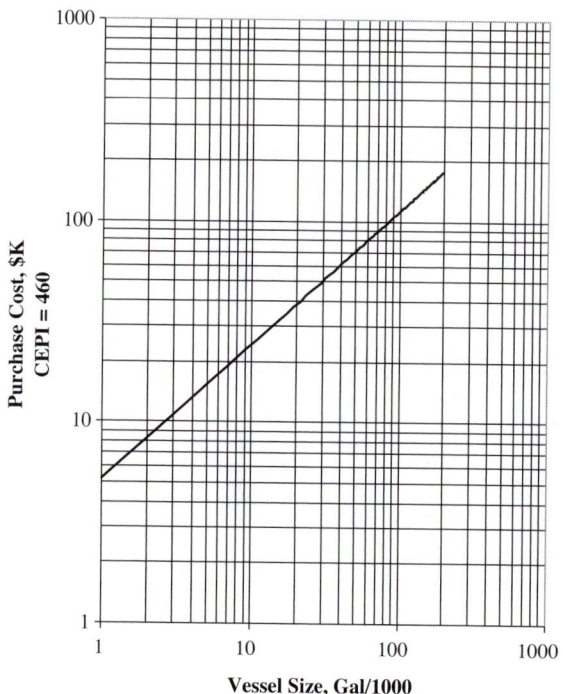

Vessel Size, Gal/1000

Equation	Material Factors	
	Carbon steel	1.0
$\$K = 5.18 \left(\dfrac{gallons}{1000}\right)^{0.67}$	Stainless steel	2.6
	Monel 400	6.7
Size Exponent: 0.67	Inconel 600	7.7
	Titanium	4.0

Pressure Factors

150 psig	1.0
Other pressures	
$F = 0.0023P + 0.66$	
Full vacuum	1.1

PUMPS

Centrifugal: ANSI and process types: With motor, cast iron, 150 psig rating
Rotary positive displacement: With motor, Cast iron, 150 psig rating
Reciprocating positive displacement: With motor, Cast iron, 150 psig rating

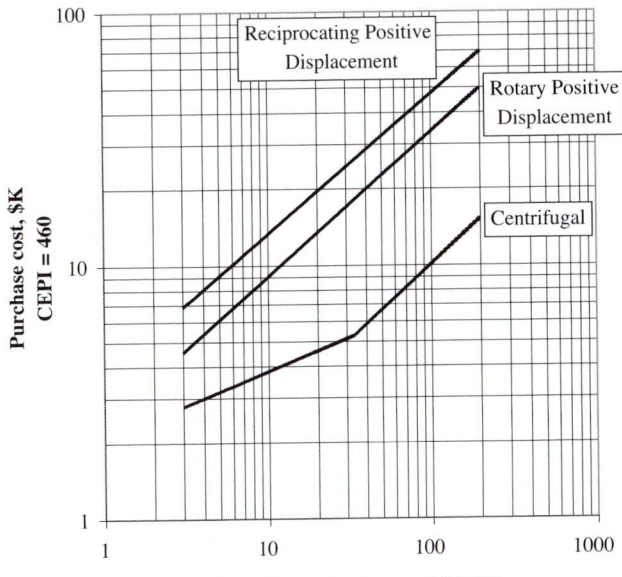

Flow * Pump Head: gpm * ft/1000

Equations

Centrifugal, (gpm * ft)/1000 ≤ 34,000

$$\$K = 2.07 \left(\frac{gpm * ft}{1000}\right)^{0.27}$$

Centrifugal, (gpm * ft)/1000 ≥ 34,000

$$\$K = 0.648 \left(\frac{gpm * ft}{1000}\right)^{0.60}$$

Rotary positive displacement

$$\$K = 2.43 \left(\frac{gpm * ft}{1000}\right)^{0.57}$$

Reciprocating positive displacement

$$\$K = 3.74 \left(\frac{gpm * ft}{1000}\right)^{0.55}$$

Size Exponents

Centrifugal, gpm * ft/100 ≤ 34,000	0.27
Centrifugal, gpm * ft/100 ≥ 34,000	0.60
Rotary positive displacement	0.57
Reciprocating positive displacement	0.55

Material Factors

Centrifugal

Cast iron	1.0
Ductile iron	1.2
Cast steel	1.4
Stainless steel	1.9
Monel	4.0
Titanium	6.0

Rotary Positive Displacement

Cast iron	1.0
Cast steel	1.4
Stainless steel	2.0

Reciprocating Positive Displacement

Cast iron	1.0
Cast steel	1.8
Stainless steel	2.4

Centrifugal Type Factors

ANSI	1.0
API	
gpm * ft	
10K	3.7
50K	2.6
100K	2.3
300K	1.6

Pump Suction Pressure Factors

Suction Pressure, psig	150	300	500	750	1000	1500
Centrifugal	1.0	1.4	1.7	2.1	2.4	—
Rotary positive displacement	1.0	—	—	2.0	—	2.5
Recip'g positive displacement	1.0	—	—	1.6	—	1.8

TANKS, STORAGE

Custom built and Off-the-shelf: Carbon steel, Atmospheric

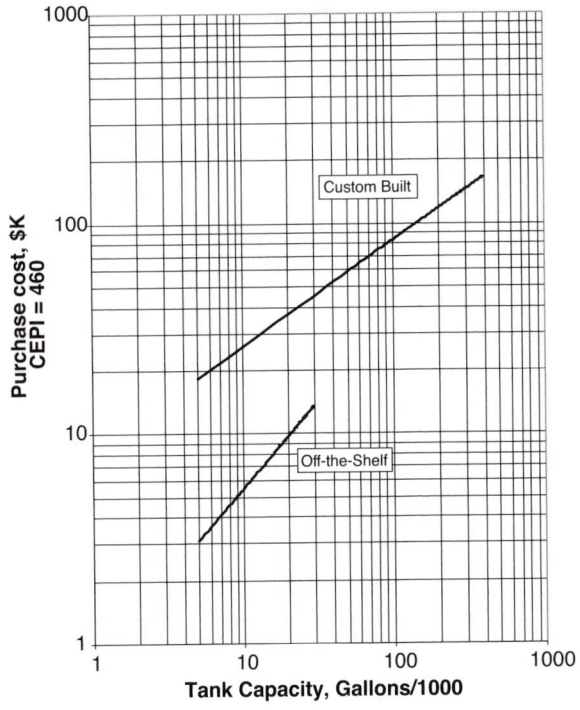

Equations

		Material Factors	
Custom built	$\$K = 8.1 \left(\dfrac{\text{gallons}}{1000}\right)^{0.5}$	Carbon steel	1.0
		Stainless steel	2.4
		Epoxy-lined	1.25
Off-the-shelf	$\$K = 0.8 \left(\dfrac{\text{gallons}}{1000}\right)^{0.83}$	Rubber-lined	1.5
		FRP (fiberglass reinforced polyester)	1.5

Size Exponents

Custom built	0.5
Off-the-shelf	0.83

Appendix V

Design Phases: Inputs and Outputs

This appendix describes what takes place in the design or project phases covered in this book — process development, feasibility, and conceptual. For each phase, the description covers:

- Its purpose
- The inputs, or the information needed before work can begin
- The outputs, or design products created

THE PROCESS DEVELOPMENT PHASE

The work in this phase is done primarily by the Process Development organization using a blend of bench-scale, pilot-plant, semi-works, or full-plant testing.

PURPOSE

The purpose of the Process Development phase is to define the key process steps and operating conditions, to develop raw material and packaging specifications, and to develop process design data (e.g., reaction kinetics, selectivity, conversion, physical properties, and so on).

A caution: Do not overspecify the process. If you do, future upgrades will be more difficult because changes to an unnecessary specification would end up having to be justified. This is a waste of time and money. For example, a process step such as heating a fluid from 120°F to 300°F would probably not need anything more than "heat oil" specified on the flow sheet. If the process developer specified a shell and tube heat exchanger (a workable option), future consideration of a plate and frame unit becomes more difficult. Alternatively, it might be critical to specify the unit operation type or equipment category for a reaction step.

INPUT

These are is the specifications for the product to be made by the process. Based on consumer research, these specifications will be different for different types of products. For example:

- Chemical specifications might cover things such as product purity, maximum levels for different impurities, color, odor, density, viscosity, particle size distribution, and so on.
- Food product specifications might be concerned with flavor, mouthfeel, texture, color, viscosity, appearance, particle size distribution, and so on.
- Pharmaceutical specifications might include drug purity, impurity levels, taste, color, viscosity, shape, rate of absorption by the body, and so on.

Specifications can also include materials to be added to the product, such as a preservative in a food item or a corrosion inhibitor in antifreeze. A good specification will explain how each product attribute fills a defined consumer need, which needs are critical, and which are just desirable. Also, the product designer may suggest raw materials, a reaction path, and separation methods for the reacted materials. These will usually be preliminary recommendations subject to verification (or change) when using commercially available materials (as opposed to high-purity laboratory materials) and subject to having economics brought into the picture.

Output

- Block flow diagram showing:
  - All major process steps in their proper sequence. When it comes to specifying the process steps or blocks, use the following guideline — specify as little as possible while ensuring the process will operate as intended. That means some process blocks will not have a unit operation type selected, some will spell out several choices of unit operation type, and some will require a certain unit operation type or equipment category. The intent of the guideline is to leave options open for the feasibility and conceptual engineers to optimize their work based on current technical and economic conditions.
  - All major flow streams — feeds, reaction products and by-products, recycle and purge streams, and emissions and waste streams.
- Raw material and packaging material specifications. When specifying materials, use the same guideline as for block flow diagrams.
- Process operation:
  - Reaction data — reaction phase; kinetic data; ranges or boundary conditions for temperature, pressure, and reactant concentrations; single-pass conversion; selectivity; and heat of reaction and catalyst requirements, if any.
  - Operating conditions, ranges, or boundaries for each process step.
  - Preliminary material balances for all major flow streams.
- Identification of potential health, safety, and environmental hazards:
  - Fire or explosion.
  - Hazardous materials used or created.
  - Emissions (planned and fugitive) and waste streams.
- Physical property data unavailable elsewhere.

THE FEASIBILITY PHASE

The work in this phase is done by the Process Development organization (using a blend of bench-scale, pilot-plant, semi-works, or plant-scale testing and verification) and by the Process or Plant Design organization (doing design calculations and process simulations).

PURPOSE

The purpose of the Feasibility phase is to determine whether or not a proposal is economically feasible. (Technical feasibility was established during Process Development.)

INPUT

The input for the Feasibility phase is the output from the Process Development phase.

OUTPUT

Process Development Output

- Product-dictated material of construction requirements, if any
- Pilot-plant material balance
- Scale-up criteria
- Small amounts of product for consumer testing

Process or Plant Design Output

- Process Flow Diagram (~75% complete), showing:
  - All major and some minor equipment. This should include utility and environmental equipment.
  - All major flow streams.
- Preliminary operating conditions for major unit operations and equipment.
- Process or plant capacity.
- Preliminary identification of health, safety, and environmental hazards plus a preliminary risk prevention and mitigation plan for the major hazards.
- Preliminary process or plant material and energy balances.
- Preliminary sizing of major equipment, including utility and environmental equipment.
- Materials of construction for major equipment.
- Preliminary selection of all major and some minor equipment. This should include utility and environmental equipment.
- Preliminary engineering and construction schedule.

THE CONCEPTUAL PHASE

The work in this phase is done primarily by the Process or Plant Design organization (sometimes supported by an engineering contractor) via design calculations, process simulations, model building (physical or computer-based), and drawings (manual or CAD-based).

PURPOSE

The purpose of the Conceptual phase is to develop the major features of the design for the selected feasible option.

INPUT

The input for the Conceptual phase is the output from the Feasibility phase.

OUTPUT

- Final issue of the Process Flow Diagram
- Preliminary health, safety, and environmental risk prevention and mitigation plan for all hazards
- First draft of the Process Description, including:
  - Operating conditions for all major equipment and some minor equipment (including health, safety, and environmental systems)
  - Important interactions among the different unit operations
  - Important design considerations
- Selection of equipment types for the major equipment and of equipment categories for most of the minor equipment
- Second or third draft of the process or plant material and energy balances
- Process control strategy
- Study models and preliminary equipment layouts
- Preliminary building and utility requirements
- Preliminary site layout
- Detailed engineering schedule plus a construction milestone schedule

Index